2013
CalDAG™

An Interpretive Manual and Checklist

Michael P. Gibbens

INTERNATIONAL
CODE COUNCIL®

CalDAG™ 2013: An Interpretive Manual and Checklist – 8th Edition

ISBN: 978-1-60983-493-7

First Printing: June 2014

Computer Graphics : Pete Karaiskos, Nicholas Karaiskos
Additional figures reproduced with permission of the International Code Council (ICC)

Published by
ICC
500 New Jersey Avenue, NW, 6th Floor,
Washington, DC 20001

The information contained herein represents the author's interpretation of the combined codes and laws regulating both the State of California and the Federal requirements for complying with disabled accessibility guidelines. Though the author has made a good faith effort to provide you with the latest and most accurate information available, the information in this book may contain technical inaccuracies or typographical errors. As the field of disabled accessibility remains to be a dynamic and constantly changing domain, changes are periodically made to the material herein, and these changes may be incorporated into new editions of the product. The information contained in this publication is subject to change without notice and does not represent a commitment on the part of Gibbens and Associates LLC. Final verification and approval for each project must be secured from regulatory authorities in their respective jurisdictions.

PRINTED IN THE U.S.A.

CalDAG 2013

California
DISABLED ACCESSIBILITY
GUIDEBOOK
INTERPRETIVE MANUAL & CHECKLIST

Eighth Edition, Printed in the United States of America

Author: **Michael P. Gibbens**

GIBBENS & ASSOCIATES, LLC
4258 Avenida Prado
Thousand Oaks, California   91360
Phone: (805) 870-0900

Computer Graphics: Pete Karaiskos, Nicholas Karaiskos
Additional figures reproduced with
permission of the International Code
Council (ICC)

Assistance: Cynthia L. Gibbens - Administration
Josh N. Shea – Information Technology

ACKNOWLEDGEMENTS

I would like to express my appreciation for the continued support, dedication and hard work from the groups and individuals who have helped to make the CalDAG the successful and trusted resource that it has become.

I want to offer sincere thanks to my publisher, the International Code Council (ICC) and the following people who work with ICC.

Mark Johnson, Executive Vice-President and Director of Business Development.

Hamid Naderi, Senior Vice-President, Business and Product Development.

Suzanne Nunes, Director of Product and Special Sales, Business and Product Development.

Cindy Rodriguez, Product Development Manager. Your patience with me and my quirks has been sorely tested over the years, but I have thoroughly appreciated your help and input.

I also want to thank:

Josh Shea, my Information Technology guru. You try to keep me from living in the dark ages, and never cease to amaze me with your computer prowess.

Pete Karaiskos, thank you for all of your assistance in formatting; and to you and your son, Nicholas Karaiskos, for providing the hundreds of computer graphics that were exclusive to the CalDag over the years.

My wife, Cynthia Gibbens, for your continued support on all fronts, and for keeping things together for me and our family, I know it's not easy to live with a writer.

And finally, I want to thank Jay Elbettar, PE, CBO, CASp, Building Official for the City of Mission Viejo and ICC Board Member, for your professional expertise in the review of this edition of the CalDAG prior to publication. Your valuable input contributed to the quality of this finished product.

Each of you has had an impact on the success of this product, and I thank you.

Sincerely,
Michael P. Gibbens, CASp
The Author

ABOUT THE AUTHOR

Michael Patrick Gibbens, CASp, ICC, is a nationally recognized author, instructor and consultant on the interpretive and technical aspects of disabled accessibility compliance in commercial and residential applications for both public and private sectors. He has developed and presented hundreds of seminars on compliance with State and Federal access laws and guidelines throughout the country and has more than 25 years experience as a disabled accessibility compliance consultant. In 1991, 6 months before the Americans with Disabilities Act (ADA) became enforceable, he had written the first nationally published manual for ADA compliance to facilitate site assessments and barrier removals, and was presenting training seminars across the country. In California, he was the first to develop and publish compilations of the more stringent provisions between State and Federal accessibility mandates for the design and construction of Public Accommodations and Commercial Facilities (Federal ADA/State Title 24 standards) and multifamily residential projects (Federal FHAA/State HCD standards) in 1991. Mr. Gibbens' fifteenth book on disabled accessibility compliance, the *CalDAG 2013 (California Disabled Accessibility Guidebook)* began national distribution in June 2013. The *CalDAG* is the only publication in existence that documents the most stringent requirements between the California State accessibility requirements and the ADA. The California Council of the American Institute of Architects (CCAIA) has recognized the *CalDAG* as the foremost subject source on combined state and federal disabled accessibility compliance in the State of California.

Mr. Gibbens' expertise in the interpretation of the legal and technical provisions of disabled accessibility standards is regularly utilized by law firms for the resolution of complaints, allegations of non-compliance and lawsuits. He is experienced with both private sector and Department of Justice actions. When the very first complaint under the newly enacted ADA was registered with the Department of Justice against the Empire State Building, Mr. Gibbens was contracted to consult on-site with the building's architects and management to evaluate the complaint and work toward a resolution. Mr. Gibbens has also been approved by the State Bar of California to provide continuing legal education (MCLE) classes to attorneys on disabled access issues, and has been utilized by the federal Office of Civil Rights (OCR) to train their attorney's and investigators on state/federal accessibility compliance. He has served as a founding member of the California Design Safety and Accessibility Advisory Board (State Access Board), having been appointed by the State Architect, and has served on the Accessibility Code Advisory Committee for the State Building Standards Commission for over 17 years (5 years as chairman). Mr. Gibbens is a California Certified Access Specialist (CASp), is certified as an Accessibility/Usability Specialist by the International Conference of Building Officials, and is an Academy Certified Expert and Diplomate by the American Academy of Certified Consultants and Experts.

Mr. Gibbens has completed facility compliance audits on virtually every category of occupancy covered under Titles II and III of the ADA and has served clients with facilities in all 50 states. He maintains continual contact with the Department of Justice and Access Board in Washington D.C. to keep abreast of new developments and proposed changes in the laws and standards. He has both designed and constructed improvements for the retrofit of facilities to provide compliance with State and Federal accessibility mandates, and is a consultant to a broad range of private businesses and governmental agencies.

ABOUT THE BOOK

For anyone who has endeavored to stay current with the constantly evolving disabled accessibility requirements contained in Part 2 of Title 24 of the California Code of Regulations (i.e., the California Building Code, or "CBC"), you know how difficult this task has been. The first edition of the CalDAG, in 1994, was the size of a phone book and contained not only a cross-referenced listing and checklist of the most stringent requirements and/or interpretations between state and federal laws and regulations, but also included the state and federal reference documents for instant cross-reference if desired. When the ADA Accessibility Standards (ADAAG) first became enforceable on January 26, 1992, there were hundreds of conflicts between California's requirements for disabled accessibility compliance and the federal requirements. Californian's were required to comply with the more stringent requirements of the federal accessibility provisions, however the state and local building officials were not legally allowed to enforce these federal provisions. Hence the reason for writing the CalDAG. The original CalDAG was the end product of many thousands of hours of research. The purpose of this research was to determine the Federal Government's position on the accuracy of California's accessibility guidelines as they compared to those of the ADA, and to establish a relevant set of standards that reflected this position. The ultimate product was to have enough background documentation to support the application of these standards if challenged in a court of law. To accomplish this task, a countless number of contacts, both written and verbal, were initiated between appropriate federal and state offices charged with the dissemination, interpretation and enforcement of disabled accessibility guidelines. The majority of these contacts were with the heads of policy and interpretation of the various technical and legal departments. In addition, the expertise of many private attorneys was solicited for their position in various simulated situations to test probable outcomes in litigation. Since the original CalDAG, California has continually made significant modifications to the disabled accessibility provisions over the years, requiring constant awareness and dedication to understanding and keeping current with these changing requirements. The dedication to providing the most current and correct information between state and federal sources in a user-friendly format has continued throughout all of the previous editions of the CalDAG, and will hopefully be apparent in this edition.

The 2010 ADA Standards are altogether different than the previous ADAAG. The new standards are in a completely new and different format from the previous standards, and a new section numbering system is being utilized. Consequently, since California has adopted the federal standards as a base document for Chapter 11B of the CBC, the California standards also have a new format and section numbering system. In previous editions of the CBC, scoping requirements and technical standards could be found together in the same location. The new format has separated scoping requirements and technical requirements in different sections; in many instances the necessary requirements for a particular use, occupancy, or improvement type are scattered throughout the code. The CalDAG 2013 has been crafted to significantly reduce this situation by listing the scoping and technical requirements for a particular use and/or improvement type in the same section heading when it was plausible, even if a particular requirement is repeated in multiple locations throughout the book.

It's hard to believe that the CalDAG is celebrating its' twentieth year in existence, and that this is the eighth edition. With California's adoption of the 2010 ADA standards into Chapter 11B of the CBC as a base document, and with the subsequent incorporation of those provisions adopted by California that are in addition to, or more stringent than, the federal accessibility requirements, the CBC disabled accessibility provisions are hopefully more stable now than they have ever been in the past. I sincerely hope that all of the information contained in the CalDAG helps the design and construction professional provide for correct disabled accessibility compliance, and that the numerous enforcement officials for the governing bodies in California can develop an increased level of consistency in the enforcement and interpretation of disabled accessibility requirements throughout the state.

TABLE OF CONTENTS

Chapter 3
GENERAL SCOPING AND TECHNICAL REQUIREMENTS OF ACCESSIBILITY APPLICABLE TO PUBLIC ACCOMMODATIONS, COMMERCIAL FACILITIES AND PUBLIC HOUSING 73

APPENDICES

HOW TO USE THIS BOOK

This book was developed to assist in the understanding and proper application of both State and Federal disabled accessibility requirements. All of the information contained herein has been cross-referenced between California's recently amended accessibility regulations (2013 California Building Code, effective January 1, 2014), selected Senate and/or Assembly Bills that contain required changes to disabled accessibility requirements, and the Code of Federal Regulations, 28 CFR Part 36, "Nondiscrimination on the Basis of Disability by Public Accommodations and in Commercial Facilities", (the ADA Title III requirements, including the ADAAG (Americans with Disabilities Act Accessibility Guidelines), and the 2010 ADA Standards for Accessible Design.

The CalDAG contains many of the State and/or Federal provisions in a checklist format. General requirements that pertain to a specific improvement or use are listed in conjunction with specific scoping provisions and technical characteristics that the feature must exhibit. Each of these applicable items includes the most stringent requirement between State and Federal sources. The listed items are then followed by a sequence of notations. The first notation, when applicable, is the California Title 24 regulation that relates to the item. This notation will be in bold print. The second notation, when applicable, is the ADA or 2010 ADA Standards provision that relates to the item. This notation is in standard italic print. The third notation, when applicable, is a reference to a corresponding diagram or figure in this book that allows the user to visualize the written requirement; this notation is also in bold print.

There are items in the book that may list a State reference but not a Federal reference (and vice-versa). In these instances the item listed is an exclusive requirement by one source and not by the other.

The example shown below is a checklist item from the book section titled "**RAMPS**" (found on page 138):

Clear Width

_____ D.   The clear width of a ramp run shall be 48 inches minimum. **11B-405.5**   *405.5* **Fig. CD-15A**

By looking up section **11B-405.2** in California's Building Code (CBC), you will find that it states: *"**Clear width**. The clear width of a ramp run shall be 48 inches (1219 mm) minimum"*. By then looking up section *405.5 of the 2010 ADA Standards*, this section will state: *"**Clear Width**. The clear width of a ramp run and, where handrails are provided, the clear width between handrails shall be 36 inches (915 mm) minimum"*. In this instance, a conflict exists between the two requirements. The 2010 ADA Standards requirement for a minimum 36 inch wide ramp run is less restrictive than California's requirement for a minimum 48 inch wide ramp run. Consequently, only the required California parameters are listed for this specific checklist item. The referenced figure will therefore show a ramp run with a minimum width requirement of 48 inches.

As previously stated, this book only lists the most stringent set of requirements between State and Federal sources. If the reader wishes to gain further clarification as to both the State and Federal requirements then they should avail themselves of the appropriate State and Federal resources.

Step 1

Chapter 1, titled "PRIVATELY-FUNDED VS. PUBLICLY-FUNDED PROJECTS", should be reviewed to determine the appropriate guidelines that you must use if your project is financed by private, state, or federal funds, or a combination of funds from these entities.

Step 2

Chapter 2, titled "APPLICATION AND ADMINISTRATION" must be thoroughly read and understood as it is relevant to both new construction and to alterations, structural repairs and additions to existing buildings or facilities, and contains foundational basics to disabled access compliance. This chapter provides the information and guidance that you will need to determine the degree of accessibility that your building or facility must exhibit, including parameters on Unreasonable Hardship, determining feasibility, and whether or not an elevator will be required.

The requirements for disabled accessibility when undertaking an Alteration or Addition to an existing building or facility are much more complicated than that for New Construction; there are a number of examples for different levels of alterations and additions provided (with diagrams) for various situations you will encounter. In addition, the information in Chapter 2 for "ACCESSIBILITY FOR EXISTING BUILDINGS AND FACILITIES", also includes the Federal ADA requirement for entities classified as "public accommodations" to remove barriers in their existing facilities when "readily achievable" as a stand-alone obligation, even though their existing buildings and/or facilities are not otherwise being altered. At the end of this chapter are the state and federal requirements for "Historic Preservation"; this section also includes the alternate standards to be allowed in those areas or improvements that qualify for this variance.

Step 3

The vast majority of the required features that individual improvements must exhibit are contained in Chapter 3 titled "REQUIREMENTS OF ACCESSIBILITY APPLICABLE TO PUBLIC ACCOMMODATIONS, COMMERCIAL FACILITIES AND PUBLIC HOUSING". While the information in this chapter will no doubt be the most continually utilized, it is important to understand all of the information in this book to provide for correct applications in various situations. The general exceptions contained in Section 1, titled "GENERAL EXCEPTIONS", and accessible route requirements contained in Section 2, ACCESSIBLE ROUTES, and Section 3, "ENTRANCES", are repeated in the remaining sections specific to a particular improvement type or use, including improvement-specific or occupancy-specific information that may be contained in other sections; this is meant to reduce the amount of bouncing around that is required under the current 2010 ADA Standards and Chapter 11B of the 2013 California Building Code.

Using The Index

The most efficient way of finding the information you are looking for on a particular scoping or technical requirement is to use the Index at the end of the book. The index section of this publication has been designed to include not only specific item location information, but also sub-item location information for certain improvement types.

DEFINITIONS & GENERAL TERMINOLOGY

GENERAL TERMINOLOGY

comply with. Meet one or more specifications of the guidelines/regulations. **202**

if, if ... then. Denotes a specification that applies only when the conditions described are present. **202**

may. Denotes an option or alternative. **202**

recommend. Does not require mandatory acceptance, but identifies a suggested action that shall be considered for the purpose of providing a greater degree of accessibility to persons with disabilities. **202**

shall. Denotes a mandatory specification or requirement. **202**

should. Denotes an advisory specification or recommendation. **202**

DEFINITIONS   11B-106   *106*

General
For the purpose of this chapter, the terms listed and defined here have the indicated meaning. **11B-106.1**   *106.1*

Terms Defined in Referenced Standards
Terms not listed in this Chapter under DEFINITIONS, and not defined in the other chapters of this publication, but specifically defined in a referenced standard, shall have the specified meaning from the referenced standard unless otherwise stated.   **11B-106.2**   *106.2*

Undefined Terms
The meaning of terms not specifically listed in this Chapter under DEFINITIONS or in referenced standards shall be as defined by collegiate dictionaries in the sense that the context implies. **11B-106.3**   *106.3*

Interchangeability
Words used in the present tense include the future; words stated in the masculine gender include the feminine and neuter; the singular number includes the plural and those used in the plural include the singular.   **CBC 201.2, 11B-106.4**   *106.4*

DEFINED TERMS   **11B-106.5**   *106.5*

A

ACCESS AISLE. [DSA-AC] An accessible pedestrian space adjacent to or between parking spaces that provides clearances in compliance with the CBC.

ACCESSIBILITY. [DSA-AC] Accessibility is the combination of various elements in a building, facility, site, or area, or portion thereof which allows access, circulation and the full use of the building and facilities by persons with disabilities in compliance with the CBC.

ACCESSIBLE. [DSA-AC] A site, building, facility, or portion thereof that is approachable and usable by persons with disabilities in compliance with the regulations.

ACCESSIBLE ELEMENT. [DSA-AC] An element specified by the regulations adopted by the Division of the State Architect-Access Compliance.

> ***ADVISORY: Definition of ACCESSIBLE ELEMENT.*** *An ACCESSIBLE ELEMENT can include a room, area, route, feature or device which provides accessibility for persons with disabilities.*

ACCESSIBLE MEANS OF EGRESS. A continuous and unobstructed way of egress travel from any accessible point in a building or facility to a public way.

ACCESSIBLE ROUTE. [DSA-AC] A continuous unobstructed path connecting accessible elements and spaces of an accessible site, building or facility that can be negotiated by a person with a disability using a wheelchair, and that is also safe for and usable by persons with other disabilities. Interior accessible routes may include corridors, hallways, floors, ramps, elevators and lifts. Exterior accessible routes may include parking access aisles, curb ramps, crosswalks at vehicular ways, walks, ramps and lifts.

ACCESSIBLE SPACE. [DSA-AC] A space that complies with the accessibility provisions of the CBC.

ADAPTABLE. [DSA-AC] Capable of being readily modified and made accessible.

> ***ADVISORY: Definition of ADAPTABLE.*** *This term means that elements can be modified or adjusted to accommodate the needs of a specific user. As part of the initial design and construction, for example, structural backing would be provided for the later installation of grab bars, base cabinets under kitchen sinks would be removable without the use of specialized tools or specialized knowledge, or countertops would be repositionable.*

ADDITION. ...
[DSA-AC] An expansion, extension or increase in the gross floor area or height of a building or facility.

ADMINISTRATIVE AUTHORITY. [DSA-AC] A governmental agency that adopts or enforces regulations and guidelines for the design, construction or alteration of buildings and facilities.

AISLE. ...
[DSA-AC] A circulation path between objects such as seats, tables, merchandise, equipment, displays, shelves, desks, etc., that provides clearances in compliance with the CBC.

ALTERATION. ...
[DSA-AC] A change, addition or modification in construction, change in occupancy or use, or structural repair to an existing building or facility. Alterations include, but are not limited to, remodeling, renovation, rehabilitation, reconstruction, historic restoration, resurfacing of circulation paths or vehicular ways, changes or rearrangement of the structural parts or elements, and changes or rearrangement in the plan configuration of walls and full-height partitions. Normal maintenance, reroofing, painting or wallpapering, or changes to mechanical and electrical systems are not alterations unless they affect the usability of the building or facility.

AMUSEMENT ATTRACTION. [DSA-AC] Any facility, or portion of a facility, located within an amusement park or theme park which provides amusement without the use of an amusement device. Amusement attractions include, but are not limited to, fun houses, barrels and other attractions without seats.

AMUSEMENT RIDE. [DSA-AC] A system that moves persons through a fixed course within a defined area for the purpose of amusement.

AMUSEMENT RIDE SEAT. [DSA-AC] A seat that is built-in or mechanically fastened to an amusement ride intended to be occupied by one or more passengers.

ANSI. [DSA-AC] The American National Standards Institute.

APPROVED. ...
[DSA-AC] "Approved" means meeting the approval of the enforcing agency, except as otherwise provided by law, when used in connection with any system, material, type of construction, fixture or appliance as the result of investigations and tests conducted by the agency, or by reason of accepted principles or tests by national authorities or technical, health or scientific organizations or agencies.

APPROVED TESTING AGENCY. [DSA-AC] Any agency, which is determined by the enforcing agency, except as otherwise provided by law, to have adequate personnel and expertise to carry out the testing of systems, materials, type of construction, fixtures or appliances.

AREA OF REFUGE. An area where persons unable to use stairways can remain temporarily to await instructions or assistance during emergency evacuation.

AREA OF SPORT ACTIVITY. [DSA-AC] That portion of a room or space where the play or practice of a sport occurs.

ASSEMBLY AREA. [DSA-AC] A building or facility, or portion thereof, used for the purpose of entertainment, educational or civic gatherings, or similar purposes. For the purposes of these requirements, assembly areas include, but are not limited to, classrooms, lecture halls, courtrooms, public meeting rooms, public hearing rooms, legislative chambers, motion picture houses, auditoria, theaters, playhouses, dinner theaters, concert halls, centers for the performing arts, amphitheaters, arenas, stadiums, grandstands or convention centers.

> *ADVISORY: Definition of ASSEMBLY AREA. The application of the accessibility provisions of the CBC is based upon the use of the space rather than the occupancy classification. For example, an assembly area may or may not be a Group A Occupancy. A large conference room in a Group B Occupancy or a multi-purpose area in a Group E Occupancy may be an assembly area.*

ASSISTIVE LISTENING SYSTEM (ALS). [DSA-AC] An amplification system utilizing transmitters, receivers and coupling devices to bypass the acoustical space between a sound source and a listener by means of induction loop, radio frequency, infrared or direct-wired equipment.

AUTOMATIC DOOR. A door equipped with a power-operated mechanism and controls that open and close the door automatically upon receipt of a momentary actuating signal. The switch that begins the automatic cycle may be a photoelectric device, floor mat or manual switch.

AUTOMATIC TELLER MACHINE (ATM). [DSA-AC] Any electronic information processing device that accepts or dispenses cash in connection with a credit, deposit or convenience account. The term does not include devices used solely to facilitate check guarantees or check authorizations, or which are used in connection with the acceptance or dispensing of cash on a person-to-person basis, such as by a store cashier.

B

BATHROOM. For the purposes of Chapters 11A and 11B of the CBC, a room which includes a water closet (toilet), a lavatory, and a bathtub and/or a shower. It does not include single-fixture facilities or those with only a water closet and lavatory. It does include a compartmented bathroom. A compartmented bathroom is one in which the fixtures are distributed among interconnected rooms. A compartmented bathroom is considered a single unit and is subject to the requirements of Chapters 11A and 11B of the CBC.

BLENDED TRANSITION. [DSA-AC] A raised pedestrian street crossing, depressed corner or similar connection between the pedestrian access route at the level of the sidewalk and the level of the pedestrian street crossing that has a grade of 5 percent or less.

BOARDING PIER. [DSA-AC] A portion of a pier where a boat is temporarily secured for the purpose of embarking or disembarking.

BOAT LAUNCH RAMP. [DSA-AC] A sloped surface designed for launching and retrieving trailered boats and other water craft to and from a body of water.

BOAT SLIP. [DSA-AC] That portion of a pier, main pier, finger pier, or float where a boat is moored for the purpose of berthing, embarking, or disembarking.

BUILDING. Any structure used or intended for supporting or sheltering any use or occupancy.

> *ADVISORY: Definition of BUILDING. The accessibility standards generally apply to buildings and facilities. Parking lots, play areas, patios, constructed trails, man-made outdoor areas are often not considered to be buildings. Rather, these elements are generally considered to be facilities. See the definition of FACILITY.*

BUILDING OFFICIAL. The officer or other designated authority charged with the administration and enforcement of the CBC, or a duly authorized representative.

C

CATCH POOL. [DSA-AC] A pool or designated section of a pool used as a terminus for water slide flumes.

CCR. [DSA-AC] The California Code of Regulations.

CHARACTERS. [DSA-AC] Letters, numbers, punctuation marks and typographic symbols.

CHILDREN'S USE. [DSA-AC] Describes spaces and elements specifically designed for use primarily by people 12 years old and younger.

CIRCULATION PATH. ...
[DSA-AC] An exterior or interior way of passage provided for pedestrian travel, including but not limited to, walks, hallways, courtyards, elevators, platform lifts, ramps, stairways and landings.

> *ADVISORY Definition of CIRCULATION PATH. A CIRCULATION PATH is a pedestrian route provided within a building, facility or site and may or may not (in the case of stairs) include an accessible route of travel. Whenever the accessible*

route diverges from the regular circulation path signage may be required to identify the departure from the regular route if not obvious.

CLEAR. [DSA-AC] Unobstructed.

CLEAR FLOOR SPACE. [DSA-AC] The minimum unobstructed floor or ground space required to accommodate a single, stationary wheelchair and occupant.

CLOSED-CIRCUIT TELEPHONE. [DSA-AC] A telephone with a dedicated line such as a house phone, courtesy phone or phone that must be used to gain entry to a facility.

COMMERCIAL FACILITIES. [DSA-AC] Facilities whose operations will affect commerce and are intended for non-residential use by a private entity. Commercial facilities shall not include (1) facilities that are covered or expressly exempted from coverage under the Fair Housing Act of 1968, as amended (42 U.S.C. 3601 - 3631); (2) aircraft; or (3) railroad locomotives, railroad freight cars, railroad cabooses, commuter or intercity passenger rail cars (including coaches, dining cars, sleeping cars, lounge cars, and food service cars), any other railroad cars described in Section 242 of the Americans With Disabilities Act or covered under Title II of the Americans With Disabilities Act, or railroad rights-of-way. For purposes of this definition, "rail" and "railroad" have the meaning given the term "railroad" in Section 202(e) of the Federal Railroad Safety Act of 1970 (45 U.S.C. 431(e)).

COMMON USE. Interior or exterior circulation paths, rooms, spaces or elements that are not for public use and are made available for the shared use of two or more people.

> *ADVISORY: Definition of COMMON USE. Employees, tenants or staff and their guests may jointly utilize common use areas where the public is not permitted general access. An example of a common use area would be a laundry room or community room within a homeless shelter. Examples of common use areas within an office building may include a break room, employee lounge, employee exercise facility or employee locker room.*

COMPLY WITH. [DSA-AC] Comply with means to meet one or more provisions of the CBC.

CROSS SLOPE. The slope that is perpendicular to the direction of travel.

CURB CUT. An interruption of a curb at a pedestrian way, which separates surfaces that are substantially at the same elevation.

CURB RAMP. A sloping pedestrian way, intended for pedestrian traffic, which provides access between a walk or sidewalk and a surface located above or below an adjacent curb face.

D

DETECTABLE WARNING. A standardized surface feature built in or applied to walking surfaces or other elements to warn of hazards on a circulation path.

> *ADVISORY: Definition of DETECTABLE WARNING. Curbs can be used by pedestrians with vision impairments to detect the boundary between a sidewalk and a vehicular way. Curb ramps remove the needed cues for persons with visual impairments; detectable warnings have been developed as a replacement cue and warning to indicate the presence of a vehicular way.*

DIRECTIONAL SIGN. [DSA-AC] A publicly displayed notice which indicates by use of words or symbols a recommended direction or route of travel.

DISABILITY. [DSA-AC] Disability is (1) a physical or mental impairment that limits one or more of the major life activities of an individual, (2) a record of such an impairment, or (3) being regarded as having such an impairment.

> **ADVISORY: Definition of DISABILITY.** *This is the definition of disability used and defined in the Americans with Disabilities Act of 1990.*

DORMITORY. A space in a building where group sleeping accommodations are provided in one room, or in a series of closely associated rooms, for persons not members of the same family group, under joint occupancy and single management, as in college dormitories or fraternity houses.

E

ELEMENT. [DSA-AC] An architectural or mechanical component of a building, facility, space or site.

ELEVATED PLAY COMPONENT. [DSA-AC] A play component that is approached above or below grade and that is part of a composite play structure consisting of two or more play components attached or functionally linked to create an integrated unit providing more than one play activity.

ELEVATOR, PASSENGER. [DSA-AC] An elevator used primarily to carry passengers.

EMPLOYEE WORK AREA. All or any portion of a space used only by employees and used only for work. Corridors, toilet rooms, kitchenettes and break rooms are not employee work areas.

ENFORCING AGENCY. [DSA-AC] Enforcing Agency is the designated department or agency as specified by statute or regulation.

ENTRANCE. Any access point to a building or portion of a building or facility used for the purpose of entering. An entrance includes the approach walk, the vertical access leading to the entrance platform, the entrance platform itself, vestibule if provided, the entry door or gate, and the hardware of the entry door or gate.

EQUIVALENT FACILITATION. The use of designs, products, or technologies as alternatives to those prescribed, resulting in substantially equivalent or greater accessibility and usability.

> **Note:** In determining equivalent facilitation, consideration shall be given to means that provide for the maximum independence of persons with disabilities while presenting the least risk of harm, injury or other hazard to such persons or others.

EXISTING BUILDING OR FACILITY. [DSA-AC] A facility in existence on any given date, without regard to whether the facility may also be considered newly constructed or altered under the CBC.

EXIT. That portion of a means of egress system between the exit access and the exit discharge or public way. Exit components include exterior exit doors at the level of exit discharge, interior exit stairways, interior exit ramps, exit passageways, exterior exit stairways and exterior exit ramps and horizontal exits.

F

FACILITY. ...
[DSA-AC] All or any portion of buildings, structures, site improvements, elements, and pedestrian routes or vehicular ways located on a site.

FUNCTIONAL AREA. [DSA-AC] A room, space or area intended or designated for a group of related activities or processes.

G

GANGWAY. [DSA-AC] A variable-sloped pedestrian walkway that links a fixed structure or land with a floating structure. Gangways that connect to vessels are not addressed by this code.

GOLF CAR PASSAGE. [DSA-AC] A continuous passage on which a motorized golf car can operate.

GRAB BAR. [DSA-AC] A bar for the purpose of being grasped by the hand for support.

> *ADVISORY: Definition of GRAB BAR. A grab bar may also provide support for a user transferring from a wheelchair onto a bench, seat or plumbing fixture.*

GRADE (Adjacent Ground Elevation). [DSA-AC] The lowest point of elevation of the finished surface of the ground, paving or sidewalk within the area between the building and the property line or, when the property line is more than 5 feet (1524 mm) from the building, between the building and a line 5 feet (1524 mm) from the building. See Health and Safety Code Section 19955.3(d).

GRADE BREAK. [DSA-AC] The line where two surface planes with different slopes meet.

GROUND FLOOR. [DSA-AC] The floor of a building with a building entrance on an accessible route. A building may have one or more ground floors.

GROUND LEVEL PLAY COMPONENT. [DSA-AC] A play component that is approached and exited at the ground level.

GUARD [DSA-AC] OR GUARDRAIL. A building component or a system of building components located at or near the open sides of elevated walking surfaces that minimizes the possibility of a fall from the walking surface to a lower level.

H

HANDRAIL. A horizontal or sloping rail intended for grasping by the hand for guidance or support.

HEALTH CARE PROVIDER. [DSA-AC] See "Professional Office of a Health Care Provider"

HISTORIC BUILDINGS. ...
[DSA-AC] See "Qualified historical building or property," C.C.R., Title 24, Part 8.

HOUSING AT A PLACE OF EDUCATION. Housing operated by or on behalf of an elementary, secondary, undergraduate, or postgraduate school, or other place of education, including dormitories, suites, apartments, or other places of residence.

I

IF, IF . . . THEN. [DSA-AC] The terms "if" and "if … then" denotes a specification that applies only when the conditions described are present.

INTERNATIONAL SYMBOL OF ACCESSIBILITY. The symbol adopted by Rehabilitation International's 11th World Congress for the purpose of indicating that buildings and facilities are accessible to persons with disabilities.

> ***ADVISORY: Definition of INTERNATIONAL SYMBOL OF ACCESSIBILITY.*** *This is also known as the "ISA." It is a graphic representation of the profile view of a wheelchair with occupant. See Chapter 11B, Figure 11B-703.7.2.1.*

K

KEY STATION. [DSA-AC] Certain rapid and light rail stations, and commuter rail stations, as defined under criteria established by the Department of Transportation in 49 CFR 37.47 and 49 CFR 37.51, respectively.

KICK PLATE. An abrasion-resistant plate affixed to the bottom portion of a door to prevent a trap condition and protect its surface.

> ***ADVISORY: Definition of KICK PLATE.*** *Although kick plates are not necessarily required by the code, they are often applied to areas of doors required to have a smooth surface and be free of abrupt surface changes. This provides a person in a wheelchair the option of opening a door by pushing with their feet or allows the front foot plates of a wheelchair to glide smoothly along the lower face of the door as a wheelchair user proceeds through a door.*

KITCHEN OR KITCHENETTE. [DSA-AC] A room, space or area with equipment for the preparation and cooking of food.

L

LAVATORY. A fixed bowl or basin with running water and drainpipe, as in a toilet or bathing facility, for washing or bathing purposes. (As differentiated from the definition of "Sink".)

M

MAIL BOXES. [DSA-AC] Receptacles for the receipt of documents, packages, or other deliverable matter. Mail boxes include, but are not limited to, post office boxes and receptacles provided by commercial mail-receiving agencies, apartment facilities or schools.

MARKED CROSSING. A crosswalk or other identified path intended for pedestrian use in crossing a vehicular way.

MAY. [DSA-AC] May denotes an option or alternative.

MEZZANINE. ...
[DSA-AC] An intermediate level or levels between the floor and ceiling of any story with an aggregate floor area of not more than one-third of the area of the room or space in which the level or levels are located. Mezzanines have sufficient elevation that space for human occupancy can be provided on the floor below.

N

NFPA. [DSA-AC] The National Fire Protection Association.

NOSING. The leading edge of treads of stairs and of landings at the top of stairway flights.

O

OCCUPANT LOAD. The number of persons for which the means of egress of a building or portion thereof is designed.

OCCUPIABLE SPACE. A room or enclosed space designed for human occupancy in which individuals congregate for amusement, educational or similar purposes or in which occupants are engaged at labor, and which is equipped with means of egress and light and ventilation facilities meeting the requirements of the CBC.

OPEN RISER. The space between two adjacent stair treads not closed by a riser.

OPERABLE PART. A component of an element used to insert or withdraw objects, or to activate, deactivate, or adjust the element.

P

PASSENGER ELEVATOR. [DSA-AC] See "Elevator, passenger"

PATH OF TRAVEL. [DSA-AC] An identifiable accessible route within an existing site, building or facility by means of which a particular area may be approached, entered and exited, and which connects a particular area with an exterior approach (including sidewalks, streets, and parking areas), an entrance to the facility, and other parts of the facility. When alterations, structural repairs or additions are made to existing buildings or facilities, the term "path of travel" also includes the toilet and bathing facilities, telephones, drinking fountains and signs serving the area of work.

> *ADVISORY: Definition of PATH OF TRAVEL. The term PATH OF TRAVEL applies only to alterations, structural repairs or additions to existing buildings or facilities. Path of travel elements may be subject to upgrade as part of the alteration to an existing building if they do not conform to current accessibility requirements.*

PEDESTRIAN. An individual who moves in walking areas with or without the use of walking assistive devices such as crutches, leg braces, wheelchairs, white cane, service animal, etc.

PEDESTRIAN WAY. A route by which a pedestrian may pass.

PERMANENT. [DSA-AC] Facilities which, are intended to be used for periods longer than those designated in this code under the definition of "Temporary."

PERMIT. An official document or certificate issued by the authority having jurisdiction which authorizes performance of a specified activity.

> ***ADVISORY: Definition of PERMIT.*** *In State-funded construction, a letter following plan review which approves the plans and allows the release of funds is equivalent to a "permit."*

PICTOGRAM. [DSA-AC] A pictorial symbol that represents activities, facilities, or concepts.

PLACE OF PUBLIC ACCOMMODATION. A facility operated by a private entity whose operations affect commerce and fall within at least one of the following categories:

1. Place of lodging, except for an establishment located within a facility that contains not more than five rooms for rent or hire and that actually is occupied by the proprietor of the establishment as the residence of the proprietor. For purposes of this code, a facility is a "place of lodging" if it is

   (i) An inn, hotel, or motel; or

   (ii) A facility that

       (A) Provides guest rooms for sleeping for stays that primarily are short-term in nature (generally 30 days or less) where the occupant does not have the right to return to a specific room or unit after the conclusion of his or her stay; and

       (B) Provides guest rooms under conditions and with amenities similar to a hotel, motel, or inn, including the following:

           (1) On- or off-site management and reservations service;

           (2) Rooms available on a walk-up or call-in basis;

           (3) Availability of housekeeping or linen service; and

           (4) Acceptance of reservations for a guest room type without guaranteeing a particular unit or room until check-in, and without a prior lease or security deposit.

2. A restaurant, bar, or other establishment serving food or drink;

3. A motion picture house, theater, concert hall, stadium, or other place of exhibition or entertainment;

4. An auditorium, convention center, lecture hall, or other place of public gathering;

5. A bakery, grocery store, clothing store, hardware store, shopping center, or other sales or rental establishment;

6. A Laundromat, dry-cleaner, bank, barber shop, beauty shop, travel service, shoe repair service, funeral parlor, gas station, office of an accountant or lawyer, pharmacy, insurance office, professional office of a health care provider, hospital, or other service establishment;

7. A terminal, depot, or other station used for specified public transportation;

8.   A museum, library, gallery, or other place of public display or collection;

9.   A park, zoo, amusement park, or other place of recreation;

10. A nursery, elementary, secondary, undergraduate, or postgraduate private school, or other place of education;

11. A day care center, senior citizen center, homeless shelter, food bank, adoption agency, or other social service center establishment;

12. A gymnasium, health spa, bowling alley, golf course, or other place of exercise or recreation;

13. A religious facility;

14. An office building; and

15. A public curb or sidewalk.

PLATFORM. A raised area within a building used for worship, the presentation of music, plays or other entertainment; the head table for special guests; the raised area for lecturers and speakers; boxing and wrestling rings; theater-in-the-round stages; and similar purposes wherein there are no overhead hanging curtains, drops, scenery or stage effects other than lighting and sound. A temporary platform is one installed for not more than 30 days.

PLATFORM (WHEELCHAIR) LIFT. A hoisting and lowering mechanism equipped with a car or platform or support that serves two landings of a building or structure and is designed to carry a passenger or passengers and/or luggage or other material a vertical distance as may be allowed.

PLAY AREA. [DSA-AC] A portion of a site containing play components designed and constructed for children.

PLAY COMPONENT. [DSA-AC] An element intended to generate specific opportunities for play, socialization, or learning. Play components are manufactured or natural; and are stand-alone or part of a composite play structure.

POINT-OF-SALE DEVICE. [DSA-AC] A device used for the purchase of a good or service where a personal identification number (PIN), zip code or signature is required.

POWDER ROOM. A room containing a water closet (toilet) and a lavatory, and which is not defined as a bathroom.

POWER-ASSISTED DOOR. [DSA-AC] A door used for human passage with a mechanism that helps to open the door, or relieves the opening resistance of a door, upon the activation of a switch or a continued force applied to the door itself.

PRIVATE BUILDING OR FACILITY. [DSA-AC] A place of public accommodation or a commercial building or facility subject to CBC Chapter 1, Section 1.9.1.2.

PROFESSIONAL OFFICE OF A HEALTH CARE PROVIDER. [DSA-AC] See CBC Chapter 11B. A location where a person or entity, regulated by the State to provide professional services related to the physical or mental health of an individual, makes such services available to the public. The facility housing the professional office of a health care provider only includes floor levels housing at least one health care provider, or any floor level designed or intended for use by at least one health care provider.

> **ADVISORY: Definition of PROFESSIONAL OFFICE OF A HEALTH CARE PROVIDER.** *The term PROFESSIONAL OFFICE OF A HEALTH CARE PROVIDER applies to the offices of doctors, psychologists, dentists, radiologists, and others certified or licensed by the State to provide physical or mental health care.*

PUBLIC BUILDING OR FACILITY. [DSA-AC] A building or facility or portion of a building or facility designed, constructed, or altered by, on behalf of, or for the use of a public entity subject to CBC Chapter 1, Section 1.9.1.1.

PUBLIC ENTITY. Any state or local government; any department, agency, special-purpose district, or other instrumentality of a state or local government.

PUBLIC ENTRANCE. An entrance that is not a service entrance or a restricted entrance.

PUBLIC HOUSING. Housing facilities owned and/or operated by, for or on behalf of a public entity including but not limited to the following:

1. Publically owned and/or operated one- or two- family dwelling units or congregate residences;

2. Publically owned and/or operated buildings or complexes with three or more residential dwellings units;

3. Publically owned and/or operated housing provided by entities subject to regulations issued by the United State Department of Housing and Urban Development under Section 504 of the Rehabilitation Act of 1973 as amended;

4. Publically owned and/or operated homeless shelters, group homes and similar social service establishments;

5. Publically owned and/or operated transient lodging, such as hotels, motels, hostels and other facilities providing accommodations of a short term nature of not more than 30 days duration;

6. Housing at a place of education owned or operated by a public entity, such as housing on or serving a public school, public college or public university campus;

7. Privately owned housing made available for public use as housing.

PUBLIC USE. [DSA-AC] Interior or exterior rooms, spaces or elements that are made available to the public. Public use may be provided at a building or facility that is privately or publicly owned. Private interior or exterior rooms, spaces or elements associated with a residential dwelling unit provided by a public housing program or in a public housing facility are not public use areas and shall not be required to be made available to the public. In the context of public housing, public use is the provision of housing programs by, for or on behalf of a public entity.

PUBLIC-USE AREAS. Interior or exterior rooms or spaces of a building that are made available to the general public and do not include common use areas. Public use areas may be provided at a building that is privately or publicly owned.

> **ADVISORY: Definition of PUBLIC-USE AREAS.** *Examples of public use areas may include a hotel lobby, movie theater, concert hall, public restroom, sales floor of a retail store, or dining room within a restaurant.*

PUBLIC WAY. A street, alley or other parcel of land open to the outside air leading to a street, that has been deeded, dedicated or otherwise permanently appropriated to the public for public use and which has a clear width and height of not less than 10 feet (3048 mm).

Q

QUALIFIED HISTORIC BUILDING OR FACILITY. [DSA-AC] A building or facility that is listed in or eligible for listing in the National Register of Historic Places, or designated as historic under an appropriate State or local law. See C.C.R. Title 24, Part 8.

R

RAMP. A walking surface that has a running slope steeper than one unit vertical in 20 units horizontal (5-percent slope).

REASONABLE PORTION. [DSA-AC] That segment of a building, facility, area, space or condition, which would normally be necessary if the activity therein is to be accessible by persons with disabilities.

> ***ADVISORY: Definition of REASONABLE PORTION.*** *The term is intended to mean that the building or facility provides equitable opportunities, advantages, and ease of use for people with disabilities as is otherwise being made available to the general public. It is not intended to mean reasonable from a cost point of view.*

RECOMMEND. [DSA-AC] Does not require mandatory acceptance, but identifies a suggested action that shall be considered for the purpose of providing a greater degree of accessibility to persons with disabilities.

REMODELING. [DSA-AC] See "Alteration."

REPAIR. The reconstruction or renewal of any part of an existing building for the purpose of its maintenance.

RESIDENTIAL DWELLING UNIT. [DSA-AC] A unit intended to be used as a residence that is primarily long-term in nature. Residential dwelling units do not include transient lodging, inpatient medical care, licensed long-term care, and detention or correctional facilities.

RESTRICTED ENTRANCE. An entrance that is made available for common use on a controlled basis but not public use and that is not a service entrance.

RISER. The upright part between two adjacent stairs treads.

RUNNING SLOPE. The slope that is parallel to the direction of travel. (As differentiated from the definition of "Cross Slope".)

S

SELF-SERVICE STORAGE. [DSA-AC] Building or facility designed and used for the purpose of renting or leasing individual storage spaces to customers for the purpose of storing and removing personal property on a self-service basis.

SERVICE ENTRANCE. An entrance intended primarily for delivery of goods or services.

SHALL. [DSA-AC] Denotes a mandatory specification or requirement.

SHOPPING CENTER (or SHOPPING MALL). [DSA-AC] One or more sales or rental establishments or stores. A shopping center may include a series of buildings on a common site, connected by a common pedestrian access route on, above or below the ground floor, that is either under common ownership or common control or developed either as one project or as a series of related projects. For the purposes of this section, "shopping center" or "shopping mall" includes a covered mall building.

> ***ADVISORY: Definition of SHOPPING CENTER (or SHOPPING MALL).** The California definition for SHOPPING CENTER is quite different from the federal definition. Federal ADA Regulations defines a shopping center or shopping mall as a building housing five or more sales or rental establishments. However, California accessibility provisions define a shopping center as only one or more sales or rental establishments or stores.*

SHOULD. [DSA-AC] Denotes an advisory specification or recommendation.

SIDEWALK. A surfaced pedestrian way contiguous to a street used by the public. (As differentiated from the definition of "Walk".)

> ***ADVISORY: Definition of SIDEWALK.** There is an important distinction between SIDEWALK and WALK and they are treated differently under the CBC. As noted in this definition, a sidewalk is contiguous to a street while a walk is not.*

SIGNAGE. [DSA-AC] Displayed verbal, symbolic, tactile, and/or pictorial information.

SINK. A fixed bowl or basin with running water and drainpipe, as in a kitchen or laundry, for washing dishes, clothing, etc. (As differentiated from the definition of "Lavatory".)

SITE. A parcel of land bounded by a lot line or a designated portion of a public right-of-way.

SLEEPING ACCOMMODATIONS. Rooms intended and designed for sleeping.

SOFT CONTAINED PLAY STRUCTURE. [DSA-AC] A play structure made up of one or more play components where the user enters a fully enclosed play environment that utilizes pliable materials, such as plastic, netting, or fabric.

SPACE. A definable area, such as, a room, toilet room, hall, assembly area, entrance, storage room, alcove, courtyard, or lobby.

SPECIFIED PUBLIC TRANSPORTATION. [DSA-AC] Transportation by bus, rail or any other conveyance (other than by aircraft) that provides the general public with general or special service (including charter service) on a regular and continuing basis.

STAGE. A space within a building utilized for entertainment or presentations, which includes overhead hanging curtains, drops, scenery or stage effects other than lighting and sound.

STAIR. A change in elevation, consisting of one or more risers.

STAIRWAY. One or more flights of stairs, either exterior or interior, with the necessary landings and platforms connecting them, to form a continuous and uninterrupted passage from one level to another.

STORY. [DSA-AC] That portion of a building or facility designed for human occupancy included between the upper surface of a floor and upper surface of the floor or roof next above. A story containing one or more mezzanines has more than one floor level. If the finished floor level directly above a basement or unused under-floor space is more than six feet (1829 mm) above grade for more than 50 percent of the total perimeter or is more than 12 feet (3658 mm) above grade at any point, the basement or unused under-floor space shall be considered as a story.

STRUCTURAL FRAME. [DSA-AC] The columns and the girders, beams and trusses having direct connections to the columns and all other members that are essential to the stability of the building or facility as a whole.

STRUCTURE. That which is built or constructed.

T

TACTILE. An object that can be perceived using the sense of touch.

TACTILE SIGN. A sign containing raised characters and/or symbols and accompanying Braille.

TECHNICALLY INFEASIBLE. An alteration of a building or a facility, that has little likelihood of being accomplished because the existing structural conditions require the removal or alteration of a load-bearing member that is an essential part of the structural frame, or because other existing physical or site constraints prohibit modification or addition of elements, spaces or features that are in full and strict compliance with the minimum requirements for new construction and which are necessary to provide accessibility.

TEEING GROUND. [DSA-AC] In golf, the starting place for the hole to be played.

TEMPORARY. [DSA-AC] Buildings and facilities intended for use at one location for not more than one year and seats intended for use at one location for not more than 90 days.

> ***ADVISORY:*** **Definition of TEMPORARY.** Temporary buildings and facilities must be accessible to the same degree as permanent facilities per CA Gov. Code §4451(e).

TEXT TELEPHONE. Machinery or equipment that employs interactive text-based communications through the transmission of coded signals across the standard telephone network. Text telephones can include, for example, devices known as TTYs (teletypewriters) or computers.

TRANSFER DEVICE. [DSA-AC] Equipment designed to facilitate the transfer of a person from a wheelchair or other mobility aid to and from an amusement ride seat.

TRANSIENT LODGING. A building or facility containing one or more guest room(s) for sleeping that provides accommodations that are primarily short-term in nature (generally 30 days or less). Transient lodging does not include residential dwelling units intended to be used as a residence, inpatient medical care facilities, licensed long-term care facilities, detention or correctional facilities, or private buildings or facilities that contain no more than five rooms for rent or hire and that are actually occupied by the proprietor as the residence of such proprietor. **[DSA-AC]** See also the definition of Place of Public Accommodation.

TRANSIT BOARDING PLATFORM. [DSA-AC] A horizontal, generally level surface, whether raised above, recessed below or level with a transit rail, from which persons embark/disembark a fixed rail vehicle.

TRANSITION PLATE. [DSA-AC] A sloping pedestrian walking surface located at the end(s) of a gangway.

TREAD. The horizontal part of a step.

TTY. An abbreviation for teletypewriter. Machinery that employs interactive text-based communication through the transmission of coded signals across the telephone network. TTYs may include, for example, devices known as TDDs (telecommunication display devices or telecommunication devices for deaf persons) or computers with special modems. TTYs are also called text telephones.

U

UNREASONABLE HARDSHIP. When the enforcing agency finds that compliance with the building standard would make the specific work of the project affected by the building standard infeasible, based on an overall evaluation of the following factors:

1. The cost of providing access.

2. The cost of all construction contemplated.

3. The impact of proposed improvements on financial feasibility of the project.

4. The nature of the accessibility which would be gained or lost.

5. The nature of the use of the facility under construction and its availability to persons with disabilities.

The details of any finding of unreasonable hardship shall be recorded and entered in the files of the enforcing agency.

USE ZONE. [DSA-AC] The ground level area beneath and immediately adjacent to a play structure or play equipment that is designated by ASTM F 1487 for unrestricted circulation around the play equipment and where it is predicted that a user would land when falling from or exiting the play equipment.

V

VALUATION THRESHOLD. [DSA-AC] An annually adjusted, dollar-amount figure used in part to determine the extent of required path of travel upgrades. The baseline valuation threshold of $50,000 is based on the January 1981, "ENR US20 Cities" Average Construction Cost Index (CCI) of 3372.02 as published in Engineering News Record, McGraw Hill Publishing Company. The current valuation threshold is determined by multiplying the baseline valuation threshold by a ratio of the current year's January CCI to the baseline January 1981 CCI.

> **ADVISORY: Definition of VALUATION THRESHOLD.** *The valuation threshold is adjusted each year in January using the Engineering News Record 20 Cities Construction Cost Index. Valuation thresholds for the current year and recent years dating back to 2000 are available on the Division of the State Architect web site at:*
> *http://www.dgs.ca.gov/dsa/Programs/progAccess/threshold.aspx.*

VEHICULAR WAY. A route provided for vehicular traffic, such as in a street, driveway, or parking facility.

W

WALK. [DSA-AC] An exterior prepared surface for pedestrian use, including pedestrian areas such as plazas and courts. (As differentiated from the definition of "Sidewalk".)

> **ADVISORY:** **Definition of WALK.** There is an important distinction between SIDEWALK and WALK and they are treated differently under the CBC. A sidewalk is contiguous to a street while a walk is not.

WET BAR. [DSA-AC] An area or space with a counter equipped with a sink and running water but without cooking facilities.

WHEELCHAIR. A chair mounted on wheels to be propelled by its occupant manually or with the aid of electric power, of a size and configuration conforming to the recognized standard models of the trade.

WHEELCHAIR SPACE. Space for a single wheelchair and its occupant.

WORKSTATION. ...
[DSA-AC] An area defined by equipment and/or work surfaces intended for use by employees only, and generally for one or a small number of employees at a time. Examples include ticket booths; the employee side of grocery store check stands; the bartender area behind a bar; the employee side of snack bars, sales counters and public counters; guardhouses; toll booths; kiosk vending stands; lifeguard stations; maintenance equipment closets; counter and equipment areas in restaurant kitchens; file rooms; storage areas; etc.

WORK AREA EQUIPMENT. [DSA-AC] Any machine, instrument, engine, motor, pump, conveyor, or other apparatus used to perform work. As used in this document, this term shall apply only to equipment that is permanently installed or built-in in employee work areas. Work area equipment does not include passenger elevators and other accessible means of vertical transportation.

Chapter 1

PRIVATELY- FUNDED VS. PUBLICLY- FUNDED PROJECTS

To establish the appropriate degree of accessibility that a specific project must exhibit, and to understand the relevant considerations that must be applied in a specific situation, the funding source for the project must first be defined. There are three principal sources of funding that may be used for a specific project; Federally funded, Publicly funded, and Privately funded. In addition, a project may utilize a combination of these types of funding. Each of these funding sources have differing degrees of responsibilities attached to them in order for the user to be in compliance with the conditions of their use. The following information makes note of these provisions and conditions, and also explains responsibilities when multiple funding sources are utilized.

FEDERALLY FUNDED

In 1968, the Architectural Barriers Act was enacted to insure that certain buildings financed with Federal funds were designed and constructed to be accessible to the physically disabled. The definition for a Federally funded project were set forth as follows:

> "Be it enacted by the Senate and House of Representatives of the United States of America in Congress assembled, that, as used in this Act, the term "building" means any building or facility (other than (A) a privately owned residential structure not leased by the Government for subsidized housing programs and (B) any building or facility on a military installation designed and constructed primarily for use by able bodied military personnel) the intended use for which either will require that such building or facility be accessible to the public, or may result in employment or residence therein of physically handicapped persons, which building or facility is -
>
> > (1) to be constructed or altered by or on behalf of the United States;
> >
> > (2) to be leased in whole or in part by the United States after August 12, 1968 (An amendment to this law states "every lease entered into on or after January 1, 1977, including any renewal of a lease entered into before such a date which renewal is on or after such date");
> >
> > (3) to be financed in whole or in part by a grant or a loan made by the United States after August 12, 1968, if such building or facility is subject to standards for design, construction, or alteration issued under authority of the law authorizing such grant or loan; or
> >
> > (4) to be constructed under authority of the National Capital Transportation Act of 1960, the National Capital Transportation Act of 1965, or title III of the Washington Metropolitan Area Transit Regulation Compact.

If the project falls into one of the above categories, it is highly unlikely that you will have to determine which standards are needed for the design and construction; the appropriate Federal agency with jurisdiction over the project will do that for you. A number of years ago I was contracted to consult on the proposed alteration of a Naval facility. The Department of Defense had provided the drawings and specifications which were formulated using the UFAS (Uniform Federal Accessibility Standards). The architects at the Naval base were concerned that they would also be required to comply with Both the ADA and California Title 24 provisions, and that the drawings did not reflect these additional provisions. I contacted a number of Federal offices regarding this situation, questioning their ultimate position in the matter. I was contacted over a month later by a representative from the Pentagon who stated that this question had been reviewed by six separate offices of the Federal Government. The Government's position was that the only set of standards that this project must comply with was the UFAS. The representative further stated that even though the UFAS was the governing standard for this project, that she *recommended* (not required) that any State standard that appeared to provide a higher degree of accessibility be *considered*, in order to provide for a higher standard of accessibility if desired.

Neither the ADA nor Title 24 apply to the Federal Government. The provisions regarding non-discrimination in Federal programs and activities are covered substantially by sections 501 and 504 of the Rehabilitation Act of 1973.

PUBLICLY FUNDED

Section 201 of the Americans with Disabilities Act defines what constitutes a "public entity" :

    (1) PUBLIC ENTITY - The term "public entity" means-

    (A)   Any State or local government;

    (B)   any department, agency, special purpose district, or other instrumentality of a State or States or local government; and

    (C)   The National Railroad Passenger Corporation, and any commuter authority (as defined in section 103(8) of the Rail Passenger Service Act).

If the project involves funding provided in whole or in part by any of the above entities, then it must comply with Title II of the ADA. Title II is intended to apply to *all* programs, activities, and services provided or operated by State or local governments. Currently, section 504 of the Rehabilitation Act only applies to programs or activities receiving Federal financial assistance. Because many State and local government operations, such as courts, licensing, and legislative facilities and proceedings do not receive Federal funds, they are beyond the reach of section 504 but must still comply with Title II. Realistically this is a moot point because Title II provides protections to individuals with disabilities that are at least equal to those provided by the nondiscrimination provisions of title V of the Rehabilitation Act. Title II may not be interpreted to provide a lesser degree of protection to individuals with disabilities than is provided under this law.

The specific applications for publicly-funded improvements under California's guidelines can be found in Section 1.9.1.1 of the CBC and state:

    1.9.1.1 Application- *See Government Code commencing with Section 4450.*

    Publicly funded buildings, structures, sidewalks, curbs and related facilities shall be accessible to and usable by persons with disabilities as follows:

> **ADVISORY:** Not only does publicly funded mean state funds, but it also means county funds, municipal funds or the funds of any political subdivision of the state. When public funds are provided by other sources and transferred to a state, county, municipality or other political subdivision of the state, the entity that collects and controls the distribution of the public funds becomes the funding source, and subject to the requirements of this section.

1.9.1.1.1 All buildings, structures, sidewalks, curbs and related facilities constructed in the state by the use of state, county or municipal funds, or the funds of any political subdivision of the state.

1.9.1.1.2 All Buildings, structures and facilities that are leased, rented contracted, sublet or hired by any municipal county, or state division of government, or by a special district.

1.9.1.1.3 All publicly funded buildings used for congregate residences or for one- or two-family dwelling unit purposes shall conform to the provisions applicable to living accommodations.

1.9.1.1.4 All existing publicly funded buildings and facilities when alterations, structural repairs or additions are made to such buildings or facilities. For detailed requirements on existing buildings see CBC Chapter 11B, Division 2, Section 11B-202.

1.9.1.1.5. With respect to buildings, structures, sidewalks, curbs and related facilities not requiring a building permit, building standards published in the California Building Standards Code relating to access for persons with disabilities and other regulations adopted pursuant to Government Code Section 4450, and in effect at the time construction is commenced, shall be applicable.

When a project involves the use of public funds as defined, The entities involved in the project have a responsibility to the public; these entities must insure that the project complies with both State and Federal requirements, not just published State requirements. Any project that does not conform to the more stringent requirements of the ADA can expose California residents to possible legal and financial consequences. State and local government agencies and their contracted service providers must endeavor to ensure that each project meets both sets of guidelines or a violation of the ADA could exist.

Title II does not disturb other Federal or State laws that provide protection for individuals with disabilities at a level greater or equal to that provided by the ADA. It does, however, prevail over any conflicting State laws.

PRIVATELY FUNDED

Both State and Federal guidelines confer a high degree of responsibility upon all entities involved in a privately funded project. The ADA considers the failure to design and construct applicable accessibility features in facilities to be discriminatory to the disabled. Under an ADA violation, the Department of Justice has been know to hold all parties involved in the project liable (owner, designer, contractor, etc.), and then let these individual parties battle out ultimate responsibilities between themselves. This situation means that each party involved in the project has an individual responsibility to ensure that all applicable State and Federal guidelines are met, at least within their scope of involvement. These entities cannot rely on a State or local government agency to accept the blame for a violation of the law simply because the entities followed an agency approved/reviewed set of design drawings. The agency, however, could be held liable if it required the entities to build features that were in direct conflict with the ADA.

The specific applications for privately funded improvements under California's guidelines can be found in Section 1.9.1.2 of the CBC and state:

> **1.9.1.2 Application-.** *See Health and Safety Code commencing with Section 19952.*
>
> *All privately funded public accommodations, as defined, and commercial facilities, as defined, shall be accessible to persons with disabilities as follows:*
>
> **EXCEPTION:** *Certain types of privately funded multistory buildings do not require installation of an elevator to provide access above and below the first floor. See CBC Chapter 11B.*
>
> **1.9.1.2.1** *Any building, structure, facility, complex, or improved area, or portions thereof, which are used by the general public.*
>
> **1.9.1.2.2** *Any sanitary facilities which are made available for the public, clients or employees in such accommodations or facilities.*
>
> **1.9.1.2.3** *Any curb or sidewalk intended for public use that is constructed in this state with private funds.*
>
> **1.9.1.2.4** *All existing privately funded public accommodations when alterations, structural repairs or additions are made to such public accommodations as set forth under Chapter 11B of the CBC..*

MULTIPLE FUNDING SOURCES

A project that will be constructed utilizing multiple funding sources requires each of the individual sources to separately ensure compliance with applicable accessibility provisions. In addition, each of the funding sources must ensure that the most stringent provisions are singled-out and utilized between the aggregated requirements of all parties.

Public entities, by definition, can never be subject to Title III of the ADA, which covers only private entities. Conversely, private entities cannot be covered by Title II, which only pertains to State and local governments. There are many situations, however, in which both public and private entities stand in very close relation to each other, with the result that certain activities may be affected, at least indirectly, by both Titles II and III. The following illustrations are directly from the Title III Technical Assistance Manual by the Department of Justice, and will explain relevant considerations:

> *ILLUSTRATION 1: A State department of parks provides a restaurant in one of its State parks. The restaurant is operated by X Corporation under a concession agreement. As a public accommodation, X Corporation is subject to title III of the ADA. The State department of parks, a public entity, is subject to title II. The parks department is obligated to ensure by contract that the restaurant will be operated in a manner that enables the parks department to meet its title II obligations, even though the restaurant is not directly subject to title II.*

ILLUSTRATION 2: The City of W owns a downtown office building occupied by W's Department of Human Resources. The first floor is leased as commercial space to a restaurant, a newsstand, and a travel agency. The City of W, as a public entity, is subject to title II in its role as landlord of the office building. As a public entity, it cannot be subject to title III, even though its tenants are public accommodations that are covered by title III.

Where public and private entities act jointly, the public entity must ensure that the relevant requirements of title II are met; and the private entity must ensure compliance with title III.

ILLUSTRATION: The City of W engages in a joint venture with T Corporation to build a new professional football stadium. The new stadium would have to be built in compliance with the accessibility guidelines of both titles II and III. In cases where the standards differ, the stadium would have to meet the standard that provides the highest degree of access to individuals with disabilities.

The moral of this story is that you should never take it for granted that all of the parties involved in a multiple funded project have met their corresponding requirements. Comparisons should be made between State, Federal and local accessibility guidelines that may or may not have an ultimate influence on the finished product. The most stringent standards between all applicable sources must be determined and applied so that all parties involved can be assured of compliance will all applicable laws and guidelines.

Chapter 2

APPLICATION & ADMINISTRATION

Purpose   **11B-101**   *101*

General
This chapter references scoping and technical requirements for accessibility to sites, facilities, buildings, and elements by individuals with disabilities. The requirements are to be applied during the design, construction, additions to, and alteration of sites, facilities, buildings, and elements to the extent required.   **11B-101.1**   *101.1*

Scope
All areas of newly designed and newly constructed buildings and facilities and altered portions of buildings and facilities shall comply as required. The requirements are to be applied during the design, construction, additions to, and alterations of sites, facilities, buildings, and elements to the extent required.   **11B-101.1, 11B-201.1**   *36.401(a)(1), 101.1, 201.1*

> *ADVISORY: **General and Scope.** In addition to these requirements, covered entities must comply with the regulations issued by the U.S. Department of Justice and the U.S. Department of Transportation under the Americans with Disabilities Act. There are issues affecting individuals with disabilities which are not addressed by these requirements, but which are covered by the U.S. Department of Justice and the U.S. Department of Transportation regulations.*
>
> *These accessibility regulations are applicable to: 1) publicly funded buildings, structures, sidewalks, curbs and related facilities; 2) privately funded public accommodations and commercial facilities; and 3) public housing and private housing available for public use.*
>
> *These requirements are to be applied to all areas of a facility unless exempted, or where scoping limits the number of multiple elements required to be accessible. For example, not all medical care patient rooms are required to be accessible; those that are not required to be accessible are not required to comply with these requirements. However, common use and public use spaces such as recovery rooms, examination rooms, and cafeterias are not exempt from these requirements and must be accessible.*
>
> *Accessible features, accommodations and elements must comply with the requirements detailed.   In some cases this requires compliance with requirements in other parts of the building code. When additional scoping or technical requirements are located in other parts of the building code, the features, accommodations and elements must comply with those regulations and these requirements.*

Dimensions for Adults and Children
The technical requirements are based on adult dimensions and anthropometrics. Technical requirements based on children's dimensions and anthropometrics for drinking fountains, water closets, toilet compartments, lavatories and sinks, dining surfaces, and work surfaces are also included.   **11B-102**   *102*

Equivalent Facilitation

Nothing in these requirements prevents the use of designs, products, or technologies as alternatives to those prescribed, provided they result in substantially equivalent or greater accessibility and usability. In determining equivalent facilitation, consideration shall be given to means that provide for the maximum independence of persons with disabilities while presenting the least risk of harm, injury or other hazard to such persons or others. **202, 11B-103** *103*

Dimensional Conventions

Dimensions that are not stated as "maximum" or "minimum" are absolute. **11B-104.1** *104.1*

Construction and Manufacturing Tolerances

All dimensions are subject to conventional building industry tolerances except where the requirement is stated as a range with specific minimum and maximum end points. **11B-104.1.1** *104.1.1*

> *ADVISORY: **Construction and Manufacturing Tolerances.** Application of conventional industry tolerances must be on a case by-case, project-by-project basis. Predetermined guidelines for construction tolerances could unnecessarily encourage contractors and others to deviate from the access regulations found in the CBC and may wrongfully be viewed by some to have the effect of law.*
>
> *Conventional industry tolerances recognized by this provision include those for field conditions and those that may be a necessary consequence of a particular manufacturing process. Recognized tolerances are not intended to apply to design work. It is a good practice when specifying dimensions to avoid specifying a tolerance where dimensions are absolute. For example if a regulation requires "1-1/2 inches", avoid specifying "1-1/2 inches plus or minus X inches".*
>
> *Where the requirement states a specified range, such as when grab bars must be installed between 33 inches and 36 inches, the range provides the adequate tolerance and therefore no tolerance outside of the range at either end point is permitted.*
>
> *Where a requirement is a minimum or a maximum dimension that does not have two specific minimum and maximum end points, tolerances may apply. Where an element is to be installed at the minimum or maximum permitted dimension, such as "15 inches minimum" or "5 pounds maximum", it would not be good practice to specify "5 pounds (plus X pounds) or "15 inches (minus X inches)." Rather, it would be good practice to specify a dimension less than the required maximum (or more than the required minimum) by the amount of the expected field or manufacturing tolerance and not to state any tolerance in conjunction with the specified dimension. An element designed to be constructed at either the maximum or minimum permitted dimensions puts the construction at risk if construction errors result in a violation of the standards. In other words, dimensions noted in accessibility provisions as "minimum" or "maximum" should not be considered dimensions for design as they represent the limits of a requirement To be sure that field tolerances result in usable construction, notes and dimensions in construction documents should anticipate expected tolerances so that a required dimensional range is not exceeded by a finish or variation in construction practice.*

> *Specifying dimensions in design in the manner above will better insure that facilities and elements accomplish the level of accessibility intended by these requirements. It will also more often produce and end result of strict and literal compliance with the stated requirements and eliminate enforcement difficulties and issues that might otherwise arise. Information on specific tolerances may be available from industry and trade organizations, code groups and building officials, and published references.*

Calculation of Percentages
Where the required number of elements or facilities to be provided is determined by calculation of ratios or percentages and remainders or fractions result, the next greater whole number of such elements or facilities shall be provided. Where the determination of the required size or dimension of an element or facility involves ratios or percentages, rounding down for values less than on e half shall be permitted. **11B-104.2** *104.2*

Figures
Unless specifically stated otherwise, figures are provided for informational purposes only. **11B-104.3** *104.3*

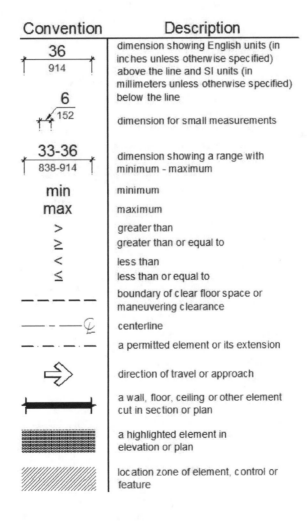

Convention	Description
36 / 914	dimension showing English units (in inches unless otherwise specified) above the line and SI units (in millimeters unless otherwise specified) below the line
6 / 152	dimension for small measurements
33-36 / 838-914	dimension showing a range with minimum - maximum
min	minimum
max	maximum
>	greater than
≥	greater than or equal to
<	less than
≤	less than or equal to
— — — —	boundary of clear floor space or maneuvering clearance
— – — ℄	centerline
— – — – —	a permitted element or its extension
⇨	direction of travel or approach
┝━━━┥	a wall, floor, ceiling or other element cut in section or plan
▨	a highlighted element in elevation or plan
▨	location zone of element, control or feature

Fig. CD-C2A
Graphic Conventions for Figures

Maintenance of Accessible Features

A public accommodation shall maintain in operable working condition those features of facilities and equipment that are required to be accessible to and usable by persons with disabilities. Isolated or temporary interruptions in service or accessibility due to maintenance or repairs shall be permitted. **11B-108** *36.211*

Application Based on Building Use

Where a site, building, facility, room or space contains more than one use, each portion shall comply with the applicable requirements for that use. **11B-201.2** *201.2*

Temporary and Permanent Structures

These requirements shall apply to temporary and permanent structures. **11B-201.3** *201.3*

The provisions of these regulations shall apply to any portable buildings leased or owned by a school district, and shall also apply to temporary and emergency buildings and facilities. Temporary buildings and facilities are not of permanent construction but are extensively used or are essential for public use for a period of time. Examples of temporary buildings or facilities include, but are not limited to: reviewing stands, temporary classrooms, bleacher areas, exhibit areas, temporary banking facilities, temporary health screening services, or temporary safe pedestrian passageways around a construction site. **1.9.1**

> **NOTE:** California defines temporary as "buildings and facilities intended for use at one location for not more than one year and seats intended for use at one location for not more than 90 days. **202**

> ***ADVISORY:** **Temporary and Permanent Structures.** Temporary buildings or facilities covered by these requirements include, but are not limited to, reviewing stands, temporary classrooms, bleacher areas, stages, platforms and daises, fixed furniture systems, wall systems, and exhibit areas, temporary banking facilities, and temporary health screening facilities. Structures and equipment directly associated with the actual processes of construction are not required to be accessible as detailed below.*

Construction Support Facilities

These requirements shall apply to temporary or permanent construction support facilities for uses and activities not directly associated with the actual processes of construction, including but not limited to offices, meeting rooms, plan rooms, other administrative or support functions. When provided, toilet and bathing facilities serving construction support facilities shall comply as required for accessible toilet facilities. When toilet and bathing facilities are provided by portable units, at least one of each type shall be accessible and connected to the construction support facilities it serves by an accessible route. **11B-201.4**

> **EXCEPTION:** During construction an accessible route shall not be required between site arrival points or the boundary of the area of construction and the entrance to the construction support facilities if the only means of access between them is a vehicular way not providing pedestrian access.

> ***ADVISORY:** **Construction Support Facilities.** This section clarifies that construction support facilities located on or adjacent to a construction site, but **not** directly associated with the actual processes of construction must comply with the accessibility provisions. Construction support facilities may include but*

> *are not limited to, offices, meeting rooms, plan rooms and other administrative and support spaces. Toilet facilities serving these construction support facilities, including portable units, must also comply with the accessibility provisions of this chapter. While an accessible route is required between construction support facilities and the toilet and parking facilities serving them, an accessible route is not required between site arrival points and construction support facilities when the only means of reaching the construction support facilities is a vehicular way.*

Religious Facilities

Religious facilities shall be accessible in accordance with the provisions of this code. Where specific areas within religious facilities contain more than one use, each portion shall comply with the applicable requirements for that use.   **11B-244.1**

> *ADVISORY: Religious Facilities. Religious facilities are exempt from federal ADA requirements, however no exception is provided for these types of facilities under the California Building Code (CBC) due to pre-existing and more stringent state regulations. Religious practice may restrict general access to specific areas, for example a raised altar area; however access for persons with disabilities to these areas may be required. The CBC requires raised areas of the facility to provide access by ramp, special access lift or elevator. Religious beliefs and practices notwithstanding, architectural barriers are not permitted at participation areas.*

Commercial Facilities Located in Private Residences

When a commercial facility is located in a private residence, that portion used exclusively in the operation of the commercial facility or that portion used both for the commercial facility and for residential purposes is covered by the new construction and alteration requirements.   **11B-245.2**

> **EXCEPTION:**   That portion of the residence used exclusively as a residence is not required to be accessible.

The portion of the residence covered extends to those elements used to enter the commercial facility, including the front sidewalk, if any, the door or entryway, and hallways; and those portions of the residence, interior or exterior, available to or used by employees or visitors of the commercial facility, including restrooms.   **11B-245.3**

Employee Work Areas

> **DEFINITION:** All or any portion of a space used only by employees and only for work. Corridors, toilet rooms, kitchenettes and break rooms are not employee work areas.

Spaces and elements within employee work areas shall only be required to comply with the requirements for an Accessible Route in Employee Work Areas, Accessible Means of Egress and Fire Alarm Systems in Employee Work Areas, and shall be designed and constructed so that individuals with disabilities can approach, enter, and exit the employee work area.   **11B-203.9**   *203.9*

Common use circulation paths within employee work areas shall be accessible routes consisting of one or more of the following components: walking surfaces with a running slope not steeper than 1:20, doorways, ramps, curb ramps excluding flared sides, elevators, and platform lifts. All components of an accessible route shall comply with the applicable requirements. **11B-206.2.8, 11B-402.2**   *206.2.8, 402.2*

EXCEPTIONS:

1. **Reserved.**

2. Common use circulation paths located within employee work areas that are an integral component of work area equipment shall not be required to provide accessible routes.

3. Common use circulation paths located within exterior employee work areas that are fully exposed to the weather shall not be required to provide accessible routes.

> *ADVISORY. **Employee Work Areas Exception 2.** Large pieces of equipment, such as electric turbines or water pumping apparatus, may have stairs and elevated walkways used for overseeing or monitoring purposes which are physically part of the turbine or pump. However, passenger elevators used for vertical transportation between stories are not considered "work area equipment" as defined.*

> *ADVISORY. **Employee Work Areas.** Although areas used exclusively by employees for work are not required to be fully accessible, consider designing such areas to include non-required turning spaces, and provide accessible elements whenever possible. Under the ADA, employees with disabilities are entitled to reasonable accommodations in the workplace; accommodations can include alterations to spaces within the facility. Designing employee work areas to be more accessible at the outset will avoid more costly retrofits when current employees become temporarily or permanently disabled., or when new employees with disabilities are hired. Contact the Equal Employment Opportunity Commission (EEOC) at www.eeoc.gov for information about title 1 of the ADA prohibiting discrimination against people with disabilities in the workplace.*

Privately-Funded Multi-Story Building Exception

For the general privately funded multistory building exception applicable to new construction and alterations, see Chapter 3, "SCOPING AND TECHNICAL REQUIREMENTS APPLICABLE TO PUBLIC BUILDINGS, PUBLIC ACCCOMMODATIONS, COMMERCIAL FACILITIES AND PUBLIC HOUSING", Section 2, "ACCESSIBLE ROUTES", Item "D".

UNREASONABLE HARDSHIP

Both State and Federal accessibility guidelines have specific provisions allowing certain features of a facility to be constructed without providing for full and strict compliance with the minimum requirements. These allowable deviations are called "Exceptions". Many of these exceptions have conditions attached that must be satisfied in order to be utilized. In many cases, California's guidelines require that a determination of "unreasonable hardship" be found in order to be eligible for a specific exception. California's definition and conditions for this unreasonable hardship can be found in Section 202 of the California Building Code (CBC) and are detailed below:

> ***UNREASONABLE HARDSHIP.*** *When the enforcing agency finds that compliance with the building standard would make the specific work of the*

project affected by the building standard infeasible, based on an overall evaluation of the following factors:

1. *The cost of providing access.*

2. *The cost of all construction contemplated.*

3. *The impact of proposed improvements on financial feasibility of the project.*

4. *The nature of the accessibility which would be gained or lost.*

5. *The nature of the use of the facility under construction and its availability to persons with disabilities.*

The details of any finding of unreasonable hardship shall be recorded and entered in the files of the enforcing agency. **202**

California presently allows the application for hardship determinations for specific exceptions in the Alteration of Existing Facilities, and does not specifically disallow hardship for New Construction. In New Construction however, the ADA has an altogether different condition that must be satisfied in order to obtain variances from "hardship".

Unreasonable Hardship in New Construction
The ADA does not allow cost to be a factor in determining whether or not a facility, or portion thereof, be made accessible in new construction. Under the ADA, the requirement that new construction be accessible does not apply where an entity can demonstrate that it is **"structurally impracticable"** to meet the requirements of the regulation. To explain structural impracticability, Section 4.1.1(5)(a) of the ADA offers the following:

(c) Exception for structural impracticability.

(1) Full compliance with the requirements of this section is not required where an entity can demonstrate that it is structurally impracticable to meet the requirements. Full compliance will be considered structurally impracticable only in those rare circumstances when the unique characteristics of terrain prevent the incorporation of accessibility features.

(2) If full compliance with this section would be structurally impracticable, compliance with this section is required to the extent it is not structurally impracticable. In that case, any portion of the facility that can be made accessible shall be made accessible to the extent that it is not structurally impracticable.

(3) If providing accessibility in conformance with this section to individuals with certain disabilities (e.g., those who use wheelchairs) would be structurally impracticable, accessibility shall nonetheless be insured to persons with other types of disabilities (e.g., those who use crutches or have sight, hearing, or mental impairments) in accordance with this section. 36.401(c)

Further explanation as to the intent of this provision can be found in the July 26, 1991 issue of the Federal Register (28 CFR Part 36, Section 36.401 preamble):

"The limited structural impracticability exception means that it is acceptable to deviate from accessibility requirements only where unique characteristics of terrain prevent the incorporation of accessibility features and where providing accessibility would destroy the physical integrity of a facility. A situation in which a building must be built on stilts because of its location in marshlands or over water is an example in which the exception for structural impracticability would apply.

This exception to accessibility requirements should not be applied to situations in which a facility is located in "hilly" terrain or on a plot of land upon which there are steep grades. In such circumstances, accessibility can be achieved without destroying the physical integrity of a structure, and is required in the construction of new facilities".

The Department of Justice was adamant that cost would not be considered a factor in providing accessibility in new construction. Prior to the development of the final Federal Guidelines, the government conducted a Preliminary Regulatory Impact Analysis (PRIA). The findings of this study indicated that if the proper incorporation of accessibility features were provided for in new construction from the conceptual design phase forward, that only an additional one percent (1%) increase in the overall cost of a project would be incurred.

Even if a determination of "unreasonable hardship" was granted for a new construction project by an enforcing agency in California, it would not preclude the possibility of future legal action for non-compliance of Federal design and construction requirements under the ADA. If a project is built incorrectly, all the parties involved in the final product could be held liable, including the owner, architect and contractor(s). The "Title III Technical Assistance Manual" by the Department of Justice illustrates this point in the following example:

ILLUSTRATION: "M owns a large piece of land on which he plans to build many facilities, including office buildings, warehouses, and stores. The eastern section of the land is fairly level, the central section of the land is extremely steep, and the western section of the land is marshland. M assumes that he only need comply with the new construction requirements in the eastern section. He notifies his architect and construction contractor to be sure that all buildings in the eastern section are built in full compliance with ADAAG. He further advises that no ADAAG requirements apply in the central and western sections.
M's advice as to two of the sections is incorrect. The central section may be extremely steep, but that is not sufficient to qualify for the "structural impracticability" exemption under the ADA. M should have advised his contractor to grade the land to provide and accessible slope at the entrance and apply all new construction requirements in the central section. M's advice as to the western section is also incorrect. Because the land is marshy, provision of an accessible grade-level entrance may be structurally impracticable. This is one of the rare situations in which the exception applies, and full compliance with ADAAG is not required. However, M should have advised his contractor to nevertheless construct the facilities in compliance with other ADAAG requirements, including provisions of features that serve individuals who use crutches or who have vision or hearing impairments. For instance, the facility needs to have stairs and railings that comply with ADAAG, and it should comply with the ADAAG signage and alarm requirements, as well.
Who is liable for violation of the ADA in the above example? *Any of the entities involved in the design and construction of the central and western sections might be liable. Thus, in any lawsuit, M (the owner), the architect, and*

the construction contractor may all be held liable in an ADA lawsuit".
(Department of Justice Title III Technical Assistance Manual, Sec. III-5.1000)

At this point it is important to note that in adoption of the 2010 ADA Standards for Accessible Design by California into the 2013 CBC, that the exception for structural impracticability was intentionally not included. This means that in a typical application to a building department a request for a determination of unreasonable hardship based on structural impracticability would not be allowed.

ACCESSIBILITY FOR EXISTING BUILDINGS AND FACILITIES

Existing Buildings and Facilities   11B-202   *202*

General
Additions and alterations to existing buildings or facilities shall comply as detailed.
11B-202.1   *202.1*

> **ADVISORY: Additions.** *An addition to site improvements, such as a new patio or playground, may require an accessible path of travel from the site arrival point to restrooms, drinking fountains, signs, public telephones (if available) serving the addition.*

Alterations
Where existing elements or spaces are altered, each altered element or space shall comply with applicable requirements as detailed, including path of travel requirements.
11B-202.3   *202.3*

> **EXCEPTIONS:**
>
> 1. **Reserved.**
>
> 2. **Technically Infeasible.**   In alterations, where the enforcing authority determines compliance with applicable requirements is technically infeasible, the alteration shall comply with the requirements to the maximum extent feasible. The details of the finding that full compliance with the requirements is technically infeasible shall be recorded and entered into the files of the enforcing agency.
>
> 3. Residential dwelling units not required to be accessible in compliance with this code shall not be required to comply with this provision.

> **ADVISORY: Alterations.** *Although covered entities are permitted to limit the scope of an alteration to individual elements, the alteration of multiple elements within a room or space may provide a cost-effective opportunity to make the entire room or space accessible. Any elements or spaces of the building or facility that are required to comply with these requirements must be made accessible within the scope of the alteration, to the maximum extent feasible. If providing accessibility in compliance with these requirements for people with one type of disability (e.g., people who use wheelchairs) is not feasible, accessibility*

must still be provided in compliance with the requirements for people with other types of disabilities (e.g., people who have hearing impairments or who have vision impairments) to the extent that such accessibility is feasible.

Prohibited Reduction in Access
An alteration that decreases or has the effect of decreasing the accessibility of a building or facility below the requirements for new construction at the time of the alteration is prohibited.   **11B-202.3.1**   *202.3.1*

Extent of Application
An alteration of an existing element, space, or area of a building or facility shall not impose a requirement for accessibility greater than required for new construction.   **11B-202.3.2** *202.3.2*

Alteration of Single Elements
If alterations of single elements, when considered together, amount to an alteration of a room or space in a building or facility, the entire room or space shall be made accessible.
11B-202.3.3

Path of Travel Requirements in Alterations, Additions and Structural Repairs
When alterations or additions are made to existing buildings or facilities, an accessible path of travel to the specific area of alteration or addition shall be provided. The primary accessible path of travel shall include: **11B-202.4**   *202.4*

1.   A primary entrance to the building or facility,

2.   Toilet and bathing facilities serving the area,

3.   Drinking fountains serving the area,

4.   Public telephones serving the area, and

5.   Signs.

   EXCEPTIONS:

   1.   Residential dwelling units shall comply with the requirements for "Alterations to Individual Dwelling Units" contained in Section 43, "RESIDENTIAL FACILITIES".

   2.   If the following elements of a path of travel have been constructed or altered in compliance with the accessibility requirements of the immediately preceding edition of the California Building Code, it shall not be required to retrofit such elements to reflect the incremental changes in this code solely because of an alteration to an area served by those elements of the path of travel:

      1.   A primary entrance to the building or facility,

      2.   Toilet and bathing facilities serving the area,

      3.   Drinking fountains serving the area,

      4.   Public telephones serving the area, and

5. Signs.

> **ADVISORY: Path of Travel.** Both the State and Federal obligations for a "path of travel" include "…a pedestrian passage by means of which the altered area may be approached , entered, and exited, and which connects the altered area with an exterior approach (including sidewalks, streets and <u>parking areas</u>), an entrance to the facility, and other parts of the facility". I have never been able to understand why California has never listed accessible parking as a primary requirement in the path of travel obligations under this section on alterations to existing facilities. Because of this, many building agencies do not require accessible parking to be a requirement of a path of travel obligation in an alteration, and as a consequence, there have been an uncountable number of lawsuits against building owners and/or their tenants for not having compliant accessible parking. In my experience, accessible parking and an accessible route from parking to an accessible entrance should be a primary concern, if not THE primary concern, in an alteration because not only is it an exterior route requirement, but non-compliant disabled parking on a site is the primary reason in alleged disabled accessibility discrimination and law suits in California.

3. Additions or alterations to meet accessibility requirements consisting of one or more of the following items shall be limited to the actual scope of work of the project and shall not be required to comply with path of travel requirements:

   1. Altering one building entrance.

   2. Altering one existing toilet facility.

   3. Altering existing elevators.

   4. Altering existing steps.

   5. Altering existing handrails.

4. Alterations solely for the purpose of barrier removal undertaken pursuant to the requirements of the Americans with Disabilities Act (Public Law 101-336, 28 C.F.R., Section 36.304) or the accessibility requirements of this code as those requirements or regulations now exist or are hereafter amended consisting of one or more of the following items shall be limited to the actual scope of work of the project and shall not be required to comply with path of travel requirements:

   1. Installing ramps.

   2. Making curb cuts in sidewalks and entrance.

   3. Repositioning shelves.

   4. Rearranging tables, chairs, vending machines, display racks, and other furniture.

   5. Repositioning telephones.

   6. Adding raised markings on elevator control buttons.

7. Installing flashing alarm lights.

8. Widening doors.

9. Installing offset hinges to widen doorways.

10. Eliminating a turnstile or providing an alternative accessible route.

11. Installing accessible door hardware.

12. Installing grab bars in toilet stalls.

13. Rearranging toilet partitions to increase maneuvering space.

14. Insulating lavatory pipes under sinks to prevent burns.

15. Installing a raised toilet seat.

16. Installing a full-length bathroom mirror.

17. Repositioning the paper towel dispenser in a bathroom.

18. Creating designated accessible parking spaces.

19. Removing high-pile, low-density carpeting.

5. Alterations of existing parking lots by resurfacing and/or restriping shall be limited to the actual scope of work of the project and shall not be required to comply with path of travel requirements.

6. The addition or replacement of signs and/or identification devices shall be limited to the actual scope of work of the project and shall not be required to comply with path of travel requirements.

7. Projects consisting only of heating, ventilation, air conditioning, reroofing, electrical work not involving placement of switches and receptacles, cosmetic work that does not affect items regulated by this code, such as painting, equipment not considered to be a part of the architecture of the building or area, such as computer terminals and office equipment shall not be required to comply with path of travel requirements, unless they affect the usability of the building or facility.

8. When the adjusted construction cost is less than or equal to the current valuation threshold, as defined, the cost of compliance with path of travel requirements shall be limited to 20 percent of the adjusted construction cost of alterations, structural repairs or additions. When the cost of full compliance with path of travel requirements would exceed 20 percent, compliance shall be provided to the greatest extent possible without exceeding 20 percent.

> **ADVISORY:** *Definition of **VALUATION THRESHOLD.** The valuation threshold is adjusted each year in January using the Engineering News Record 20 Cities Construction Cost Index. Valuation thresholds for the current year and recent years dating back to 2000 are available on the Division of the State Architect web site at:*
> *http://www.dgs.ca.gov/dsa/Programs/progAccess/threshold.aspx.*

When the adjusted construction cost exceeds the current valuation threshold, as defined, and the enforcing agency determines the cost of compliance with path of travel requirements is an unreasonable hardship, as defined, full compliance with path of travel requirements shall not be required. Compliance shall be provided by equivalent facilitation or to the greatest extent possible without creating an unreasonable hardship; but in no case shall the cost of compliance be less than 20 percent of the adjusted construction cost of alterations, structural repairs or additions. The details of the finding of unreasonable hardship shall be recorded and entered into the files of the enforcing agency and shall be subject to CBC Chapter 1, Section 1.9.1.5, Special Conditions for Persons with Disabilities Requiring Appeals Action Ratification.

For the purposes of this exception, the adjusted construction cost of alterations, structural repairs or additions shall not include the cost of alterations to path of travel elements required to comply with path of travel requirements.

In choosing which accessible elements to provide, priority should be given to those elements that will provide the greatest access in the following order:

1. An accessible entrance;

2. An accessible route to the altered area;

3. At least one accessible restroom for each sex;

4. Accessible telephones;

5. Accessible drinking fountains; and

6. When possible, additional accessible elements such as parking, storage and alarms.

If an area has been altered without providing an accessible path of travel to that area, and subsequent alterations of that area or a different area on the same path of travel are undertaken within three years of the original alteration, the total cost of alterations to the areas on that path of travel during the preceding three-year period shall be considered in determining whether the cost of making that path of travel accessible is disproportionate.

9. Certain types of privately funded, multistory buildings and facilities were formerly exempt from accessibility requirements above and below the first floor under the CBC, but as of, April 1, 1994, are no longer exempt due to more restrictive provisions in

the federal Americans with Disabilities Act. In alteration projects involving buildings and facilities previously approved and built without elevators, areas above and below the ground floor are subject to the 20-percent disproportionality provisions described in Exception 8, above, even if the value of the project exceeds the valuation threshold in Exception 8. The types of buildings and facilities are:

1. Office buildings and passenger vehicle service stations of three stories or more and 3,000 or more square feet (279 m^2) per floor.

2. Offices of physicians and surgeons.

3. Shopping centers.

4. Other buildings and facilities three stories or more and 3,000 or more square feet (279 m^2) per floor if a reasonable portion of services sought and used by the public is available on the accessible level.

For the general privately funded multistory building exception applicable to new construction and alterations, see Chapter 3, "SCOPING AND TECHNICAL REQUIREMENTS APPLICABLE TO PUBLIC BUILDINGS, PUBLIC ACCCOMMODATIONS, COMMERCIAL FACILITIES AND PUBLIC HOUSING", Section 2, "ACCESSIBLE ROUTES", Item "D".

The elevator exception set forth in this section does not obviate or limit in any way the obligation to comply with the other accessibility requirements in this code. For example, floors above or below the accessible ground floor must meet the requirements of this section except for elevator service. If toilet or bathing facilities are provided on a level not served by an elevator, then toilet or bathing facilities must be provided on the accessible ground floor.

> **ADVISORY: Path of travel requirements in alterations, additions and structural repairs. New access regulations made effective after the date of last construction may require owners to upgrade their facility to comply with the current regulations during alterations, additions or structural repair.**

Alterations to Qualified Historic Buildings and Facilities
Alterations to a qualified historic building or facility shall comply with the State Historical Building Code, Part 8, Title 24, of the California Code of Regulations (See "HISTORIC PRESERVATION", in this Chapter. **11B-202.5** *202.5*

> **EXCEPTION: Reserved.**

DETERMINING PRIMARY OBLIGATIONS IN ALTERATIONS
It is extremely important to understand the difference between what is required in the specific area of alteration, structural repair or addition; and what additional modifications may be required to alter the primary path of travel to this area of alteration including the

toilet and bathing facilities, drinking fountains, public telephones and signs serving this area. The requirement to make the *specific area of alteration* accessible to the maximum extent feasible should not be confused with the requirement to provide modifications to the *primary path of travel, toilet and bathing facilities, drinking fountains, public telephones and signs serving the area of alteration.* These are totally separate issues and must be dealt with as such.

Specific area of alteration, addition, or structural repair-The first step in the process is to determine the extent of the proposed construction in the specific area of alteration, addition, or structural repair, and the total cost associated with providing all of these desired improvements. For an alteration that has existing toilet and bathing facilities, drinking fountain(s), public telephone(s) or signs within the specific area of alteration, or some or any of these improvements outside the specific area of alteration, this means all construction that is planned <u>without</u> including the toilet and bathing facilities, drinking fountains, public telephones and signs, or the primary path of travel from the exterior approaches through the entrance of the building or facility to the actual boundaries of the alteration. This total cost must include every item necessary to complete the desired alteration, including electrical, HVAC, carpeting, paint, ceiling tiles, lever-handled door hardware, walls, partitions, etc.

Primary path of travel, toilet and bathing facilities, drinking fountains, public telephones and signs -The second step in the process is to determine what modifications will be necessary to provide an accessible path of travel <u>to</u> the specific area of alteration; and what the associated costs will be to make these modifications. To determine the extent of what constitutes a required path of travel in an alteration, the following definition must be understood:

> **PATH OF TRAVEL. [DSA-AC]** *An identifiable accessible route within an existing site, building or facility by means of which a particular area may be approached, entered and exited, and which connects a particular area with an exterior approach (including sidewalks, streets, and parking areas), an entrance to the facility, and other parts of the facility. When alterations, structural repairs or additions are made to existing buildings or facilities, the term "path of travel" also includes the toilet and bathing facilities, telephones, drinking fountains and signs serving the area of work.*

Note should be taken that it is only necessary to provide a single, primary path of travel to the specific area of alteration. This path of travel should be the most direct common route from the specific area of alteration to the primary entrance and out to the exterior approaches; and must include access to sanitary facilities, drinking fountains, public telephones and signs serving the specific area of alteration. The following examples should help clarify the differences between these two areas:

<u>EXAMPLE #1</u>
This example is an existing privately funded single-story office building with multiple tenants. The tenant in space "H" wishes to totally remodel their leased area by reconfiguring interior rooms, adding a reception area, new ceiling, carpet, wallpaper, and adding a fire sprinkler system. The "specific area of alteration" is delineated as shown on the following page:

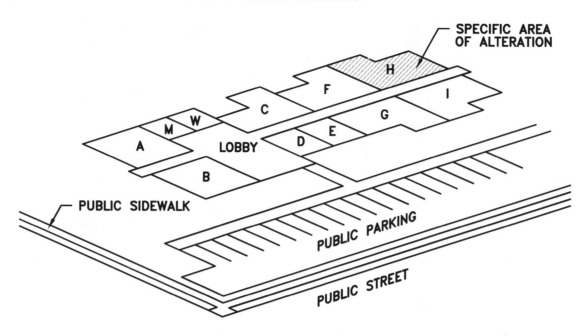

All of the improvements within this specific area of alteration must be made "readily accessible" to the maximum extent feasible. This will require providing accessible doors and hardware, appropriate corridor widths, light switches and electrical controls within reach ranges, an accessible counter at the reception area, appropriate signage, etc.

The tenant in this example must now determine what improvements and/or modifications are necessary to provide an accessible primary path of travel to the specific area of alteration. The next diagram delineates the minimum required considerations:

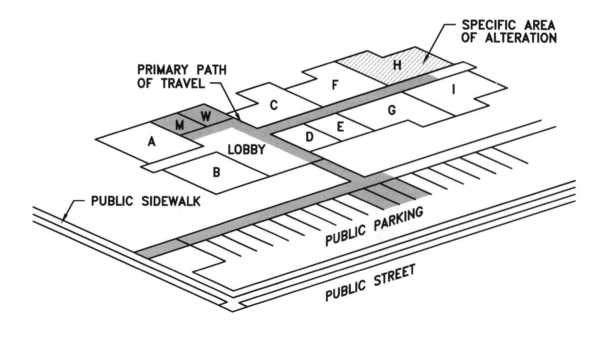

Note that this path of travel includes the existing sanitary facilities and access through the lobby to the primary entrance of the building. The exterior portion includes access to public parking and sidewalk. In this example, the tenant will be required to extend the existing frontage sidewalk out to their property line to join the existing public sidewalk and provide accessible parking stalls. The tenant will also be required to modify any other deficiencies, including the restrooms, drinking fountains, signage and public telephones in these areas to the maximum extent feasible.

EXAMPLE #2

This example is an existing privately funded two-story office building with multiple tenants. Vertical access between floors is provided by means of a full passenger elevator. The tenant in space "2E" wishes to completely remodel their leased area.

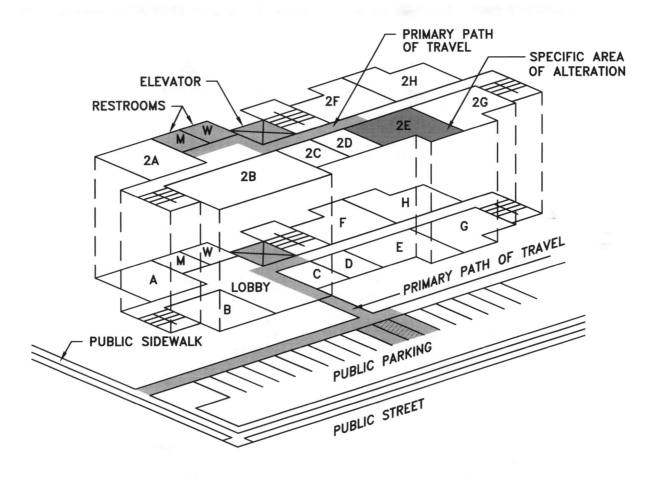

All construction in the tenant's specific area of alteration must be "readily accessible" to the maximum extent feasible. A primary path of travel must be provided from the specific area of alteration to the primary entrance of the building and out to the exterior approaches. Since this building has a full passenger elevator that provides the primary path of travel between floors, the stairs do not need to be accessible. The elevator itself, however, must meet all of the requirements and be "readily accessible" to the maximum extent feasible. The restrooms, drinking fountains, signs and telephones serving the specific area of alteration must be modified to the maximum extent feasible if deficiencies from required accessibility features exist.

EXAMPLE #3

This example is an existing privately funded two-story office building with multiple tenants. Vertical access between floors is provided by means of a full passenger elevator. The tenant in space "2E" wishes to remodel a specific portion of their office by removing some existing walls and then installing partitioned cubicles in the widened area. Upon evaluating the restrooms down the hall from the office, it is determined that physical and legal constraints will prevent these restrooms from being made fully accessible. The restrooms on the first floor, however, can be made accessible by widening the rooms out further into the lobby area.

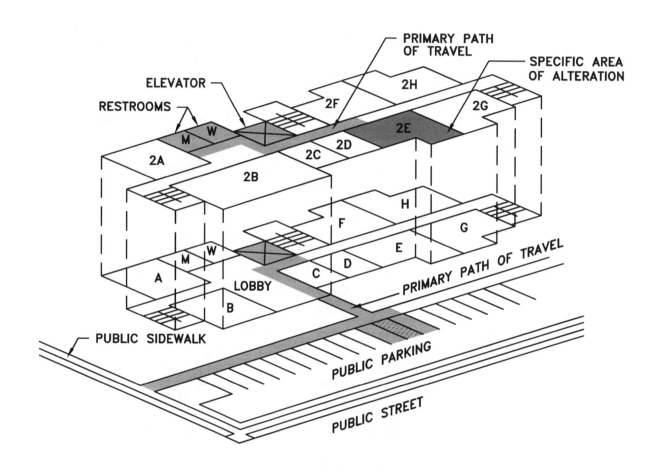

The specific area of alteration must be made "readily accessible" to the maximum extent feasible. An accessible path of travel from this specific area of alteration must be provided through the unaltered office areas, through the building and out to the exterior approaches. Due to the fact that this building has a full passenger elevator that provides the primary path of travel between floors, the stairs do not need to be accessible. The elevator itself, however, must be made accessible to the "maximum extent feasible". Since the restrooms serving the second floor cannot be made accessible, and the first floor restrooms are within a reasonable distance and can be made accessible, these first floor restrooms will require any necessary modifications. A sign must be placed at the restrooms on the second floor explaining that accessible restrooms are located in the lobby on the first floor.

Additional Considerations - Each of the previous examples only extended the primary path of travel obligation to the existing public sidewalk; this was intentional. As was stated earlier, any alteration must comply with the requirements for new construction, to the maximum extent feasible. Both State and Federal guidelines require an accessible route to accessible building entrances and between the public way. Neither the State nor the Federal guidelines, however, require this route to extend beyond the boundaries of the site. The actual wording in both sets of guidelines is the same:

> ### Site Arrival Points.
> *"At least one accessible route shall be provided <u>within the site</u> from accessible parking spaces and accessible passenger loading zones; public streets and sidewalks; and public transportation stops, to the accessible building or facility entrance they serve."* **11B-206.2.1** *206.2.1*

Neither State nor Federal law requires a private entity to extend their accessibility obligations into public right-of-way outside of the actual boundaries of their property. This has always caused consternation between building owners and local government jurisdictions. In some instances, local building agencies will require a permit applicant to include modifications to improvements in the public right-of-way as a "condition" to obtaining a building permit. Although the State accessibility guidelines do not provide further guidance regarding this situation, the ADA does; the following passage is from the "Section-by-Section Analysis and Response to Comments" section of the July 26, 1991 issue of the Federal Register, 28 CFR Part 36, Part III, and provides the Federal interpretation to this issue:

> *"Ensuring access to a newly constructed facility will include providing access to the facility from the street or parking lot, to the extent the responsible entity has control over the route from those locations. In some cases, the private entity will have no control over access at the point where streets, curbs, or sidewalks already exist, and in those instances the entity is encouraged to request modifications to a sidewalk, including installation of curb cuts, from a public entity responsible for them. However, as some commenters pointed out, there is no obligation for a private entity subject to title III of the ADA to seek or ensure compliance by a public entity with title II. Thus, although a locality may have an obligation under title II of the Act to install curb cuts at a particular location, that responsibility is separate from the private entity's title III obligation, and any involvement by a private entity in seeking cooperation from a public entity is purely voluntary in this context."*

This brings up another previously regularly encountered situation, which we will call "The Path of Travel to No Where".

"The Path of Travel to No Where"- California building agencies under previous access requirements frequently required existing facilities to provide a path of travel from a building or facility through vehicular ways to other buildings on the site and to the nearest adjacent public street regardless of excessive elevation differentials, sometimes phenomenal distances between site improvements through vehicular ways, streets without sidewalks, etc. Because of new accessible route requirements, this often challenging situation is no longer an issue:

> ### Within a Site
> *At least one accessible route shall connect accessible buildings, accessible facilities, accessible elements, and accessible spaces that are on the same site.*

> **EXCEPTION:** An accessible route shall not be required between accessible buildings, accessible facilities, accessible elements, and accessible spaces if the only means of access between them is a vehicular way not providing pedestrian access. **11B-206.2.2** *206.2.2*

Unreasonable Hardship in Alterations

Both State and Federal accessibility guidelines have specific provisions allowing certain features of a facility to be constructed without providing for full and strict compliance with the minimum requirements. These allowable deviations are called "Exceptions". Many of these exceptions have conditions attached that must be satisfied in order to be utilized. In many cases, California's guidelines require that a determination of "unreasonable hardship" be found in order to be eligible for a specific exception.

California's definition and conditions for this unreasonable hardship can be found in Section 202 of the California Building Code (CBC) and are detailed below:

UNREASONABLE HARDSHIP. When the enforcing agency finds that compliance with the building standard would make the specific work of the project affected by the building standard infeasible, based on an overall evaluation of the following factors:

> 1. The cost of providing access.
> 2. The cost of all construction contemplated.
> 3. The impact of proposed improvements on financial feasibility of the project.
> 4. The nature of the accessibility which would be gained or lost.
> 5. The nature of the use of the facility under construction and its availability to persons with disabilities.

The details of any finding of unreasonable hardship shall be recorded and entered in the files of the enforcing agency. **202**

Hardship when alteration costs <u>do not exceed</u> the ENR Average Construction Cost Index

In the past, California had allowed builders to apply for an "Unreasonable Hardship", if the cost of a proposed alteration did not exceed a specific valuation threshold known as the "ENR US20 Cities, average construction cost index of 3372.02" (Engineering News Record, McGraw Hill Publishing Company). In January of 2014, this threshold limit was $143,303.00. If approved, this hardship allowed the builder to limit accessibility modifications to the actual work of the project (the specific area of alteration) and did not require additional modifications to the entrance, path of travel, restrooms, drinking fountains, signs and public telephones. Since April 1, 1994, however, any builder that applies for and receives a determination of unreasonable hardship is required to spend up to twenty percent (20%)of the cost of the alteration making additional modifications to the entrance, path of travel, etc. In other words, since April 1, 1994, there were no more total exemptions from path of travel requirements with a hardship determination.

To explain this requirement, lets look at a proposed tenant improvement that will cost $50,000.00. This $50,000.00 is the cost to remodel the specific tenant's space alone. Upon reviewing the entrance, path of travel, restrooms, drinking fountains and public telephones that service the proposed alteration, it is determined that it will cost an additional $20,000.00 to make these features accessible. If the builder applies for a determination of unreasonable hardship from the local building agency, the builder will still have to spend up to $10,000.00, (20% of $50,000.00 is $10,000.00) making these additional modifications before the cost of the *overall alteration* is considered *disproportionate*. Under previous hardship allowances, the builder would only have been required to complete the original $50,000.00 remodel. Under the current regulations, the builder must spend up to $60,000.00 even with a hardship determination in this example.

If the facilities public and common use improvements already comply with the combined State and Federal access guidelines, this additional 20% cost will not likely be an issue. To be safe however, the budgets prepared for future alterations should include a minimum additional factor of 20% over and above the specific alteration cost.

Hardship when alteration costs <u>exceed</u> the ENR Average Construction Cost Index
An entity can still apply for a determination of unreasonable hardship when the alteration cost exceeds the allowable ENR cost index limit. The major difference is that the obligation to spend 20% of the alteration cost on the path of travel becomes a minimum limit instead of a maximum one. To explain this situation, lets look at a proposed tenant improvement that will cost $200,000.00. This $200,000.00 is the cost to remodel the specific tenants space alone. Upon reviewing the entrance, path of travel, restrooms, drinking fountains and public telephones that service the proposed alteration, it is determined that it will cost an additional $65,000.00 to make these improvements accessible as required. Unless the tenant applies for and receives a determination of unreasonable hardship from the local building agency, the tenant will be required to make all of the necessary modifications. If the tenant applies for and is successful in receiving a hardship determination, the <u>minimum</u> additional path of travel expense will be $40,000.00 (20% of $200,000.00 is $40,000.00). It is at the building agencies discretion to determine how much <u>over</u> this 20% constitutes a hardship upon reviewing the particular circumstances involved. There is a caveat to this standard however; If the building is a privately funded multistory building that does not have an elevator, the standards place a 20% limit on the path of travel obligations on floors above and below the ground floor even when the ENR threshold is exceeded for most buildings (see CBC Section 11B-202.4, Exception 9 for specifics.

Priority for completing path of travel requirements when unreasonable hardship occurs
Whenever an entity obtains a determination of unreasonable hardship on a proposed alteration, it means that all of the improvements that are necessary to make the primary path of travel, restrooms, drinking fountains and public telephones "readily accessible" will not be accomplished. In this situation, both State and Federal guidelines request that an entity prioritize the required modifications so that the greatest degree of access can be provided until the money runs out. In choosing which accessible elements to provide, priority should be given in the following order:

    A. An accessible entrance,
    B. An accessible route to the altered area,
    C. At least one accessible restroom for each sex,
    D. Accessible telephones,
    E. Accessible drinking fountains, and
    F. When possible, additional accessible elements such as parking, storage and alarms. **11B-202.4, Exception 8**

Following this list of priorities is not required (unless a building agency tells you it is) however it is highly recommended to do so. As a final note, although item "F" mentions parking as a last priority, this is for <u>additional</u> accessible parking if some accessible parking is already provided. Item "B" which asks for "an accessible route to the altered area" includes providing accessible parking and a path of travel. An accessible building is of no use to the disabled if they can't even get to the front door.

Legal and physical constraints that prohibit full accessibility compliance
Besides the cost limitations that may prevent full compliance with applicable guidelines in alterations, certain legal or physical constraints on a site may also prevent full compliance. One of these physical constraints is called "technical infeasibility":

> *Technically Infeasible. Means, with respect to an alteration of a building or a facility, that it has little likelihood of being accomplished because existing structural conditions would require removing or altering a load-bearing member which is an essential part of the structural frame; or because other existing physical or site constraints prohibit modification or addition of elements, spaces, or features which are in full and strict compliance with the minimum requirements for new construction and which are necessary to provide accessibility".* 106.5 **202**

Some examples of this would be a structural column that prohibits widening a toilet stall to the proper dimensions; or the significant structural modifications that would be necessary in order to install an elevator in an existing building. Some examples of legal constraints could include the inability to widen a toilet stall because it would require encroachment into another entities leased space; or because widening the stall would require removing an adjacent toilet that would cause the required fixture count for the occupancy load of the building to be deficient.

The fact that it may not be feasible to make a particular improvement "readily accessible" does not excuse the entity from providing partial accessibility to the improvement if possible, or from implementing other accessibility standards that are not affected by the subject constraints. The law requires an entity to ensure that, to the maximum extent feasible, the altered portions of the facility are readily accessible to and usable by individuals with disabilities, including individuals who use wheelchairs:

> *"To the maximum extent feasible. The phrase "to the maximum extent feasible," as used in this section, applies to the occasional case where the nature of an existing facility makes it virtually impossible to comply fully with applicable accessibility standards through a planned alteration. In these circumstances, the alteration shall provide the maximum physical accessibility feasible. Any altered features of the facility that can be made accessible shall be made accessible. If providing accessibility in conformance with this section to individuals with certain disabilities (e.g., those who use wheelchairs) would not be feasible, the facility shall be made accessible to persons with other types of disabilities (e.g., those who use crutches, those who have impaired vision or hearing, or those who have other impairments)." ADA §36.402(c)*

Evading path of travel requirements by performing a series of small alterations
In some situations, the cost to provide a specific improvement to the path of travel will be disproportionate in all but the most expensive alterations (e.g., elevators). Accordingly, some building owners will attempt to perform a series of smaller, separately permitted alterations to a specific area in order to avoid having to provide an expensive improvement. To address this possible "loop-hole" in alterations of existing buildings, the State guidelines have adopted the ADA's requirement for "Aggregation". Aggregation involves adding together the total cost of all alterations on the same path of travel over a three-year period to determine ultimate obligations:

> *"The obligation to provide access may not be evaded by performing a series of small alterations to the area served by a single path of travel if those alterations could have been performed as a single undertaking. If an area has been altered without providing an accessible path of travel to that area, and subsequent alterations of that area or a different area on the same path of travel are undertaken within three years of the original alteration, the total cost of alterations to the areas on that path of travel during the preceding three-year period shall be considered in determining whether the cost of making that path of travel accessible is disproportionate. Only alterations undertaken after January*

26, 1992, shall be considered in determining if the cost of providing an accessible path of travel is disproportionate to the overall alteration."
11B-202.4, Exception 8   *36.403(h)*

EXAMPLE #1

This example shows the first floor of an existing office building that is owner occupied. A 50-foot long concrete sidewalk ramp extends from the parking area to the main entry door threshold. The difference in elevation between the door threshold and the parking area is six feet, and the existing ramp exhibits a slope of 12%. All of the public and common use areas inside the building are "readily accessible", however the exterior entry ramp is seriously out of compliance. To make this entrance accessible, a "switch-back" ramp with a correct door landing will have to be constructed. A feasibility study determines that this ramp and landing will cost $20,000.00. In 2013, three separately permitted alterations were completed to areas "A", "E" and "G" that each cost $25,000.00. In each alteration, the owner was successful in obtaining a determination of unreasonable hardship and did not have to make path of travel modifications to the ramp.

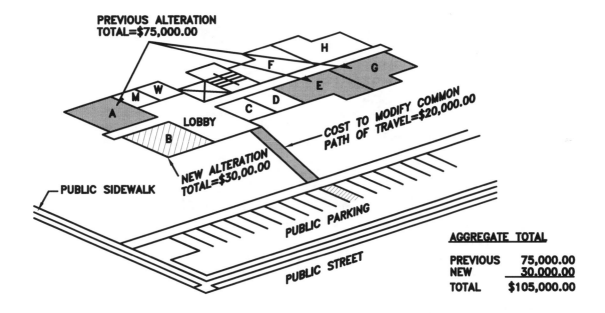

In April of 2014, the owner applies for a permit to construct a $30,000.00 alteration to space "B". The owner again applies for unreasonable hardship, but upon reviewing the total cost of all of the previous improvements that are serviced by this exterior common path of travel, the building agency discovers that $75,000.00 has been spent within the prior 3 years without providing for any path of travel modifications. By combining the previous alteration costs of $75,000.00 with the proposed new alteration cost of $30,000.00, a total of $105,000.00 is now the determining amount. By taking 20% of the "aggregated" alteration costs, the building agency determines that the owner is obligated to spend up to an additional $21,000.00 in path of travel costs (20% of $105,000.00 is $21,000.00). Since the total cost of altering the ramp and landing is under this amount, the agency requires the owner to construct the improvements as a condition of obtaining a permit.

EXAMPLE #2

This example shows the first floor of an existing office building that is owner occupied. A 50-foot long concrete sidewalk ramp extends from the parking area to the main entry door threshold. The difference in elevation between the door threshold and the parking area is six feet, and the existing ramp exhibits a slope of 12%. All of the public and common use areas inside the building are "readily accessible" with the exception of the restrooms, which will require significant modifications. To make the exterior entrance accessible, a "switch-back" ramp with a correct door landing will have to be constructed. A feasibility study determines that the exterior ramp and landing will cost $20,000.00, and each interior restroom will cost $4,000.00. In 2013, three separately permitted alterations were completed to areas "A", "E" and "G" that each cost $25,000.00. In each alteration, the owner was successful in obtaining a determination of unreasonable hardship and did not have to make path of travel modifications to the ramp or restrooms.

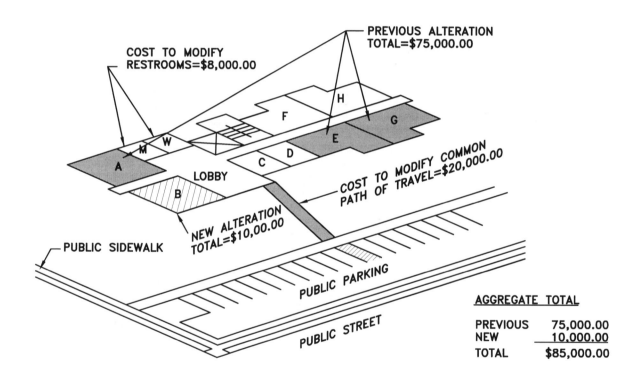

AGGREGATE TOTAL	
PREVIOUS	75,000.00
NEW	10,000.00
TOTAL	$85,000.00

In February of 2014 the owner applies for a permit to construct a $10,000.00 alteration to space "B". The owner again applies for unreasonable hardship, but upon reviewing the total cost of all of the previous improvements that are serviced by this exterior common path of travel, the building agency discovers that $75,000.00 has been spent within the prior 3 years without providing for any path of travel modifications. By combining the previous alteration costs of $75,000.00 with the proposed new alteration cost of $10,000.00, a total of $85,000.00 is now the determining amount. By taking 20% of the "aggregated" alteration costs, the building agency determines that the owner is obligated to spend up to an additional $17,000.00 in path of travel costs (20% of $85,000.00 is

$17,000.00) with a determination of unreasonable hardship. Since the total cost of altering the ramp and landing is over this amount, the agency requires the owner to spend $8,000.00 and modify both restrooms as a condition of obtaining a permit.

If the owner in example #2 chose to make additional alterations in a three-year period, the total cost of all the alterations would again be aggregated to determine the total 20% obligation, however the $8,000.00 that was spent on the restrooms would have to be subtracted from this 20% total.

In both of the above examples, the building was owner-occupied. If the building consisted of multiple separate tenants, the principal of aggregation would not apply between these tenants. You cannot aggregate the total costs of all alterations on a particular path of travel if different tenants permit each of the alterations separately. You cannot make one tenant responsible for another tenant's obligations. This situation also applies to the landlord of the facility. If a tenant is making alterations that would trigger the path of travel requirement outside of the areas that only the tenant occupies, this alteration does not trigger a path of travel obligation upon the landlord. This obligation is the responsibility of the tenant to work out if they want to complete their alteration as planned. The building agency is neither obligated nor required to determine whether the tenant, owner or another party is responsible for paying for the improvements; their only obligation is to enforce the requirements of the code upon the applicant. If an individual tenant undertakes multiple alterations to their leased space however, the principal of aggregation would apply to the areas of the facility under their control. Due to the fact that you cannot aggregate the individual tenants alterations in a multi-tenant facility, very few owners will choose to build out these tenant areas themselves; most will instead provide the funding directly to the tenant through rent reductions or other means to avoid having to deal with aggregated path of travel obligations. Regardless of your personal feelings as to whether or not this is appropriate, this situation is legal and allowable under the law, as it presently exists.

As a final note, the principal of aggregation applies only to alterations in areas that are serviced by the same specific path of travel. When multiple alterations are completed to different locations in a facility, which are serviced by different paths of travel, only those alteration costs applicable to that specific area can be aggregated. If portions of the path of travel were shared however, such as when different interior paths of travel converge into a single building lobby to the exterior arrival points, the total cost of all the alterations would be aggregated to determine the extent of obligations to the lobby and exterior.

Elevator requirements in alterations
One of the most significant changes between previous Title 24 guidelines and the current guidelines is in regards to elevator requirements in alterations. Under old Title 24 guidelines, certain types of privately funded, multistory buildings and facilities were formerly exempt from accessibility requirements above and below the first floor. As of April 1, 1994, these buildings and facilities were no longer exempt from providing accessibility features on non-accessible floors; and may additionally be required to install an elevator and/or ramps to provide for vertical access between floors. In alteration projects involving buildings and facilities that were previously approved and built without elevators, the areas above and below the ground floor must now be made "readily accessible" to the maximum extent feasible. In addition, the installation of an elevator must be considered as a possible "path of travel" improvement if the cost of the alteration is significant enough to pay for it, among other factors. The determination as to whether or not an elevator must be installed in an existing building or facility can be a complex one. Considerations include the facilities use, square footage, number of stories, structural feasibility, previous alterations and most significantly, the cost. If the alteration

project will be limited to an accessible level of the facility, you do not need to consider the installation of an elevator. In other words, undertaking an alteration to an accessible first floor will not trigger a path of travel requirement to other inaccessible floors, regardless of the cost of the alteration.

DETERMINATION OF DISPROPORTIONALITY IN ALTERATIONS

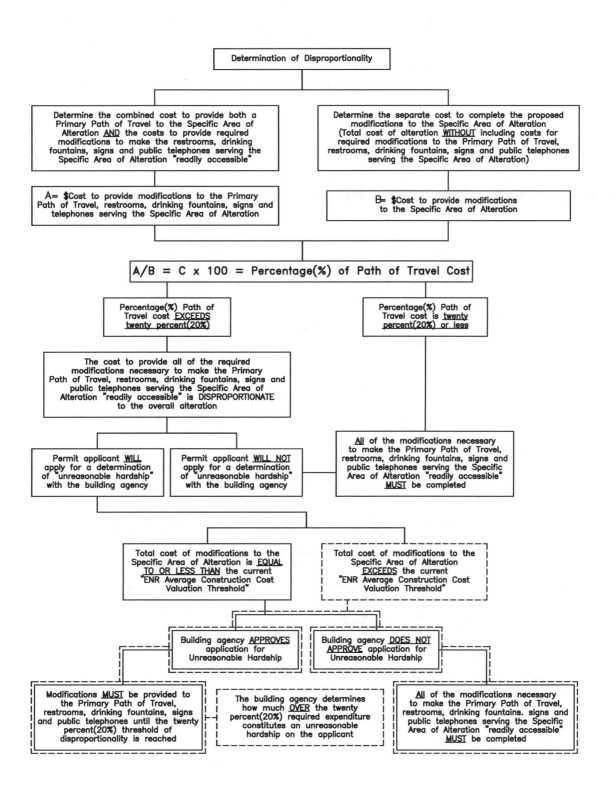

EXAMPLE #1

This example is an existing office building that went into financial receivership. The new owner has been involved in completely refurbishing the building as medical offices. In 2013, the owner pulled a permit and completely renovated the left half of the building including the ground floor lobby. Since the second floor was not accessible, the building agency did not require the owner to provide accessible features on this floor. The ground floor level was constructed to be "readily accessible", and included exterior path of travel modifications. The total cost for this construction was as follows:

Second Floor	$100,000.00
First Floor	$125,000.00

In May of 2014, the owner applied for a permit to refurbish the right half of the building. The proposed alteration costs were as follows:

Second Floor	$125,000.00
First Floor	$100,000.00

This time the building agency requires the owner to provide accessibility modifications to the second floor, which includes the four restrooms in this half of the building; and additionally requires the installation of an elevator. A feasibility study determines that an elevator can be installed by removing the existing main stairway for a cost of $70,000.00. The modifications to the restrooms will cost $8,000.00. The owner applies for a determination of unreasonable hardship, wishing only to modify the restrooms and upgrade the existing stairway for full compliance.

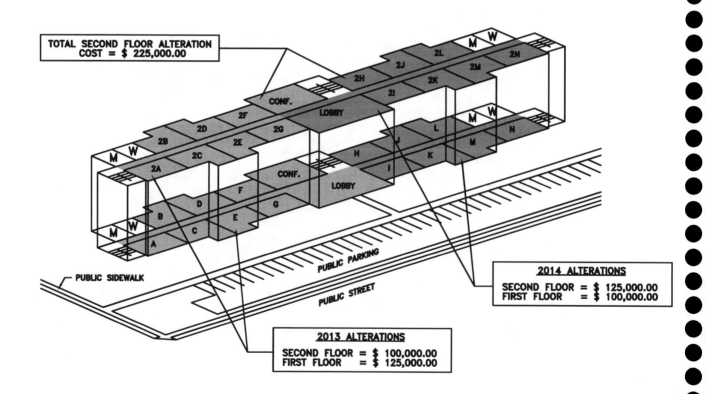

TOTAL SECOND FLOOR ALTERATION COST = $ 225,000.00

PUBLIC SIDEWALK

PUBLIC PARKING

PUBLIC STREET

2013 ALTERATIONS
SECOND FLOOR = $ 100,000.00
FIRST FLOOR = $ 125,000.00

2014 ALTERATIONS
SECOND FLOOR = $ 125,000.00
FIRST FLOOR = $ 100,000.00

In determining the path of travel obligations, the building agency must aggregate the costs for both halves of the <u>second floor</u> remodel. Because both halves will be serviced by the same vertical path of travel (i.e. the main stairway), all of the alterations on levels served by this path of travel that are constructed within a three year period must be totaled to determine if the path of travel costs are disproportionate.

Total cost for second story alterations = $225,000.00 **x 20% = $45,000.00**

As is shown, $45,000.00 is the 20% limit at which the path of travel costs are disproportionate in cost and scope to the overall alteration. As this is an office building and is less than 3 stories in height, 20% is the maximum obligation, even though the ENR cost threshold has been exceeded. At this point the building agency only has the ability to require the bathrooms and main stairs to be made accessible. The building agency cannot require an elevator because the 20% limit is not enough to pay for the elevator. Additionally, if the stairwells on either side of the building deviate from existing access standards, these are not on the single primary path of travel from the second floor to the main lobby and neither state nor federal law requires these to be modified to New Construction standards. Remember, any modifications that can be made to the primary path of travel, restrooms, drinking fountains, telephones and signs within this 20% threshold limit <u>MUST</u> be completed regardless of whether or not a determination of unreasonable hardship is granted.

To provide further awareness regarding the vertical access requirements in this example, it should be understood that the obligation to determine the feasibility of providing an elevator in this building is due to the fact that the second floor will harbor "the professional office of a health care provider". Since this facility is a multi-story office building that is less than three stories in height, it does not require the consideration of vertical accessibility unless it will house a shopping center, shopping mall, the professional office of a health care provider, etc.

EXAMPLE #2

This example is an existing office building that went into financial receivership. The building, originally constructed in the early 1960's, does not presently contain any features for disabled access. The new owner wishes to completely renovate the building for lease to various tenants, and additionally desires to open a snack shop on the second floor for the building occupants. As the building is a two-story office complex, the owner does not intend to install an elevator but instead will modify the existing stairs for full compliance. The construction plans include full accessibility modifications to all restrooms and exterior improvements.

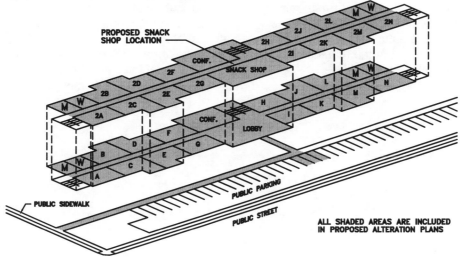

Upon review of the plans, the building agency determines that the proposed snack shop on the second floor of the facility will categorize this second floor as a "Shopping Center" under California's definitions ("Shopping Center" = "One or more sales or rental establishments or stores", **Section 202**). The building agency requires the owner to place the shop on the accessible first floor, remove it altogether, or install an elevator to the second floor if 20% of the cost of the second story alterations will pay for it. The building agency further "conditions" the applicant that within a 5-year period, no "sales establishments or stores" or "offices of physicians or surgeons" be placed on the second floor if the elevator exemption is to be maintained. At this point, the owner has the ability take his/her chances and apply for a determination of unreasonable hardship on the elevator, or comply/contest with the agency's conditions.

At this point, you may be wondering about the legality of conditioning an applicant to restrict the use of their facility. This example was meant to open the "Pandora's box" regarding the obligation to install an elevator in various situations. The owner in this example was renovating the entire facility and bringing in an entirely new group of tenants. In retrospect, the owner will probably wish that he/she had not provided details regarding the snack shop on the construction drawings, but had rather waited until the facility construction was completed and approved and then applied for the snack shop occupancy as a separate issue to try to avoid the elevator requirements. With the principal of aggregation however, the total alteration costs for the improvements on the second floor within a three-year period are utilized to determine the feasibility of an elevator, and the owner will therefore have to deal with this issue now or in the future.

It is not within a building agency's obligation to be a "watchdog" for builders that try to slip around the access requirements. It is within a building agency's discretion however to protect itself from future claims of complicity in providing a permit to a builder that ultimately constructs inaccessible facilities. Both agencies and owners should be aware that it is not uncommon for a complaint to be filed that alleges an *"intentional misrepresentation of the facility's proposed usage to avoid disabled accessibility requirements"*.

It is <u>not</u> illegal for a "health care provider" or a "sales establishment" to locate their business in an existing multi-story facility that does not have an elevator. It is also not illegal for a tenant or leasing agent to lease an inaccessible facility. Whether or not the owner deliberately misrepresented the ultimate use of the facility will likely be an issue of fact subject to trial. The determination of liability for a complaint of this sort would involve investigating whether or not the facility was "designed or intended for use" by a non-exempt entity. In determining whether an inaccessible floor was intended for such use, factors to be considered include the types of establishments that first occupied the floor, the nature of the developer's marketing strategy, i.e., what types of establishments were sought in the marketing strategy, and inclusion of any design features particular to the non-exempt entities, such as special plumbing, electrical or other features needed by health care providers, or any design features particular to sales establishments.

Additions

Additions
Each addition to an existing building or facility shall comply with the requirements for new construction and shall comply with path of travel requirements.
11B-202.2   *202.2*

Each addition to an existing building or facility shall be regarded as an alteration. Each space or element added to the existing building or facility must meet the requirements for New Construction and be designed and constructed to be "readily accessible".

The requirements for the addition of new buildings or facilities to an existing building or facility is fairly straightforward; you must meet the guidelines for New Construction with the added structures, AND you must provide modifications to the "path of travel, restrooms, drinking fountains, telephones and signs", that service the new addition(s), just as you would with an alteration. The principal difference between an alteration and an addition is in the calculation of cost obligations for these modifications. In an addition, every feature that is constructed must be "readily accessible" just as in a new building or facility. This includes any restrooms, drinking fountains, telephones, signs or path of travel improvements <u>within the scope of the addition</u>. Unlike an alteration, however, the TOTAL costs for this entire addition are taken as a whole to determine the amount of the 20% obligation to provide additional modifications to the path of travel, restrooms, drinking fountains and telephones that adjoin and serve the addition, if necessary. The following examples will hopefully clarify the necessary considerations:

EXAMPLE #1
This example is an existing single-story office building. The owner submits plans to lengthen the building and add six new offices. All portions of the new addition must meet the guidelines for New Construction and be "readily accessible". Since the new addition will not contain an "Area for Evacuation Assistance" or a "Supervised Automatic Sprinkler System", an accessible and egressible path of travel must be provided from the exit-only door at the end of the addition to the nearest accessible route. The total cost for the new addition will be $250,000.00.

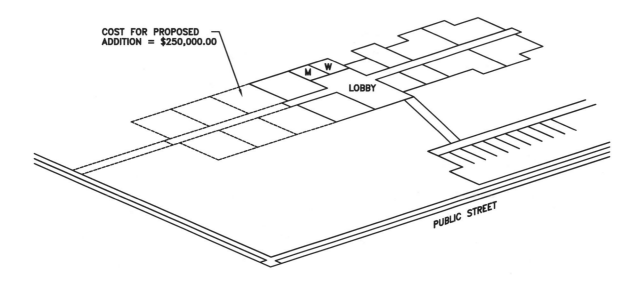

COST FOR PROPOSED
ADDITION = $250,000.00

M   W

LOBBY

PUBLIC STREET

Upon reviewing the interior and exterior improvements of the existing building, the building agency finds that a 50-foot long concrete sidewalk ramp extends from the parking area to the main entry door threshold. The difference in elevation between the door threshold and the parking area is six feet, and the existing ramp exhibits a slope of 12%. The site does not presently provide for disabled parking stalls or an accessible route to existing public sidewalk. All of the public and common use areas inside the existing building are "readily accessible" with the exception of the restroom facilities which will need to be accessible and enlarged to accommodate the new occupancy load of the building. To make the exterior entrance accessible, a "switch-back" ramp with a correct door landing will have to be constructed. In addition, the correct number of disabled parking stalls and an accessible path of travel to the sidewalk will have to be constructed. A feasibility study determines that the ramp and parking improvements will cost $25,000.00, the route to public sidewalk will cost $7,000.00, and modifications to both restrooms will cost $25,000.00 (total = $57,000.00).

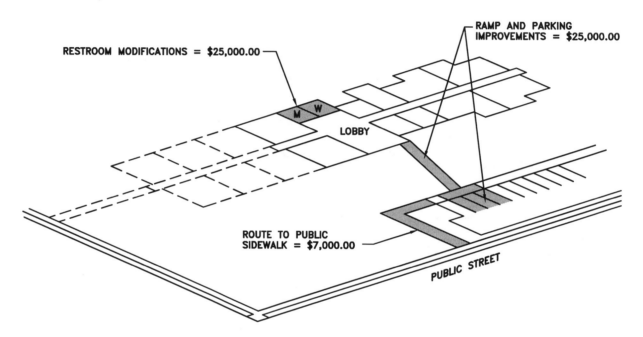

The owner applies for a determination of unreasonable hardship, wishing only to modify the interior restrooms for $25,000.00. The building agency informs the owner that since the addition will cost $250,000.00, the owner is obligated to spend 20% of this amount, or $50,000.00, even with a determination of unreasonable hardship. The building agency grants the hardship for $7,000.00 and requires the owner to construct all improvements except for the route to the public sidewalk.

EXAMPLE #2
This example is an existing single-story office building. The owner submits plans to lengthen the building and add five new offices and two restrooms. The owner also plans to build a new parking lot and accessible entrance to this addition. The new parking lot will be separate from the existing parking lot. Since all portions of the addition must meet the requirements for New Construction and be "readily accessible", this entrance must also provide an accessible path of travel to the adjacent public sidewalk. The owner determines that it will cost $220,000.00 to complete the basic addition, $25,000.00 to provide new restrooms in the addition, and $35,000.00 to build the new parking lot and sidewalk (total cost = $280,000.00).

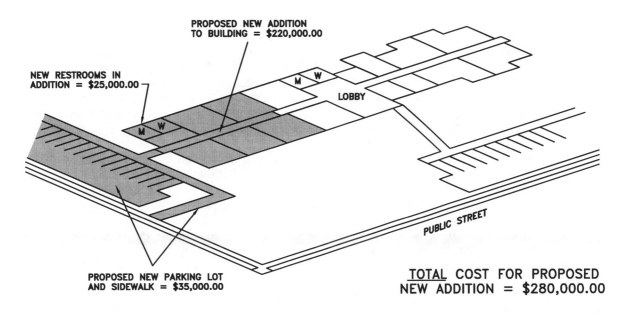

PROPOSED NEW ADDITION TO BUILDING = $220,000.00

NEW RESTROOMS IN ADDITION = $25,000.00

PROPOSED NEW PARKING LOT AND SIDEWALK = $35,000.00

TOTAL COST FOR PROPOSED NEW ADDITION = $280,000.00

Upon reviewing the <u>existing</u> building, the building agency finds that the existing restrooms are inaccessible. In addition, the existing parking lot and path of travel to the existing entrance are seriously out of compliance. A feasibility study indicates that modifications to the restrooms and exterior improvements of the existing building will cost $45,000.00. Since California requires that all primary entrances be accessible, and since patrons of the facility will be able to park in either of the two available lots, the owner is obligated to make path of travel modifications from the existing improvements to the new addition. The building agency directs the owner to make these additional modifications to the existing building as a condition to performing the new addition. The diagram below illustrates the additional areas that the owner is required to modify:

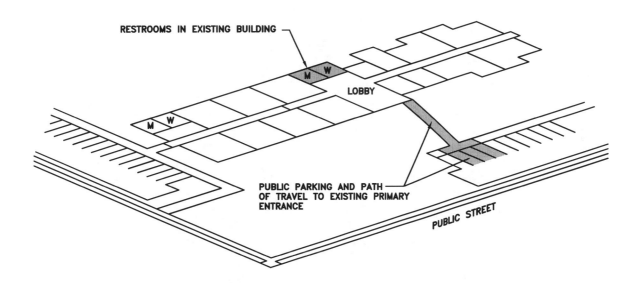

RESTROOMS IN EXISTING BUILDING

PUBLIC PARKING AND PATH OF TRAVEL TO EXISTING PRIMARY ENTRANCE

TOTAL COST FOR REQUIRED MODIFICATIONS TO <u>EXISTING</u> BUILDING = $45,000.00

The owner contests the agency's determination, stating that the construction of the new addition without the restrooms or parking lot will only cost $220,000.00; and that 20% of this amount is $44,000.00. The owner further states that he/she is already spending $60,000.00 on the restrooms and exterior improvements for the new addition, an amount well over that that would be required in an alteration. The owner does not believe that he/she has any further obligations. The building agency then informs the owner that his/her interpretation of the requirements for an addition are incorrect. To correctly determine the 20% obligation, the TOTAL cost of the full scope of work involved in the construction of the addition is utilized to determine the 20% disproportionality limit. This amount would be 20% of $280,000.00, a total of $56,000.00. In this situation, this path of travel obligation must be applied to modify the improvements within the route from the exterior portions of the old building all the way to the new addition. Since the existing restrooms are on this path of travel, they must also be included. The owner cannot include the costs for the restrooms or parking lot of the new addition to fulfill path of travel obligations, as these were required to be accessible anyway.

BARRIER REMOVAL IN EXISTING BUILDINGS AND FACILITIES

Both the ADA and the CBC have similar provisions when an entity is undertaking an alteration to an existing building or facility. The ADA however, has a requirement that neither the CBC nor state laws contain that requires an entity that can be classified as a "public accommodation" to remove structural barriers to disabled access in existing buildings and facilities that are not otherwise being altered, when such removal is considered "readily achievable". The readily achievable obligation to remove barriers in existing facilities is found in Section 36.304 of the ADA Title III regulations and states:

> ### Section 36.304 Removal of Barriers
> **(a)   General.** *A public accommodation shall remove architectural barriers in existing facilities, including communication barriers that are structural in nature, where such removal is readily achievable, i.e., easily accomplishable and able to be carried out without much difficulty or expense.*

As stated, a *public accommodation* is required to remove architectural and communications barriers to the disabled in existing *public accommodation* facilities. Non-compliance with this mandate is considered discrimination against the disabled and is a civil rights violation. Consequently, there are no "grandfather" clauses under this portion of the law; the obligations to remove existing barriers became an instant and on-going consideration for all *public accommodations* when the ADA was signed into law in 1990.

What IS a public accommodation? A public accommodation means *"a private entity that owns, leases (or leases to), or operates a place of public accommodation"*. In order to be a *Place of Public Accommodation,* a facility must be operated by a private entity, its operations must affect commerce, and it must fall into one of the following 12 categories:

1) An inn, hotel, motel, or other place of lodging, except for an establishment located within a building that contains not more than five rooms for rent or hire and that is actually occupied by the proprietor of the establishment as the residence of the proprietor;
2) A restaurant, bar, or other establishment serving food or drink;
3) A motion picture house, theater, concert hall, stadium, or other place of exhibition or entertainment;
4) An auditorium, convention center, lecture hall, or other place of public gathering;
5) A bakery, grocery store, clothing store, hardware store, shopping center, or other sales or rental establishment;
6) A Laundromat, dry-cleaner, bank, barber shop, beauty shop, travel service, shoe repair service, funeral parlor, gas station, office of an accountant or lawyer, pharmacy, insurance office, professional office of a health care provider, hospital, or other service establishment;
7) A terminal, depot, or other station used for specified public transportation;
8) A museum, library, gallery, or other place of public display or collection;
9) A park, zoo, amusement park, or other place of recreation;
10) A nursery, elementary, secondary, under-graduate, or postgraduate private school, or other place of education;
11) A day care center, senior citizen center, homeless shelter, food bank, adoption agency, or other social service center establishment; and
12) A gymnasium, health spa, bowling alley, golf course, or other place of exercise or recreation.

If you have now determined that you or your client are indeed a "public accommodation" and must now consider removing barriers in your "place of public accommodation", the "readily achievable" standard is your next likely concern. The determination as to whether or not a specific barrier removal action is readily achievable depends on a number of considerations. The definition of readily achievable is found in Section 36.104 of the ADA Title III regulations and states:

> *Readily achievable means easily accomplishable and able to be carried out without much difficulty or expense. In determining whether an action is readily achievable, factors to be considered include—*
>
> *(1) The nature and cost of the action needed under this part;*
> *(2) The overall financial resources of the site or sites involved in the action; the number of persons employed at the site; the effect on expenses and resources; legitimate safety requirements that are necessary for safe operation; including crime prevention measures; or the impact otherwise of the action upon the operation of the site;*
> *(3) The geographic separateness, and the administrative or fiscal relationship of the site or sites in question to any parent corporation or entity;*
> *(4) If applicable, the overall financial resources of any parent corporation or entity; the overall size of the parent corporation or entity with respect to the number of its employees; the number, type, and location of its facilities; and*
> *(5) If applicable, the type of operation or operations of any parent corporation or entity, including the composition, structure, and functions of the workforce of the parent corporation or entity.*

As stated, the definition of "readily achievable" is easily accomplishable and able to be carried out without much difficulty or expense. One of these issues, as stated, is expense. The ultimate cost, however, to remove a specific barrier or barriers, is not the sole consideration as to whether or not a barrier removal is readily achievable; the standard requires consideration of other factors. One of these factors is the extent of the difficulty that would be involved in removing the barrier. As defined, a readily achievable barrier removal is one that is easily accomplishable, and can be accomplished **without** much difficulty. It is the Preamble to the Title III regulations in the Federal Register that contains the "Section-By-Section Analysis and Response to Comments"; this text contains definitive information on how the regulations were designed to be applied and the extent of obligation that is/is not required under specific provisions. This text under Section 36.304 contains explanatory language that limits the level of modifications required under the readily achievable standard absent of cost:

> *Section 36.304(b) provides a wide-ranging list of the types of <u>modest measures</u> that may be taken to remove barriers and that are likely to be readily achievable. The list includes examples of measures, such as adding raised letter markings on elevator control buttons and installing flashing alarm lights, that would be used to remove communications barriers that are structural in nature. It is not an exhaustive list, but merely an <u>illustrative</u> one. Moreover, the inclusion of a measure on this list does not mean that it is readily achievable in all cases. Whether or not any of these measures is readily achievable is to be determined on a case-by-case basis in light of the particular circumstances presented and the factors listed in the definition of readily achievable (§36.104).*
>
> *A public accommodation generally would not be required to remove a barrier to physical access posed by a flight of steps, if removal would require extensive ramping or an elevator. Ramping a single step, however, will likely be readily achievable, and ramping several steps will in many circumstances also be readily achievable. <u>The readily achievable standard **does not require barrier removal that requires extensive restructuring** or burdensome expense</u>. Thus, where it is not readily achievable to do, the ADA would not require a restaurant to provide access to a restroom reachable only by a flight of stairs.*

As the highlighted text above illustrates, the readily achievable standard *"does not require barrier removal that requires extensive restructuring or burdensome expense"*. The text also explains that the list of barrier removal measures that are likely to be readily achievable under the standard are modest measures, and that these measures are illustrative of the types of measures required under the standard. This list is found in the text referenced Section 36.304(b) and is presented below:

> *(b) Examples. Examples of steps to remove barriers include, but are not limited to, the following actions--*
> *(1) Installing ramps;*
> *(2) Making curb cuts in sidewalks and entrances;*
> *(3) Repositioning shelves;*
> *(4) Rearranging tables, chairs, vending machines, display racks, and other furniture;*
> *(5) Repositioning telephones;*
> *(6) Adding raised markings on elevator control buttons;*
> *(7) Installing flashing alarm lights;*
> *(8) Widening doors;*
> *(9) Installing offset hinges to widen doorways;*
> *(10) Eliminating a turnstile or providing an alternative accessible path;*
> *(11) Installing accessible door hardware;*
> *(12) Installing grab bars in toilet stalls;*
> *(13) Rearranging toilet partitions to increase maneuvering space;*
> *(14) Insulating lavatory pipes under sinks to prevent burns;*
> *(15) Installing a raised toilet seat;*
> *(16) Installing a full-length bathroom mirror;*
> *(17) Repositioning the paper towel dispenser in a bathroom;*
> *(18) Creating designated accessible parking spaces;*
> *(19) Installing an accessible paper cup dispenser at an existing inaccessible water fountain;*
> *(20) Removing high pile, low density carpeting; or*
> *(21) Installing vehicle hand controls.*

In addition to limitations regarding extensive restructuring and difficulty under the readily achievable standard, the DOJ also specifically addressed the issue as to whether or not the profitability of various businesses or other private entities should represent a factor in consideration of what is or is not readily achievable. As the following language under the preamble to Section 36.304 explains, the notion of profitability was specifically addressed and purposely excluded from consideration.

> *Section 36.304(f)(1) of the proposed rule, which stated that "barrier removal is not readily achievable if it would result in significant loss of profit or significant loss of efficiency of operation," has been **deleted** from the final rule. Many commenters objected to this provision **because it impermissibly introduced the notion of profit** into a statutory standard that **did not include it**. Concern was expressed that, in order for an action not to be considered readily achievable, a public accommodation would **inappropriately** have to show, for example, **not only that the action could not be done without "much difficulty or expense", but that a significant loss of profit would result as well.** In addition, some commenters asserted use of the word "significant" which is used in the definition of undue hardship under title I (the standard for interpreting the meaning of undue burden as a defense to title III's auxiliary aids requirements) (see §§ 36.104, 36.304(f)), blurs the fact that the readily achievable standard requires a lower level of effort on the part of a public accommodation than does the undue burden standard.*

BARRIER REMOVALS DO NOT TRIGGER PATH-OF-TRAVEL OBLIGATIONS
Although "readily achievable" barrier removals, in some cases, could be classified as an alteration of an existing facility, the same conditions for additional path-of-travel obligations do NOT apply as with regular alterations. When an entity undertakes an alteration, structural repair or addition with the sole purpose of removing barriers to the disabled, then the scope of the alteration is limited to the actual work of the project. The entity performing the alteration is NOT required to engage in additional modifications to the path-of-travel, restrooms, drinking fountains, telephones or signs that may serve the specific area of alteration. This situation applies even when a permit is required to perform the alteration.

Although both state and federal sources list specific examples of the types of modifications that may be undertaken to remove barriers to the disabled, these lists are far from exhaustive. It is sufficient to state than any modification undertaken with the sole purpose of barrier removal does not trigger any additional requirements beyond the scope of the specific barrier removal action.

HISTORIC PRESERVATION

DEFINITION - ADAAG:
A **qualified historic building or facility** is a building or facility that is:
(i) Listed in or eligible for listing in the National Register of Historic Places; or
(ii) Designated as historic under an appropriate State or local law.

DEFINITION – CBC:
QUALIFIED HISTORICAL BUILDING OR PROPERTY.
Any building, site, object place, location, district or collection of structures, and their associated sites, deemed of importance to the history, architecture or culture of an area by an appropriate local, state or federal governmental jurisdiction. This shall include historical buildings or properties on, or determined eligible for, national, state or local historical registers or inventories, such as the National Register of Historic Places, California Register of Historical Resources, State Historical Landmarks, State Points of Historical Interest, and city or county registers, inventories, or surveys of historical or architecturally significant sites, places or landmarks. **8-201**

General Rule- Alterations to a qualified historic building or facility shall comply with all applicable accessibility requirements for Alterations unless it is determined in accordance with the procedures listed below that compliance with the requirements for accessible routes (exterior and interior), ramps, entrances or toilets would threaten or destroy the historic significance of the building or facility. If it is determined that any of the accessibility requirements would threaten or destroy the historic significance or character-defining features of a building or facility, the alternative modifications listed below may be utilized for the feature. **8-602.1** *202.5*

ADVISORY: There are a number of differences between state and federal requirements in regards to the alteration of historic buildings and facilities. California allows the building official to determine whether or not a specific alteration will threaten or destroy the historical significance or character-defining features of the building or facility in question, and then allows this official to determine what alternative modifications can be utilized. Under federal provisions, a building official may not make these same determinations unless they are proceeding in consultation with the State Historic Preservation Officer. A local government body or official may only make decisions regarding alternative modifications to eligible historic structures after the State Historic Preservation Officer has delegated responsibility to a federally certified local historic preservation program. At this point, the local historic preservation program can designate appropriate responsibilities to be carried out by the appropriate local government body or official.

As stated above in the General Rule, any time an alteration to a qualified historic structure is undertaken you must ensure that any modifications comply with the full range of accessible features UNLESS the modification, or portion thereof, will threaten or destroy the historic significance of the building or facility. It is only after the appropriate steps have been taken to determine whether or not the historical significance of the building may be compromised that alternative modifications or deviations from established guidelines may be utilized.

It is appropriate at this point to also mention that while certain buildings or facilities may be historic in nature in their entirety, there are many buildings or facilities of which only a portion of the structures qualify as historic. In these situations, you may only apply qualifying alternative modifications to the specific

> historic portion of the structure(s), and must utilize standard accessible features in all remaining non-historic areas.
>
> The following text titled "PROCEDURES" is taken directly from the ADAAG and delineates the required procedures for dealing with federal, state, and local historic structures. In addition, the sections of federal text that are referred to in these procedures are included at the end of this text under "IMPLEMENTING REGULATIONS" for your further use.
>
> Immediately following the section on implementing regulations are the current California Historical Building Code provisions. It is recommended to become familiar with both sets of standards to insure the proper application of both.

FEDERAL STANDARDS

(A) PROCEDURES

(1) Alterations to Qualified Historic Buildings and Facilities Subject to Section 106 of the National Historic Preservation Act:

(a) **Section 106 Process.** Section 106 of the National Historic Preservation Act (16 U.S.C. 470 f) requires that a Federal agency with jurisdiction over a Federal, federally assisted, or federally licensed undertaking consider the effects of the agency's undertaking on buildings and facilities listed in or eligible for listing in the National Register of Historic Places and give the Advisory Council on Historic Preservation a reasonable opportunity to comment on the undertaking prior to approval of the undertaking.

(b) **ADA Application.** Where alterations are undertaken to a qualified historic building or facility that is subject to section 106 of the National Historic Preservation Act, the Federal agency with jurisdiction over the undertaking shall follow the section 106 process. If the State Historic Preservation Officer or Advisory Council on Historic Preservation agrees that compliance with the requirements for accessible routes (exterior and interior), ramps, entrances, toilets, parking or signage would threaten or destroy the historic significance of the building or facility, the alternative modifications listed may be used for the feature (see below). *206.2.3, Exception 7, 36.405(b)*

(2) Alterations to Qualified Historic Buildings and Facilities Not Subject to Section 106 of the National Historic Preservation Act.
Where alterations are undertaken to a qualified historic building or facility that is not subject to section 106 of the National Historic Preservation Act, if the entity undertaking the alterations believes that compliance with the requirements for accessible routes (exterior and interior), ramps, entrances, toilets, parking or signage would threaten or destroy the historic significance of the building or facility and that the alternative modifications listed should be used for the feature, the entity should consult with the State Historic Preservation Officer. If the State Historic Preservation Officer agrees that compliance with the accessibility requirements for accessible routes (exterior and interior), ramps, entrances, toilets, parking or signage would threaten or destroy the historical significance of the building or facility, the alternative modifications listed may be used (see below). *202.5, Exception*

Alterations to Qualified Historic Buildings and Facilities

Alterations to a qualified historic building or facility shall comply with the State Historical Building Code, Part 8, Title 24, of the California Code of Regulations. **11B-202.5**

Exception: *Reserved.*

ADVISORY: Alterations to Qualified Historic Buildings and Facilities. The basic criteria for designating a building or property as a qualified historic building or facility are administered by the National Register program of the National Park Service. In California, those responsibilities are delegated to the Office of Historic Preservation. The Office of Historic Preservation administers the National Register and has created and administers the California Register, a similar listing. The Office of Historic Preservation delegates authority to cities and counties as "certified local governments" to apply National Register criteria for creating local lists of historic buildings and properties. There are also local governments and agencies that designate buildings and properties as historic outside of the Office of Historic Preservation program.

The State Historical Building Code provides a definition of qualified historic building or property, and can be used to determine if a building or facility is qualified. The State Historical Building Code is used in conjunction with the accessibility requirements of Chapter 11B. In general, alteration projects in qualified historic buildings and facilities must comply with the new construction requirements of Chapter 11B, however, the State Historical Building Code provides alternative accessibility provisions when an entity believes that compliance with the Chapter 11B requirements for specific elements would threaten or destroy the historical significance of the building or facility. Alternative provisions are provided for entrances, doors, power-assisted doors, toilet rooms, and exterior and interior ramps and lifts. Equivalent facilitation is permitted in specific cases when the alternative provisions themselves would threaten or destroy the historical significance or character defining features of the historic building or facility. Alternative provisions and equivalent facilitation are applied on a case-by-case basis only.

IMPLEMENTING REGULATIONS

Section 470f. Effect of Federal undertakings upon property listed in National Register: comment by Advisory Council on Historic Preservation

SOURCE- Section 106 of the National Historic Preservation Act (16 U.S.C. 470f)
The head of any Federal agency having direct or indirect jurisdiction over a proposed Federal or federally assisted undertaking in any State and the head of any Federal department or independent agency having authority to license any undertaking shall, prior to the approval of the expenditure of any Federal funds on the undertaking or prior to the issuance of any license, as the case may be, take into account the effect of the undertaking on any district, site, building, structure, or object that is included in or eligible for inclusion in the National Register. The head of any such Federal agency shall afford the Advisory Council on Historic Preservation established under sections 470I to 470v of this title a reasonable opportunity to comment with regard to such undertaking.

(Pub. L. 89-665, title I, Section 106, Oct. 15, 1966, 80 State. 917; Pub. L 94-422, title II, Section 201(3), Sept. 28, 1976, 90 State. 1320.)

Section 470a(c). Certification of local governments by State Historic Preservation Officer; transfer of portion of grants; certification by Secretary; nomination of properties by local governments for inclusion on National Register

SOURCE- Section 101(c) of the National Historic Preservation Act of 1966 (16 U.S.C. 470a (c))

(1) Any State program approved under this section shall provide a mechanism for the certification by the State Historic Preservation Officer of local governments to carry out the purposes of this subchapter and provide for the transfer, in accordance with section 470c(c) of this title, of a portion of the grants received by the States under this subchapter, to such local governments. Any local government shall be certified to participate under the provisions of this section if the applicable State Historic Preservation Officer, and the Secretary, certifies that the local government

(A) enforces appropriate state or local legislation for the designation and protection of historic properties;

(B) has established an adequate and qualified historic preservation review commission by State or local legislation;

(C) maintains a system for the survey and inventory of historic properties that furthers the purposes of subsection (b) of this section;

(D) provides for adequate public participation in the local historic preservation program, including the process of recommending properties for nomination to the National Register; and

(E) satisfactorily performs the responsibilities delegated to it under this subchapter.

Where there is no approved State program, a local government may be certified by the secretary if he determines that such local government meets the requirements of subparagraphs (A) through (E); and in any such case the Secretary may make grants-in-aid to the local government for purposes of this section.

(2)(A) Before a property within the jurisdiction of the certified local government may be considered by the State to be nominated to the Secretary for inclusion on the National Register, the State Historic Preservation Officer shall notify the owner, the applicable chief local elected official, and the local historic preservation commission. The commission, after reasonable opportunity for public comment, shall prepare a report as to whether or not such property, in its opinion, meets the criteria of the National Register. Within sixty days of notice from the State Historic Preservation Officer, the chief local elected official shall transmit the report of the commission and his recommendation to the State Historic Preservation Officer. Except as provided in subparagraph (B), after receipt of such report and recommendation, or if no such report and recommendation are received within sixty days, the State shall make the nomination pursuant to subsection (a) of this section. The State may expedite such process with the concurrence of the certified local government.

(B) If both the commission and the chief local elected official recommend that a property not be nominated to the National Register, the State Historic Preservation Officer shall take no further action, unless within thirty days of the receipt of such recommendation by the State Historic Preservation Officer an appeal is filed with the State. If such an appeal is filed, the State shall follow the procedures for making a nomination pursuant to subsection (a) of this section. Any report and recommendations made under this section shall be included with any nomination submitted by the State to the Secretary.

(3) Any local government certified under this section or which is making efforts to become so certified shall be eligible for funds under the provisions of section 470c(c)

of this title, and shall carry out any responsibilities delegated to it in accordance with such terms and conditions as the Secretary deems necessary or advisable.

Section 61.5 Approved Local Programs.

SOURCE- Section 61.5 of 36 CFR Ch. 1 (7-1-93 Edition)

(a) All approved State programs shall provide a mechanism for certifying local governments to participate in the National program.

(b) All approved State historic preservation programs shall develop, for approval by the Secretary, procedures for the certification of local governments. Procedures also shall be defined for removal of CLG status for cause. States shall indicate specific requirements for certification, specific responsibilities that will be delegated to certified local governments (CLG's), and the schedule for the certification process. The requirements outlined in paragraph (c) of this section must be incorporated into the State's process for local certification that is submitted to the Secretary for approval. Beyond the minimum delegations of authority that must be made to all CLG's. These delegations may not include the State's overall responsibility derived from the *National Historic Preservation Act, as amended,* or where specified by law or regulation (e.g., the operations of the Review Board and nominations to the National Register). Regulations and standards governing performance of State functions (e.g., rules relating to conflict of interest) are to be enforced by States when the functions are delegated.

(c) States shall require local governments to satisfy the following minimum requirements:

(1) Enforce appropriate State or local legislation for the designation and protection of historic properties. The State shall define what constitutes appropriate legislation so long as it is consistent with the purposes of the Act. Where State enabling legislation or home rule authority permits local historic preservation ordinances, a State may require adoption of an ordinance and indicate specific provisions that must be included in the ordinance.

(2) Establish by State or local law an adequate and qualified historic preservation review commission (Commission) composed of professional and lay members. All Commission members shall have a demonstrated interest, competence, or knowledge in historic preservation. To the extent available in the community, the local government shall appoint professional members from the disciplines of architecture, history, architectural history, planning, archeology, or other historic preservation related disciplines, such as urban planning, American Studies, American Civilization, Cultural Geography, or Cultural Anthropology.

(I) States may specify the minimum number and type of professional members that the local government shall appoint to the Commission and indicate how additional expertise can be obtained. Requirements set by the State for local Commissions shall not be more stringent or comprehensive that its requirements for the State Review Board. Local governments may be certified without the minimum number or types of disciplines if they can demonstrate that they have made a reasonable effort to fill those positions. When a discipline is not represented in the Commission membership, the Commission shall be required to seek expertise in this area when considering National Register nominations and other actions that will impact properties which are normally evaluated by a professional in such discipline. This can be accomplished through consulting (e.g., universities, private preservation organizations, or regional planning commissions) or by other means that the State determines appropriate.

(ii) States shall specify the role and responsibilities of the local government's Commission in local preservation decisions. These responsibilities must be

---

complementary to and carried out in coordination with those of the State as outlined in Section 61.4(b) of these rules.

(iii) States shall make available orientation materials and training to all local Commissions. The orientation and training shall be designed to provide a working knowledge of the roles and operations of Federal, State, and local preservation programs.

(3) Maintain a system for the survey and inventory of historic properties. States shall formulate guidelines for local survey and inventory systems which ensure that such systems and the data they produce can be readily integrated into statewide comprehensive historic preservation planning and other appropriate planning processes. Local government survey and inventory efforts shall be coordinated with and complementary to those of the State. The State also shall require that local survey data be in a format that is consistent with the planning processes noted above.

(4) Provide for adequate public participation in the historic preservation program, including the process of recommending properties to the National Register. States shall require adequate public participation in relation to all responsibilities that are delegated to CLG's. States shall outline specific mechanisms to ensure adequate public participation in local preservation programs. These may include requirements for open meeting, published minutes, and the publication of procedures by which assessments of potential National Register nominations, design review, etc. will be carried out as well as compliance with appropriate regulations. National Register notification requirements may be found in 36 CFR Part 60.

(5) Satisfactorily perform the responsibilities delegated to it under the Act. States shall monitor and evaluate the performance of CLG's in program operation and administration will be evaluated. Written records shall be maintained for all State evaluations of CLG's so that results are available for the Secretary's performance evaluations of States. If a State evaluation of a CLG's performance indicates that, in the State's judgment, such performance is inadequate, the State shall suggest ways to improve performance. If, after a period of time stipulated by the State, the State determines that there has not been sufficient improvement, it may recommend decertification of the local government to the Secretary for his concurrence. This recommendation shall cite the specific reasons why decertification is proposed. If the Secretary does not object within 30 working days of receipt, the decertification shall be considered approved by the Secretary. Appropriately documented State recommendations for decertification ordinarily will be accepted by the Secretary. When a local government is decertified, the State shall conduct financial assistance closeout procedures as specified in *The National Register Programs Manual*.

(d) Effects of certification:

(1) Inclusion in the process of nominating properties to the National Register of Historic Places in accordance with Sections 101 (c)(2)(A) and (c)(2)(B) of the Act. The State may delegate to a CLG any of the responsibilities of the SHPO and the State Review Board in processing National Register nominations as specified in 36 CFR Part 60, except for the authority to review and nominate properties directly to the National Register. CLG's may make nominations directly to the National Park Service only when the State does not have an approved program. States shall ensure that CLG performance of these responsibilities is consistent and coordinated with the identification, evaluation, and preservation priorities of the comprehensive State historic preservation planning process.

(2) Eligibility to apply for a portion of the State's annual HPF grant. At least 10 percent of the State's annual HPF apportionment shall be set aside for transfer to CLG's. All CLG's in the State shall be eligible to receive funds from the designated CLG share of the State's annual HPF grant; no government, however, is automatically entitled to receive funds. Local governments that receive these monies shall be considered subgrantees of the State.

(3) The requirements set forth in paragraph (c) of this section may be amplified by the Secretary and/or the States as necessary to reflect particular State and or local government program concerns.

(e) States shall submit, within 180 days of publication of the final rule for local certification and funds transfer, their proposed local certification processes to the Secretary for review and approval. In developing the submission, the State shall consult with local governments, local historic preservation commissions, and all other parties likely to be interested in the CLG process; consider local preservation needs and capabilities; and invite comments on the proposed process from local governments, commissions, and parties in the State likely to be interested. The State's proposal shall review the results of this consultation process. States shall keep a record of their consultation processes and make them available to the Secretary upon request.

(f) States shall establish procedures to ensure that all parties likely to be interested are notified and provide a 60-day period for public comment on the proposal before it is submitted to the Secretary. Records of all comments received during the commenting period shall be kept by the State and shall be made available to the Secretary upon request. The State should be able to respond to all suggestions that it does not adopt.

(g) The Secretary shall review State proposals and within 90 days of receipt issue an approval or disapproval. This review will be based on compliance with all requirements set forth in this section.

(h) If a State proposal is disapproved, the Secretary will recommend changes that would make the proposed process acceptable and, in consultation with the State, will designate a date by which the revision must be submitted. Final approval by the Secretary must be achieved by October 1, 1985, or States will be ineligible to continue their approved State program status beyond that time.

(I) A State may begin certification of local governments as soon as the State's proposed certification process is approved by the Secretary. When a local government certification request has been approved in accordance with the State's approved certification process, the State shall prepare a written certification agreement that lists the specific responsibilities of the local government when certified. The State shall forward to the Secretary a cop of the approved request and the certification agreement. If the Secretary does not take exception to the request within 15 working days of receipt, the local government shall be regarded as certified by the Secretary.

(j) A State may agree with a CLG to change the delegation of responsibilities by amending the certification agreement. The State must submit the amendment to the National Park Service for review to ensure that it is in conformance with the approved State process, this rule, and the Act. If the National Park Service does not object within 15 working days, the amendment shall be considered approved.

(k) States may amend their approved State certification and funds transfer processes. In developing the amendment, the State shall follow to the extent appropriate the same consultation procedures outlined in Section 61.5(e)/(f) and Section 61.7(g)(h). The State shall submit the proposed amendment to the National ParService. The National Park Service shall review the proposed amendment for conformance with this rule and the Act, and, within 45 working days of receipt, issue an approval or disapproval notice.

(l) State administration of its local certification process shall be reviewed by the Secretary during performance evaluations and audits of State programs as required by section 101 (b)(2) of the *National Historic Preservation Act, as amended.* Local governments may appeal to the Secretary State decisions to deny certification or to decertify. Appealed actions shall be examined for conformance with approved State procedures for CLG's, these regulations, and the Act.

(m) The district of Columbia shall be exempted from the requirements of section 61.5 because there are no subordinated local governments in the District. If a territory believes that its political subdivisions lack authorities similar to those of local governments in other States and hence cannot satisfy the requirements for local certification, it may apply to the Secretary for exemption from the requirements of Section 61.5.

(n) *Procedures for direct certification by the Secretary where there is no approved State program.* (1) When there is no approved State program, local governments wishing to be certified must apply directly to the Secretary.

(2) The application must demonstrate that the local government meets the specifications for certification set forth in paragraph (c) of this section.

(3) The Secretary shall review certification applications under this subsection and take action within 90 days of receipt.

(4) To the extent feasible, the Secretary will ensure that there is consistency and continuity in the CLG program of a State that does not have an approved historic preservation program. Therefore, if a now disapproved State program had an approved local government certification process and had already certified local governments, the Secretary will consider the process in his review of any applications for local government certification from within the State.

STATE STANDARDS

ALTERNATIVE ACCESSIBILITY PROVISIONS

SECTION 8-601 -- PURPOSE, INTENT AND SCOPE

8.601.1 Purpose. The purpose of the California Historic Building Code (CHBC) is to provide alternative regulations to facilitate access and use by persons with disabilities to and throughout facilities designated as qualified historical buildings or properties. These regulations require enforcing agencies to accept alternatives to regular code when dealing with qualified historical buildings or properties.

8-601.2 Intent. It is the intent of this chapter to preserve the integrity of qualified historical buildings and properties while providing access to and use by persons with disabilities.

8-601.3 Scope. The CBHC shall apply to every qualified historical building or property that is required to provide access to persons with disabilities.

1. Provisions of this chapter do not apply to new construction or reconstruction/replicas of historical buildings.

2. Where provisions of this chapter apply to alteration of qualified historical buildings or properties, alteration is defined in the California Building Code (CBC) as:

> **ALTERATION. ...**
> **[DSA-AC]** A change, addition or modification in construction, change in occupancy or use, or structural repair to an existing building or facility. Alterations include, but are not limited to, remodeling, renovation, rehabilitation, reconstruction, historic restoration, resurfacing of circulation paths or vehicular ways, changes or rearrangement of the structural parts or elements, and changes or rearrangement in the plan configuration of walls and full-height partitions. Normal maintenance, reroofing, painting or wallpapering, or changes to mechanical and

electrical systems are not alterations unless they affect the usability of the building or facility. **202**

8-601.4 General Application. The provisions in the CHBC apply to local and state governments (Title II entities); alteration of commercial facilities and places of public accommodation (Title III entities); and barrier removal in places of public accommodation (Title III entities). Except as noted in this chapter

SECTION 8-602 – BASIC PROVISIONS

8-602.1 Regular Code. The regular code for access for persons with disabilities shall be applied to qualified historical buildings or properties unless strict compliance with the regular code will threaten or destroy the historical significance or character-defining features of the building or property.

8-602.2 Alternative Provisions. If the historical significance or character-defining features are threatened, alternative provisions for access may be applied pursuant to this chapter provided the following conditions are met:

1. These provisions shall be applied only on an item-by-item or a case-by-case basis.

2. Documentation is provided, including meeting minutes or letters, stating the reasons for the application of the alternative provisions. Such documentation shall be maintained in the permanent file of the enforcing agency.

SECTION 8-603 – ALTERNATIVES

8-603.1 Alternative minimum standards. The alternative minimum standards for alterations of qualified historical buildings or facilities are contained in Section 4.1.7(3) of ADA Standards for Accessible Design, as incorporated and set forth in federal regulation 28 C.F.R. Part 36 (Included in this Chapter under Federal Standards, Section B, "Historic Preservation: Minimum Requirements").

8-603.2 Entry. These alternatives do not allow exceptions for the requirement of level landings in front of doors, except as provided in Section 8-603.4. Alternatives listed in order of priority are:

1. Access to any entrance used by the general public and no further than 200 feet from the primary entrance.

2. Access at any entrance not used by the general public but open and unlocked with directional signs at the primary entrance and as close as possible to, but no further than 200 feet from, the primary entrance.
3. The accessible entrance shall have a notification system. Where security is a problem, remote monitoring may be used.

8-603.3 Doors. Alternatives listed in order of priority are:

1. Single-leaf door which provides a minimum 30 inches of clear opening.

2. Single-leaf door which provides a minimum 29-1/2 inches clear opening.

3. Double door, one leaf of which provides a minimum 29-1/2 inches clear opening.

4. Double doors operable with a power-assist device to provide a minimum 29-1/2 inches clear opening when both doors are in the open position.

8-603.4 Power-assisted Doors. A power-assisted door or doors may be considered an equivalent alternative to level landings, strikeside clearance and door-opening forces required by the regular code.

8-603.5 Toilet Rooms. In lieu of separate-gender toilet facilities as required in the regular code, an accessible unisex toilet facility may be designated.

8-603.6 Exterior and Interior Ramps and Lifts. Alternatives listed in order of priority are:

1. A lift or a ramp of greater than standard slope but no greater than 1:10, for horizontal distances not to exceed 5 feet. Signs shall be posted at upper and lower levels to indicate steepness of the slope.

2. Access by ramps of 1:6 slope for horizontal distance not to exceed 13 inches. Signs shall be posted at upper and lower levels to indicate steepness of the slope.

SECTION 8-604 – EQUIVALENT FACILITATION

Use of other designs and technologies, or deviation from particular technical and scoping requirements, are permitted if the application of the alternative provisions contained in Section 8-603 would threaten or destroy the historical significance or character-defining features of the historical building or property.

Alternatives to Section 8-604 are permitted only where the following conditions are met:

1. Such alternatives shall be applied only on an item-by-item or case-by-case basis.

2. Access provided by experiences, services, functions, materials and resources through methods including, but not limited to, maps, plans, videos. Virtual reality and related equipment, at accessible levels. The alternative design and/or technologies used will provide substantially equivalent or greater accessibility to, and usability of, the facility.

3. The official charged with enforcement of the standards shall document the reasons for the application of the alternative design and/or technologies and their effect on the historical significance or character-defining features. Such documentation shall be in accordance with Section 8-602.2, Item 2, and shall include the opinions and comments of state or local accessibility officials, and the opinions and comments of representative local groups of people with disabilities. Such documentation shall retained in the permanent file of the enforcing agency. Copies of the required documentation should be available at the facility upon request.

   Note: For commercial facilities and places of public accommodation (Title III entities).
   Equivalent facilitation for an element of a building or property when applied as a waiver of an ADA accessibility requirement will not be entitled to the Federal Department of Justice certification of this code as rebuttable evidence of compliance for that element.

Chapter 3

SCOPING AND TECHNICAL REQUIREMENTS OF ACCESSIBILITY APPLICABLE TO PUBLIC ACCOMMODATIONS, COMMERCIAL FACILITIES, AND PUBLIC HOUSING

The CalDAG - California Disabled Accessibility Guidebook

1. <u>GENERAL EXCEPTIONS</u>

NOTE: The general exceptions specific to a particular improvement type or use included in this section are also repeated in the section specific to that particular improvement type or use.

General
Sites, buildings, facilities, and elements are exempt from these requirements to the extent specified. **11B-203.1** *203.1*

Construction Sites
Structures and sites directly associated with the actual processes of construction, including but not limited to, scaffolding, bridging, material hosts, material storage, and construction trailers shall not be required to comply with these requirements or to be on an accessible route. Portable toilet units provided for use exclusively by construction personnel on a construction site shall not be required to comply with the requirements for accessible toilet facilities and bathing facilities, or to be on an accessible route. **11B-203.2** *203.2*

> ***ADVISORY: Construction Sites.*** *This section provides a general exception for structures and sites directly associated with the actual processes of construction from the accessibility provisions of Chapter 11B. Construction associated structures and sites may include, but are not limited to,* ***scaffolding, bridging, materials hoists, materials storage and construction trailers.*** *Portable toilet units provided exclusively for use by construction personnel on a construction site are also exempted from the accessibility provisions of this chapter.*

Raised Area
Areas raised primarily for purposes of security, life safety, or fire safety, including but not limited to, observation or lookout galleries, prison guard towers, fire towers, or life guard stands shall not be required to comply with these requirements or to be on an accessible route. **11B-203.3** *203.3*

Limited Access Spaces
Spaces not customarily occupied and accessed only by ladders, catwalks, crawl spaces, or very narrow passageways shall not be required to comply wit these requirements or to be on an accessible route. **11B-203.4** *203.4*

Machinery Spaces
Spaces frequented only by service personnel for maintenance, repair or occasional monitoring of equipment shall not be required to comply with these requirements or to be on an accessible route. Machinery spaces include, but are not limited to, elevator pits or elevator penthouses; mechanical, electrical or communications equipment rooms; piping or equipment catwalks; water or sewage treatment pump rooms and stations; electric substations and transformer vaults; and highway and tunnel utility facilities. **11B-203.5** *203.5*

Single Occupant Structures
Single occupant structures accessed only by passageways below grade or elevated above standard curb height, including but not limited to, toll booths that are accessed only by underground tunnels, shall not be required to comply with these requirements or to be on an accessible route. **11B-203.6** *203.6*

Detention and Correctional Facilities

In detention and correctional facilities, common use areas that are used only by inmates and detainees and security personnel and that do not serve holding cells or housing cells required to comply, shall not be required to comply with these requirements or to be on an accessible route. **11B-203.7** *203.7*

Residential Facilities

In residential facilities, common use areas that do not serve residential dwelling units required to provide mobility features, or adaptable features complying with Chapter 11A, Division IV, shall not be required to comply with these requirements or to be on an accessible route. **11B-203.8** *203.8*

Employee Work Areas

Spaces and elements within employee work areas shall only be required to comply with the requirements for an accessible route in Employee Work Areas, Accessible Means of Egress and Fire Alarm Systems in Employee Work Areas, and shall be designed and constructed so that individuals with disabilities can approach, enter, and exit the employee work area. **11B-203.9** *203.9*

> *ADVISORY. Employee Work Areas. Although areas used exclusively by employees for work are not required to be fully accessible, consider designing such areas to include non-required turning spaces, and provide accessible elements whenever possible. Under the ADA, employees with disabilities are entitled to reasonable accommodations in the workplace; accommodations can include alterations to spaces within the facility. Designing employee work areas to be more accessible at the outset will avoid more costly retrofits when current employees become temporarily or permanently disabled., or when new employees with disabilities are hired. Contact the Equal Employment Opportunity Commission (EEOC) at www.eeoc.gov for information about title 1 of the ADA prohibiting discrimination against people with disabilities in the workplace.*

Raised Refereeing, Judging, and Scoring Areas

Raised structures used solely for refereeing, judging, or scoring a sport shall not be required to comply with these requirements or to be on an accessible route. A compliant accessible route shall be provided to the ground- or floor-level entry points, where provided, of stairs, ladders, or other means of reaching the raised elements or areas. **11B-203.10** *203.10*

Water Slides

Water slides shall not be required to comply with these requirements or to be on an accessible route. A compliant accessible route shall be provided to the ground- or floor-level entry points, where provided, of stairs, ladders or other means of reaching raised elements or areas. **11B-203.11** *203.11*

Animal Containment Areas

Animal containment areas that are not for public use shall not be required to comply with these requirements or to be on an accessible route. Animal containment areas for public use shall be on an accessible route. **11B.203.12** *203.12*

> *ADVISORY. Animal Containment Areas. Public circulation routes where animals may travel, such as in petting zoos and passageways alongside animal pens in State fairs, are not eligible for the exception.*

Raised Boxing or Wrestling Rings

Raised boxing or wrestling rings shall not be required to comply with these requirements or to be on an accessible route. A compliant accessible route shall be provided to the ground- or floor-level entry points, where provided, of stairs, ladders or other means of reaching raised elements or areas. **11B-203.13** *2103.13*

Raised Diving Boards and Diving Platforms

Raised diving boards and diving platforms shall not be required to comply with these requirements or to be on an accessible route. A compliant accessible route shall be provided to the ground- or floor-level entry points, where provided, of stairs, ladders or other means of reaching raised elements or areas. **11B-203.14** *203.14*

2. ACCESSIBLE ROUTES

NOTE: The requirements for accessible routes specific to a particular improvement type or use included in this section are also repeated in the section specific to that particular improvement type or use.

General

Accessible Routes shall comply as specified herein. **11B-206.1** *206.1*

Where Required

Accessible routes shall be provided where required as specified herein.
11B-206.2 *206.2*

Components

Accessible routes shall consist of one or more of the following components: walking surfaces with a running slope not steeper than 1:20, doorways, ramps, curb ramps excluding the flared sides, elevators, and platform lifts. All components of an accessible route shall comply as required. **11B-404.2** *404.2*

Site Arrival Points

_____A.   At least one accessible route shall be provided within the site is provided to an accessible building or facility entrance from:

_____ accessible parking spaces;

_____ accessible passenger loading zones;

_____ public streets and sidewalks;

_____ public transportation stops.   **11B-206.2.1**   *206.2.1*

_____B.   When more than one route is provided, all routes shall be accessible.
11B-206.2.1   *206.2.1*

> **EXCEPTIONS:**
>
> 1.   **Reserved.**
>
> 2.   An accessible route shall not be required between site arrival points and the building or facility entrance if the only means of access between them is a vehicular way not providing pedestrian access.

3.   General circulation paths shall be permitted when located in close proximity to an accessible route.

> **ADVISORY. *Site Arrival Point.*** *Each site arrival point must be connected by an accessible route to the accessible building entrance or entrances served. Where two or more similar site arrival points, such as bus stops, serve the same accessible entrance or entrances, both bus stops must be on accessible routes. In addition, the accessible routes must serve all of the accessible entrances on the site.*

> **ADVISORY.** *Access from site arrival points may include vehicular ways.*
> *Where a vehicular way, or portion of a vehicular way, is provided for pedestrian travel, such as within a shopping mall parking lot, Exception 2, above, does not apply.*

Within a Site

_____C.   At least one accessible route connects accessible buildings, accessible facilities, accessible elements, and accessible spaces that are on the same site.
11B-206.2.2   *206.2.2*

     EXCEPTION: An accessible route shall not be required between accessible buildings, accessible facilities, accessible elements, and accessible spaces if the only means of access between them is a vehicular way not providing pedestrian access.

> **ADVISORY. *Within a Site.*** *An accessible route is required to connect to the boundary of each area of sport activity. Examples of areas of sport activity include: soccer fields, basketball courts baseball fields, running tracks, skating rinks, ad the area surrounding a piece of gymnastic equipment. While the size of the sport activity may vary from sport to sport, each includes only the space needed to play. Where multiple sports fields or courts are provided, an accessible route is required to each field or area of sport activity.*

Multi-Story Buildings and Facilities

_____D.   At least one accessible route shall connect each story and mezzanine in multi-story buildings and facilities. **11B-206.2.3**   *206.2.3*

     EXCEPTIONS:

         1.   The following types of privately funded multistory buildings do not require a ramp or elevator above and below the first floor:

            1.1   Multistoried office buildings (other than the professional office of a health care provider) and passenger vehicle service stations less than three stories high or less than 3,000 square feet per story.

            1.2   Any other privately funded multistoried building that is not a shopping center, shopping mall or the professional office of a health care provider, or a terminal, depot or other station used for specified public transportation, or an airport passenger terminal and that is less than three stories high or less than 3,000 square feet per story if a reasonable portion of all facilities and accommodations normally sought and used by the public in such a building are accessible to and usable by persons with disabilities.

2.   **Reserved.**

3.   In detention and correctional facilities, an accessible route shall not be required to connect stories where cells with mobility features, all common use areas serving cells with mobility features, and all public use areas are on an accessible route.   **11B-206.2.3, Exception 3**, *206.2.3, Exception 3*

4.   In residential facilities, an accessible route shall not be required to connect stories where residential dwelling units with required mobility features and also adaptable features complying with Chapter 11A, Division IV, and all common use areas serving residential dwelling units with required mobility features and also adaptable features complying with Chapter 11A, Division IV, and public use areas serving residential dwelling units are on an accessible route. **11B-206.2.3, Exception 4**, *206.2.3, Exception 4*

5.   Within multi-story transient lodging guest rooms with mobility features, an accessible route shall not be required to connect stories provided that the living and dining areas, exterior spaces and sleeping areas are on an accessible route and sleeping accommodations for two persons minimum are provided on a story served by an accessible route. **11B-206.2.3, Exception 5**, *206.2.3, Exception 5*

6.   In air traffic control towers, an accessible route shall not be required to serve the cab and the equipment areas on the floor immediately below the cab. **11B-206.2.3, Exception 6**, *206.2.3, Exception 6*

7.   **Reserved.**

*ADVISORY. **Multi-Story Buildings and Facilities.** Space and elements located on a level not required to be served by an accessible route must fully comply with these requirements. While a mezzanine may be a change in level, it is not a story. If an accessible route is required to connect stories within a building or facility, the accessible route must serve all mezzanines.*

*ADVISORY. **Multi-Story Buildings and Facilities Exceptions 1.1 and 1.2.** These exceptions are only available to privately-funded buildings and do not include a waiver of other access features required on upper or lower floors. In other words, the exception is only for the elevator; everything else must comply. Many people with non-mobility (for example sight or hearing impairments) or semi-ambulatory conditions are served by the remaining access features required by this code. Many wheelchair users can get to upper floors through the use of crutches and other assistance, and can use their*

> *wheelchair brought to that floor where access to accessible restrooms, hallways, and accommodations are important.*

> ***ADVISORY. Multi-Story Buildings and Facilities Exception 1.2.*** *What is a reasonable portion? Typically, one of each type of accommodation that is normally sought or used by the general public which is provided on inaccessible floors must be provided must be provided on the ground floor or an accessible floor; for example, equivalent meeting rooms, classrooms, etc.*
>
> *In facilities that house a shopping center or shopping mall, or a professional office of a health care provider on an accessible ground floor, the floors that are above or below the ground floor and that do not house sales or rental establishments, or a professional office of a health care provider must still meet the accessibility requirements of this code, except for elevator service.*

> ***ADVISORY. Multi-Story Buildings and Facilities Exception 4.*** *Where common use areas are provided for the use of residents, it is presumed that all such common use areas "serve" accessible dwelling units unless use is restricted to residents occupying certain dwelling units. For example, if all residents are permitted to use all laundry rooms, then all laundry rooms "serve" accessible dwelling units. However, if the laundry room on the first floor is restricted for use by residents on the first floor, and the second floor laundry room is for use by occupants of the second floor, then first floor accessible units are "served" only by laundry rooms on the first floor. In this example, an accessible route is not required to the second floor provided that all accessible units and all common use areas serving them are on the first floor.*

Stairs and Escalators in Existing Buildings

In alterations and additions, where an escalator or stair is provided where none existed previously and major structural modifications are necessary for the installation, an accessible route shall be provided between the levels served by the escalator or stair unless exempted by the Multi-Story Buildings and Facilities Exceptions 1 through 7, above. **11B-206.2.3.1** 206.2.3.1

Distance to Elevators

In new construction of buildings where elevators are required for multi-story buildings and facilities, and which exceed 10,000 square feet on any floor, an accessible means of vertical access via ramp, elevator or lift shall be provided within 200 feet of travel of each stair and each escalator. In existing buildings that exceed 10,000 square feet on any floor and in which elevators are required, whenever a newly constructed means of vertical access is provided via stairs or an escalator, an accessible means of vertical access via ramp, elevator or lift shall be provided within 200 feet of travel of each new stair or escalator. **11B-206.2.3.2**

Spaces and Elements

At least one accessible route shall connect accessible building or facility entrances with all accessible spaces and elements within the building or facility, including mezzanines, which are otherwise connected by a circulation path unless exempted by the Multi-Story Buildings and Facilities Exceptions 1 through 7, above. **11B-206.2.4** 206.2.4

> **EXCEPTIONS:**
>
> 1. **Reserved.**
>
> 2. In assembly areas with fixed seating required to comply with the requirements for accessible Assembly Areas, an accessible route shall not be required to served fixed seating where

---

wheelchair spaces required to be on an accessible route are not provided.

3. **Reserved**

> **ADVISORY.** **Spaces and Elements.** *Accessible Routes must connect all spaces and elements required to be accessible, including, but not limited to, raised areas and speaker platforms.*

Restaurants, Cafeterias, Banquet Facilities and Bars

In restaurants, cafeterias, banquet facilities, bars, and similar facilities, an accessible route shall be provided to all functional areas, including raised or sunken dining areas, and outdoor dining areas. **11B-206.2.5** *206.2.5*

EXCEPTIONS:

1. In alterations of buildings or facilities not required to provide an accessible route between stories, an accessible route shall not be required to a mezzanine dining area where the mezzanine contains less than 25 percent of the total combined area for seating and dining and where the same décor and services are provided in the accessible area.

2. **Reserved.**

3. In sports facilities, tiered dining areas providing seating required to comply with the requirements for accessible Assembly Areas shall be required to have accessible routes serving at least 25 percent of the dining area provided that accessible routes serve seating required to comply with the requirements for accessible Assembly Areas, and each tier is provided with the same services.

Performance Areas

Where a circulation path directly connects a performance area to an assembly seating area, an accessible route shall directly connect the assembly seating area with the performance area. An accessible route shall be provided from performance areas to ancillary areas or facilities used by performers unless exempted by the Multi-Story Buildings and Facilities Exceptions 1 through 7, above. **11B-206.2.6** *206.2.6*

> **ADVISORY.** **Performance Areas.** *Performance areas, including but not limited to, stages, platforms and orchestra pits, are treated as raised or lowered areas within a given story and all are required to be accessible by a ramp, elevator or, when allowed, by platform lift. Wheelchair lifts are allowed to be used to access stages, platforms and orchestra pits used as performing areas. These areas are required to be accessible whether temporary or not.*
>
> *Generally, the accessible route to the stage shall coincide with the route for the general public, to the maximum extent feasible. For example, requiring persons with disabilities to go outside the building and reenter the building to gain access to the stage when others have a direct route would not be considered coinciding.*

Press Boxes

Press boxes in assembly areas shall be on an accessible route. **11B-206.2.7** *206.2.7*

EXCEPTIONS:

1.   An accessible route shall not be required to press boxes in bleachers that have points of entry at only one level provided that the aggregate area of all press boxes if 500 square feet maximum.

2.   An accessible route shall not be required to free-standing press boxes that are elevated above grade 12 feet minimum provided that the aggregate area of all press boxes is 500 square feet maximum.

Employee Work Areas

Common use circulation paths within employee work areas shall be accessible routes consisting of one or more of the following components: walking surfaces with a running slope not steeper than 1:20, doorways, ramps, curb ramps excluding flared sides, elevators, and platform lifts. All components of an accessible route shall comply with the applicable requirements.
11B-206.2.8, 11B-402.2   *206.2.8, 402.2*

EXCEPTIONS:

1.   **Reserved.**

2.   Common use circulation paths located within employee work areas that are an integral component of work area equipment shall not be required to provide accessible routes.

3.   Common use circulation paths located within exterior employee work areas that are fully exposed to the weather shall not be required to provide accessible routes.

> *ADVISORY. Employee Work Areas Exception 2. Large pieces of equipment, such as electric turbines or water pumping apparatus, may have stairs and elevated walkways used for overseeing or monitoring purposes which are physically part of the turbine or pump. However, passenger elevators used for vertical transportation between stories are not considered "work area equipment" as defined.*

Amusement Rides

Amusement rides required to comply shall provide accessible routes as detailed below. Accessible routes serving amusement rides shall comply except as modified.
11B-206.2.9   *206.2.9*

Load and Unload Areas

Load and unload areas shall be on an accessible route. Where load and unload areas have more than one loading or unloading position, at least one loading and unloading position shall be on an accessible route. **11B-206.2.9.1**   *206.2.9.1*

Wheelchair Spaces, Ride Seats Designed for Transfer, and Transfer Devices

When amusement rides are in the load and unload position, wheelchair spaces, amusement ride seats designed for transfer, and transfer devices shall be on an accessible route. **11B-206.2.9.2**   *206.2.9.2*

Recreational Boating Facilities

Boat slips required to comply, and boarding piers at boat launch ramps required to comply, shall be on an accessible route. Accessible routes serving recreational boating facilities shall comply except as modified. **11B-206.2.10**   *206.2.10*

Bowling Lanes

Where bowling lanes are provided, at least 5 percent, but no fewer than one of each type of bowling lane, shall be on an accessible route. **11B-206.2.11** *206.2.11*

Court Sports

In court sports, at least one accessible route shall directly connect both sides of the court. **11B-206.2.12** *206.2.12*

Exercise Machines and Equipment

Exercise machines and equipment required to comply shall be on an accessible route. **11B-206.2.13** *206.2.13*

Fishing Piers and Platforms

Fishing piers and platforms shall be on an accessible route. Accessible routes serving fishing piers and platforms shall comply except as modified. **11B-206.2.14** *206.2.14*

Golf Facilities

At least one accessible route shall connect accessible elements and spaces within the boundary of the golf course. In addition, accessible routes serving golf car rental areas; bag drop areas; course weather shelters, course toilet rooms and practice putting greens, practice teeing grounds, and teeing stations at driving ranges shall comply except as modified. **11B-206.2.15** *206.2.15*

> **EXCEPTION:** Golf car passages that comply shall be permitted to be used for all or part of accessible routes required.

Miniature Golf Facilities

Holes required to comply, including the start of play, shall be on an accessible route. Accessible routes serving miniature golf facilities shall comply except as modified. **11B-206.2.16** *206.2.16*

Play Areas

Play areas shall provide accessible routes as required. Accessible routes serving play areas shall comply except as modified. **11B-206.2.17** *206.2.17*

Ground Level and Elevated Play Components

At least one accessible route shall be provided within the play area. The accessible route shall connect ground level play components and elevated play components required to comply, including entry and exit points of the play components. **11B-206.2.17.1** *206.2.17.1*

Soft Contained Play Structures

Where three or fewer entry points are provided for soft contained play structures, at least one entry point shall be on an accessible route. Where four or more entry points are provided for soft contained play structures, at least two entry points shall be on an accessible route. **11B-206.2.17.2** *206.2.17.2*

Area of Sport Activity

An accessible route shall be provided to the boundary of each area of sport activity. **11B-206.2.18**

Location

Accessible routes shall coincide with or be located in the same area as general circulation paths. Where circulation paths are interior, required accessible routes shall also be interior. An accessible route shall not pass through kitchens, storage rooms, restrooms, closets or other spaces used for similar purposes, except as permitted by CBC Chapter 10. **11B-206.3** *206.3*

> **ADVISORY. Location.** *The accessible route must be in the same area as the general circulation path. This means that circulation paths, such as vehicular ways designed for pedestrian traffic, walks, and unpaved paths that are designed to be routinely used by pedestrians must be accessible or have an accessible route nearby. Additionally, accessible vertical interior circulation must be in the same area as stairs and escalators, not isolated in the back of the facility.*

Entrances

Entrances shall be provided as required. Entrance doors, doorways, and gates shall comply and shall be on an accessible route. **11B-206.4** *206.4*

EXCEPTIONS:

1. **Reserved.**

2. **Reserved.**

Entrances and Exterior Ground Floor Exits

All entrances and exterior ground floor exit doors to buildings and facilities shall comply as detailed herein. **11B-206.4.1** *206.4.1*

EXCEPTIONS:

1. Exterior ground-floor exits serving smoke-proof enclosures, stairwells, and exit doors servicing stairs only shall not be required to comply with these provisions.

2. Exits in excess of those required by Chapter 10 of the CBC, and which are more than 24 inches above grade shall not be required to comply with these provisions. Such doors shall have warning signs that comply with the requirements for accessible Visual Characters (see Section 30, SIGNS", stating that they are not accessible.

> **ADVISORY 11B-206.4.1 Entrances and Exterior Ground Floor Exits Exception.** *Exits in excess of those required by Chapter 10 and which are more than 24 inches above grade are exempted from the requirement for an accessible route complying with Section 404. The 2013 CBC Section 1101.4 requires such a door to be identified by a tactile exit sign with the following words "EXIT STAIR DOWN" or "EXIT STAIR UP". In an emergency situation much time can be wasted if persons with disabilities must travel the full distance to the inaccessible exit door before there is a sign posted indicating that the exit door is inaccessible. Placing a directional sign or signs along the main accessible route or routes leading to the inaccessible exit door indicating the location of the nearest accessible exit door can save valuable travel time for persons with disabilities attempting to exit a building. The directional signs must comply with the requirements of Section 11B-703.5 for visual characters.*

Parking Structure Entrances

Where direct access is provided for pedestrians from a parking structure to a building on facility entrance, each direct access to the building or facility entrance shall comply as detailed. **11B-206.4.2** *206.4.2*

Entrances from Tunnels or Elevated Walkways

Where direct access is provided for pedestrians from a pedestrian tunnel or elevated walkway to a building or facility all entrances to the building or facility from each tunnel or walkway shall comply with the requirements of Section 14, "DOORS, DOORWAYS AND GATES", and Section 14A, "MANEUVERING CLEARANCES AT DOORWAYS AND GATES". **11B-206.4.3** *206.4.3*

Transportation Facilities

In addition to the requirements for Parking Structures Entrances, Entrances from Tunnels or Elevated Walkways, Tenant Spaces, Residential Dwelling Unit Primary Entrance, Restricted Entrances. Service Entrances and Entrances for Inmates or Detainees, transportation facilities shall provide entrances as detailed. **11B-206.4.4** *206.4.4*

Location

In transportation facilities, where different entrances serve different transportation fixed routes or groups of fixed routes, entrances serving each fixed route or group of fixed routes shall comply with the requirements of Section 14, "DOORS, DOORWAYS AND GATES", and Section 14A, "MANEUVERING CLEARANCES AT DOORWAYS AND GATES". **11B-206.4.4.1** *206.4.4.1*

> **EXCEPTION:** Entrances to key stations or existing intercity rail stations retrofitted in accordance with 49 CFR 37.49 or 49 CFR 37.51 shall not be required to comply.

Direct Connections

Direct connections to the other facilities shall provide an accessible route which complies with the requirements of Section 14, "DOORS, DOORWAYS AND GATES", and Section 14A, "MANEUVERING CLEARANCES AT DOORWAYS AND GATES" from the point of connection to boarding platforms and all transportation system elements required to be accessible. Any elements provided to facilitate future direct connections shall be on an accessible route connecting boarding platforms and all transportation system elements required to be accessible. **11B-206.4.4.2** *206.4.4.2*

> **EXCEPTION:** In key stations and existing intercity rail stations, existing direct connections shall not be required to comply.

Key Stations and Intercity Rail Stations

Key stations and existing intercity rail stations required by Subpart C of 49 CFR part 37 to be altered, shall have entrances that comply with the requirements of Section 14, "DOORS, DOORWAYS AND GATES", and Section 14A, "MANEUVERING CLEARANCES AT DOORWAYS AND GATES". **11B-206.4.4.3** *206.4.4.3*

Tenant Spaces

All entrances to each tenancy in a facility shall comply with the requirements for Doors, Doorways and Gates. **11B-206.4.5** *206.4.5*

> **EXCEPTION:** Self-service storage facilities not required to comply shall not be required to be on an accessible route.

Residential Dwelling Unit Primary Entrance

In residential dwelling units, at least one primary entrance shall comply with the requirements of Section 14, "DOORS, DOORWAYS AND GATES", and Section 14A, "MANEUVERING CLEARANCES AT DOORWAYS AND GATES". The primary entrance to a residential dwelling unit shall not be to a bedroom. **11B-206.4.6** *206.4.6*

Restricted Entrances

Where restricted entrances are provided to a building or facility, all restricted entrances to the building or facility shall comply with the requirements of Section 14, "DOORS, DOORWAYS AND GATES", and Section 14A, "MANEUVERING CLEARANCES AT DOORWAYS AND GATES". **11B-206.4.7** *206.4.7*

Service Entrances

If a service entrance is the only entrance to a building or to a tenancy in a facility, that entrance shall comply with the requirements of Section 14, "DOORS, DOORWAYS AND GATES", and Section 14A, "MANEUVERING CLEARANCES AT DOORWAYS AND GATES". In existing buildings and facilities, a service entrance shall not be the sole accessible entrance unless it is the only entrance to a building or facility. **11B-206.4.8** *206.4.8*

Entrance for Inmates or Detainees

Where entrances used only by inmates or detainees and security personnel are provided at judicial facilities, detention facilities, or correctional facilities, at least one such entrance shall comply with the requirements of Section 14, "DOORS, DOORWAYS AND GATES", and Section 14A, "MANEUVERING CLEARANCES AT DOORWAYS AND GATES". **11B-206.4.9** *206.4.9*

Medical Care and Long-Term Care Facilities

Weather protection by a canopy or roof overhang shall be provided at a minimum of one accessible entrance to licensed medical care and licensed long-term care facilities where the period of stay may exceed twenty-four hours. The area of weather protection shall include the accessible passenger loading zone and the accessible route from the passenger loading zone to the accessible entrance it serves. **11B-206.4.10**

Doors, Doorways and Gates

Doors, doorways and gates providing user passage shall be provided as detailed herein. **11B-206.5** *206.5*

Entrances

Each entrance to a building or facility required to be accessible shall comply as detailed herein. **11B-206.5.1** *206.5.1*

Rooms and Spaces

Within a building or facility, every door, doorway or gate serving rooms and spaces complying with the California Building Code shall comply with the requirements of Section 14, "DOORS, DOORWAYS AND GATES", and Section 14A, "MANEUVERING CLEARANCES AT DOORWAYS AND GATES". **11B-206.5.2** *206.5.2*

Transient Lodging Facilities

In transient lodging facilities, entrances, doors, and doorways providing user passage into and within guest rooms that are not required to provide mobility features shall provide compliant clear width. **11B-206.5.3** *206.5.3*

> **EXCEPTION:** Shower and sauna doors in guest rooms that are not required to provide mobility features shall not be required to provide compliant clear width.

Residential Dwelling Units
In residential dwelling units required to provide mobility features, all doors and doorways providing user passage shall comply as detailed herein. **11B-206.5.4** *206.5.4*

Elevators
Elevators provided for passengers shall comply as detailed herein. Where multiple elevators are provided, each elevator shall comply as detailed. **11B-206.6** *206.6*

EXCEPTIONS:

1. In a building or facility permitted to use the exceptions for Multi-Story Buildings and Facilities or permitted to use a platform lift, elevators complying with the requirements for accessible Limited-Use/Limited-Activation Elevators shall be permitted.

2. Elevators complying with the requirements for accessible Limited-Use/Limited-Activation Elevators or Private Residence Elevators shall be permitted in multi-story residential dwelling units. Elevators provided as a means of access within a private residence shall be installed so that they are not accessible to the general public or to the other occupants of the building.

Existing Elevators
Where elements of existing elevators are altered, the same element shall also be altered in all elevators that are programmed to respond to the same hall call control as the altered elevator and shall comply with the requirements for Elevators for the altered element. **11B-206.6.1** *206.6.1*

Platform Lifts
Platform lifts shall comply as required. Platform lifts shall be permitted as a component of an accessible route in new construction as detailed herein. Platform lifts shall be permitted as a component of an accessible route in an existing building or facility. **11B-206.7** *206.7*

Performance Areas and Speakers' Platforms
Platform lifts shall be permitted to provide accessible routes to performance areas and speakers' platforms. **11B-206.7.1** *206.7.1*

Wheelchair Spaces
Platform lifts shall be permitted to provide an accessible route to comply with the wheelchair space dispersion and line-of-sight requirements for Assembly Areas, Wheelchair Spaces, Companion Seats, Designated Aisle Seats and Semi-Ambulant Seats. **11B-206.7.2** *206.7.2*

Incidental Spaces
Platform lifts shall be permitted to provide an accessible route to incidental spaces which are not public use spaces and which are occupied by five persons maximum. **11B-206.7.3** *206.7.3*

Judicial Spaces
Platform lifts shall be permitted to provide an accessible route to: jury boxes and witness stands; raised courtroom stations including, judges' benches, clerks'

stations, bailiffs' stations, deputy clerks' stations, and court reporters' stations; and to depressed areas such as the well of the court. **11B-206.7.4** *206.7.4*

Existing Site Constraints
Platform lifts shall be permitted where existing exterior site constraints make use of a ramp or elevator infeasible. **11B-206.7.5** *206.7.5*

> ***ADVISORY. Existing Site Constraints.*** *The exception applies where topography or other similar existing site constraints necessitate the use of a platform lift as the only feasible alternative. While the site constraint must reflect exterior conditions, the lift can be installed in the interior of a building. For example, a new building constructed between and connected to two existing buildings may have insufficient space to coordinate floor levels and also to provide ramped entry from the public way. In this example, an exterior or interior platform lift could be used to provide an accessible entrance or to coordinate one or more interior floor levels.*

Guest Rooms and Residential Dwelling Units
Platform lifts shall be permitted to connect levels within transient lodging guest rooms required to provide mobility features complying with Section 39, "TRANSIENT LODGING GUEST ROOMS", or residential dwelling units required to provide mobility features complying with Section 43, "RESIDENTIAL DWELLING UNITS" and adaptable features complying with Chapter 11A, Division IV. **11B-206.7.6** *206.7.6*

Amusement Rides
Platform lifts shall be permitted to provide accessible routes to load and unload areas serving amusement rides. **11B-206.7.7** *206.7.7*

Play Areas
Platform lifts shall be permitted to provide accessible routes to play components or soft contained play structures. **11B-206.7.8** *206.7.8*

Team or Player Seating
Platform lifts shall be permitted to provide accessible routes to team or player seating areas serving areas of sport activity. **11B-206.7.9** *206.7.9*

Recreational Boating Facilities and Fishing Piers and Platforms
Platform lifts shall be permitted to be used instead of gangways that are part of accessible routes serving recreational boating facilities and fishing piers and platforms. **11B-206.7.10** *206.7.10*

Security Barriers
Security barriers, including, but not limited to, security bollards and security check points. Shall not obstruct a required accessible route or accessible means of egress. **11B-206.8** *206.8*

> **EXCEPTION:** Where security barriers incorporate elements that cannot comply with these requirements such as certain metal detectors, fluoroscopes, or other similar devices, the accessible route shall be permitted to be located adjacent to security screening devices. The accessible route shall permit persons with disabilities passing around security barriers to maintain visual contact with their personal items to the same extent provided others passing through the security barrier.

3. ENTRANCES

NOTE: The requirements for entrances specific to a particular improvement type or use included in this section are also repeated in the section specific to that particular improvement type or use.

Entrances

Entrances shall be provided as required. Entrance doors, doorways, and gates shall comply as detailed herein and shall be on an accessible route. **11B-206.4** *206.4*

> **EXCEPTIONS:**
>
> 1. **Reserved.**
>
> 2. **Reserved.**

Entrances and Exterior Ground Floor Exits

All entrances and exterior ground floor exit doors to buildings and facilities shall comply as detailed herein. **11B-206.4.1** *206.4.1*

> **EXCEPTIONS:**
>
> 1. Exterior ground-floor exits serving smoke-proof enclosures, stairwells, and exit doors servicing stairs only shall not be required to comply with these provisions.
>
> 2. Exits in excess of those required by Chapter 10 of the CBC, and which are more than 24 inches above grade shall not be required to comply with these provisions. Such doors shall have warning signs that comply with the requirements for accessible Visual Characters (see Checklist Section on "Signs"), stating that they are not accessible.

> *ADVISORY 11B-206.4.1 Entrances and Exterior Ground Floor Exits Exception. Exits in excess of those required by Chapter 10 and which are more than 24 inches above grade are exempted from the requirement for an accessible route complying with Section 404. The 2013 CBC Section 1101.4 requires such a door to be identified by a tactile exit sign with the following words "EXIT STAIR DOWN" or "EXIT STAIR UP". In an emergency situation much time can be wasted if persons with disabilities must travel the full distance to the inaccessible exit door before there is a sign posted indicating that the exit door is inaccessible. Placing a directional sign or signs along the main accessible route or routes leading to the inaccessible exit door indicating the location of the nearest accessible exit door can save valuable travel time for persons with disabilities attempting to exit a building. The directional signs must comply with the requirements of Section 11B-703.5 for visual characters.*

Parking Structure Entrances

Where direct access is provided for pedestrians from a parking structure to a building on facility entrance, each direct access to the building or facility entrance shall comply as detailed. **11B-206.4.2** *206.4.2*

Entrances from Tunnels or Elevated Walkways

Where direct access is provided for pedestrians from a pedestrian tunnel or elevated walkway to a building or facility all entrances to the building or facility from each tunnel or walkway shall comply with the requirements of Section 14, "DOORS, DOORWAYS AND GATES", and Section 14A, "MANEUVERING CLEARANCES AT DOORWAYS AND GATES". **11B-206.4.3** *206.4.3*

Transportation Facilities

In addition to the requirements for Parking Structures Entrances, Entrances from Tunnels or Elevated Walkways, Tenant Spaces, Residential Dwelling Unit Primary Entrance, Restricted Entrances. Service Entrances and Entrances for Inmates or Detainees, transportation facilities shall provide entrances as detailed. **11B-206.4.4** *206.4.4*

Location

In transportation facilities, where different entrances serve different transportation fixed routes or groups of fixed routes, entrances serving each fixed route or group of fixed routes shall comply with the requirements of Section 14, "DOORS, DOORWAYS AND GATES", and Section 14A, "MANEUVERING CLEARANCES AT DOORWAYS AND GATES". **11B-206.4.4.1** 206.4.4.1

> **EXCEPTION:** Entrances to key stations or existing intercity rail stations retrofitted in accordance with 49 CFR 37.49 or 49 CFR 37.51 shall not be required to comply.

Direct Connections

Direct connections to the other facilities shall provide an accessible route which complies with the requirements of Section 14, "DOORS, DOORWAYS AND GATES", and Section 14A, "MANEUVERING CLEARANCES AT DOORWAYS AND GATES", from the point of connection to boarding platforms and all transportation system elements required to be accessible. Any elements provided to facilitate future direct connections shall be on an accessible route connecting boarding platforms and all transportation system elements required to be accessible. **11B-206.4.4.2** *206.4.4.2*

> **EXCEPTION:** In key stations and existing intercity rail stations, existing direct connections shall not be required to comply.

Key Stations and Intercity Rail Stations

Key stations and existing intercity rail stations required by Subpart C of 49 CFR part 37 to be altered, shall have entrances that comply with the requirements of Section 14, "DOORS, DOORWAYS AND GATES", and Section 14A, "MANEUVERING CLEARANCES AT DOORWAYS AND GATES". **11B-206.4.4.3** *206.4.4.3*

Tenant Spaces

All entrances to each tenancy in a facility shall comply with the requirements of Section 14, "DOORS, DOORWAYS AND GATES", and Section 14A, "MANEUVERING CLEARANCES AT DOORWAYS AND GATES". **11B-206.4.5** *206.4.5*

> **EXCEPTION:** Self-service storage facilities not required to comply shall not be required to be on an accessible route.

Residential Dwelling Unit Primary Entrance

In residential dwelling units, at least one primary entrance shall comply with the requirements of Section 14, "DOORS, DOORWAYS AND GATES", and Section 14A, "MANEUVERING CLEARANCES AT DOORWAYS AND GATES". The primary entrance to a residential dwelling unit shall not be to a bedroom. **11B-206.4.6** *206.4.6*

Restricted Entrances

Where restricted entrances are provided to a building or facility, all restricted entrances to the building or facility shall comply with the requirements of Section 14, "DOORS, DOORWAYS AND GATES", and Section 14A, "MANEUVERING CLEARANCES AT DOORWAYS AND GATES". **11B-206.4.7** *206.4.7*

Service Entrances

If a service entrance is the only entrance to a building or to a tenancy in a facility, that entrance shall comply with the requirements of Section 14, "DOORS, DOORWAYS AND GATES", and Section 14A, "MANEUVERING CLEARANCES AT DOORWAYS AND GATES". In existing buildings and facilities, a service entrance shall not be the sole accessible entrance unless it is the only entrance to a building or facility. **11B-206.4.8** *206.4.8*

Entrance for Inmates or Detainees

Where entrances used only by inmates or detainees and security personnel are provided at judicial facilities, detention facilities, or correctional facilities, at least one such entrance shall comply with the requirements of Section 14, "DOORS, DOORWAYS AND GATES", and Section 14A, "MANEUVERING CLEARANCES AT DOORWAYS AND GATES". **11B-206.4.9** *206.4.9*

Medical Care and Long-Term Care Facilities

Weather protection by a canopy or roof overhang shall be provided at a minimum of one accessible entrance to licensed medical care and licensed long-term care facilities where the period of stay may exceed twenty-four hours. The area of weather protection shall include the accessible passenger loading zone and the accessible route from the passenger loading zone to the accessible entrance it serves. **11B-206.4.10**

Signage for Entrances in Existing Buildings and Facilities

In existing buildings and facilities where not all entrances comply with Section 14, "DOORS, DOORWAYS AND GATES", and Section 14A, "MANEUVERING CLEARANCES AT DOORWAYS AND GATES", entrances that comply with Section 14, "DOORS, DOORWAYS AND GATES", and Section 14A, "MANEUVERING CLEARANCES AT DOORWAYS AND GATES", shall be identified by the International Symbol of Accessibility. Directional signs complying with the requirements for accessible Visual Characters shall be provided at entrances that do not comply. Directional signs with accessible Visual Characters, Including the International Symbol of Accessibility, indicating the accessible route to the nearest accessible entrance shall be provided at junctions when the accessible route diverges from the regular circulation path. (See Section 30, "SIGNS"). **11B-216.6** *216.6* **Fig. CD-3A**

EXCEPTIONS:

1. An International Symbol of Accessibility is not required at entrances to individual rooms, suites, offices, sales or rental establishments, or other such spaces when all entrances to the building or facility are accessible and persons entering the building or facility have passed through one or more entrances with signage complying with this section.

2. An International Symbol of Accessibility is not required at entrances to machinery spaces frequented only by service personnel for maintenance, repair, or occasional

monitoring of equipment; for example, elevator pits or elevator penthouses; mechanical, electrical or communications equipment rooms; piping or equipment catwalks; electric substations and transformer vaults; and highway and tunnel utility facilities.

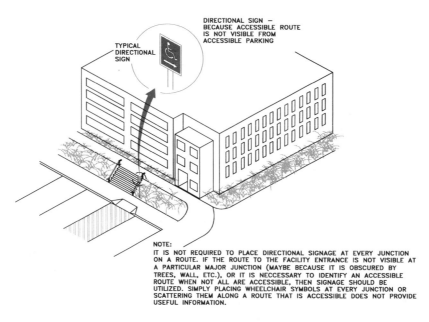

NOTE:
IT IS NOT REQUIRED TO PLACE DIRECTIONAL SIGNAGE AT EVERY JUNCTION ON A ROUTE. IF THE ROUTE TO THE FACILITY ENTRANCE IS NOT VISIBLE AT A PARTICULAR MAJOR JUNCTION (MAYBE BECAUSE IT IS OBSCURED BY TREES, WALL, ETC.), OR IT IS NECCESSARY TO IDENTIFY AN ACCESSIBLE ROUTE WHEN NOT ALL ARE ACCESSIBLE, THEN SIGNAGE SHOULD BE UTILIZED. SIMPLY PLACING WHEELCHAIR SYMBOLS AT EVERY JUNCTION OR SCATTERING THEM ALONG A ROUTE THAT IS ACCESSIBLE DOES NOT PROVIDE USEFUL INFORMATION.

Fig. CD-3A
Directional Signage

4. ACCESSIBLE MEANS OF EGRESS

General
Means of egress shall comply with CBC Chapter 10, Section 1007. **11B-207.1** *207.1*

> **EXCEPTIONS:**
>
> 1. Where means of egress are permitted by local building or life safety codes to share a common path of egress travel, accessible means of egress shall be permitted to share a common path of egress travel.
>
> 2. Areas of refuge shall not be required in detention and correctional facilities.

Platform Lifts
Standby power shall be provided for platform lifts permitted by CBC Chapter 10, Section 1007.5 to serve as part of an accessible means of egress. To ensure continued operation in case of primary power loss, platform lifts shall be provided with standby power or with self-rechargeable battery power that provides sufficient power to operate all platform lift functions for a minimum of five upward and downward trips. **11B-207.2** *207.2*

Security Barriers
Security barriers, including but not limited to, security bollards and security check points, shall not obstruct a required accessible route or accessible means of egress. **11B-206.8** *206.8*

> **EXCEPTION:** Where security barriers incorporate elements that cannot comply with these requirements such as certain metal detectors, fluoroscopes, or other similar devices, the accessible route shall be permitted to be located adjacent to security screening devices. The accessible route shall permit persons with disabilities passing around security barriers to maintain visual contact with their personal items to the same extent provided others passing through the security barrier.

Applicability
The general requirements specified in CBC Sections 1003 through 1013 shall apply to all three elements of the means of egress system, in addition to those specific requirements for the exit access, the exit and the exit discharge detailed elsewhere in CBC Chapter 10. **1003.1**

[DSA-AC] In addition to the requirement of this chapter, means of egress, which provide access to, or egress from, buildings or facilities where accessibility is required for applications listed in Section 1.8.2.1.2 regulated by the Department of Housing and Community Development, or Section 1.9.1 regulated by the Division of the State Architect-Access Compliance, shall also comply with CBC Chapter 11A or Chapter 11B, as applicable.

Accessible Means of Egress Required
Accessible means of egress shall comply with this section. Accessible spaces shall be provided with not less than one accessible means of egress. Where more than one means of egress are required by Section 1015.1 or 1021.1 from any accessible space, each accessible portion of the space shall be served by accessible means of egress in at least the same number as required by Section 1015.1 or 1021.1. In addition to the requirements of this chapter, means of egress, which provide access to, or egress from, buildings for persons with disabilities, shall also comply with the requirements of CBC Chapter 11A or 11B as applicable. **1007.1**

EXCEPTIONS:

1. Accessible means of egress are not required in alterations to existing buildings.

2. One accessible means of egress is required from an accessible mezzanine level in accordance with Section 1007.3, 1007.4 or 1007.5 and Chapter 11A or 11B, as applicable.

3. In assembly areas with sloped or stepped aisles, one accessible means of egress is permitted where the common path of travel is accessible and meets the requirements in Section 1028.8, and Chapter 11A or 11B, as applicable.

Continuity and Components

Each required accessible means of egress shall be continuous to a public way and shall consist of one or more of the following components: **1007.2**

1. Accessible routes complying with Chapter 11A, Sections 1110A.1 and 1120A, or Chapter 11B, Sections 11B-206 and 11B-402, as applicable.

2. Interior exit stairways complying with Sections 1007.3 and 1022, Chapter 11A, Section 1123A, or Chapter 11B, Sections 11B-210 and 11B-504, as applicable.

3. Interior exit access stairways complying with Sections 1007.3 and 1009.3, Chapter 11A, Section 1123A, or Chapter 11B, Sections 11B-210 and 11B-504, as applicable.

4. Exterior exit stairways complying with Sections 1007.3, 1026, and Chapter 11A, Section 1115A, or Chapter 11B, Sections 11B-210 and 11B-504, as applicable.

5. Elevators complying with Section 1007.4, and Chapter 11A, Section 1124A, or Chapter 11B, Sections 11B-206.6 and 11B-407, as applicable.

6. Platform lifts complying with Section 1007.5 and Chapter 11A, Section 1124A, or Chapter 11B, Sections 11B-206.7, 11B-207.2 and 11B-410 as applicable.

7. Horizontal exits complying with Section 1025.

8. Ramps complying with Section 1010, and Chapter 11A, Sections 1114A and 1122A, or Chapter 11B, Section 11B-405, as applicable.

9. Areas of refuge complying with Section 1007.6.

10. Exterior area for assisted rescue complying with Section 1007.7.

Elevators Required

In buildings where a required accessible floor is four or more stories above or below a level of exit discharge, at least one required accessible means of egress shall be an elevator complying with Section 1007.4.   **1007.2.1**

EXCEPTIONS:

1. In buildings equipped throughout with an automatic sprinkler system installed in accordance with Section 903.3.1.1 or 903.3.1.2, the elevator shall not be required on floors provided with a horizontal exit and located at or above the levels of exit discharge.

2. In buildings equipped throughout with an automatic sprinkler system installed in accordance with Section 903.3.1.1 or 903.3.1.2, the elevator shall not be required on floors provided with a ramp conforming to the provisions of Section 1010.

Stairways

In order to be considered part of an accessible means of egress, a stairway between stories shall have a clear width of 48 inches (1219 mm) minimum between handrails and shall either incorporate an area of refuge within an enlarged floor-level landing or shall be accessed from either an area of refuge complying with Section 1007.6 or a horizontal exit. Exit access stairways that connect levels in the same story are not permitted as part an accessible means of egress. **[DSA-AC]** In addition, exit stairways shall comply with Chapter 11A, Section 1115A and 1123A, or Chapter 11B, Sections 11B-210 and 11B-504, as applicable. **1007.3**

EXCEPTIONS:

1. The clear width of 48 inches (1219 mm) between handrails is not required in buildings equipped throughout with an automatic sprinkler system installed in accordance with Section 903.3.1.1 or 903.3.1.2.

2. Areas of refuge are not required at stairways in buildings equipped throughout by an automatic sprinkler system installed in accordance with Section 903.3.1.1 or 903.3.1.2.

3. The clear width of 48 inches (1219 mm) between handrails is not required for stairways accessed from a horizontal exit.

4. Areas of refuge are not required at stairways serving open parking garages.

5. Areas of refuge are not required for smoke protected seating areas complying with Section 1028.6.2.

6. The areas of refuge are not required in Group R-2 occupancies.

Elevators

In order to be considered part of an accessible means of egress, an elevator shall comply with the emergency operation and signaling device requirements of California Code of Regulations, Title 8, Division 1, Chapter 4, Subchapter 6, Elevator Safety Orders. Standby power shall be provided in accordance with Chapter 27 and Section 3003. The elevator shall be accessed from either an area of refuge complying with Section 1007.6 or a horizontal exit. **1007.4**

EXCEPTIONS:

1. Elevators are not required to be accessed from an area of refuge or horizontal exit in open parking garages.

2. Elevators are not required to be accessed from an area of refuge or horizontal exit in buildings and facilities equipped throughout with an automatic sprinkler system installed in accordance with Section 903.3.1.1 or 903.3.1.2.

3. Elevators not required to be located in a shaft in accordance with Section 712 are not required to be accessed from an area of refuge or horizontal exit.

4. Elevators are not required to be accessed from an area of refuge or horizontal exit for smoke protected seating areas complying with Section 1028.6.2.

Platform Lifts

Platform (wheelchair) lifts shall not serve as part of an accessible means of egress, except where allowed as part of a required accessible route in Chapter 11A, Sections 1121A and 1124A.11, or Chapter 11B, Sections 11B-206.7.1 through 11B-206.7.10, as applicable. Standby power shall be provided in accordance with Chapter 27 for platform lifts permitted to serve as part of a means of egress. **1007.5**

> **[DSA-AC]** See Chapter 11B, Section 11B-207.2 for additional accessible means of egress requirements at platform lifts.

Openness
Platform lifts on an accessible means of egress shall not be installed in a fully enclosed hoistway. **1007.5.1**

Areas of Refuge

Every required area of refuge shall be accessible from the space it serves by an accessible means of egress. The maximum travel distance from any accessible space to an area of refuge shall not exceed the travel distance permitted for the occupancy in accordance with Section 1016.1. Every required area of refuge shall have direct access to a stairway complying with Sections 1007.3 or an elevator complying with Section 1007.4. Where an elevator lobby is used as an area of refuge, the shaft and lobby shall comply with Section 1022.10 for smokeproof enclosures except where the elevators are in an area of refuge formed by a horizontal exit or smoke barrier. **[DSA-AC]** Areas of refuge shall comply with the requirements of this code and shall adjoin an accessible route complying with Sections 11B-206 and 11B-402. **1007.6**

Size
Each area of refuge shall be sized to accommodate *two* wheelchair space*s that are not less than* 30 inches by 48 inches (762 mm by 1219 mm). The total number of such 30-inch by 48-inch (762 mm by 1219 mm) spaces per story shall be not less than one for every 200 persons of calculated occupant load served by the area of refuge. Such wheelchair spaces shall not reduce the required means of egress width. Access to any of the required wheelchair spaces in an area of refuge shall not be obstructed by more than one adjoining wheelchair space. **1007.6.1**

EXCEPTION: The enforcing agency may reduce the size of each required area of refuge to accommodate one wheelchair space that is not less than 30 inches by 48 inches (762 mm by 1219 mm) on floors where the occupant load is less than 200.

Separation
Each area of refuge shall be separated from the remainder of the story by a smoke barrier complying with Section 709 or a horizontal exit complying with Section 1025. Each area of refuge shall be designed to minimize the intrusion of smoke. **1007.6.2**

> **EXCEPTION:** Areas of refuge located within an enclosure for exit access stairways or interior exit stairways.

Two-Way Communication
Areas of refuge shall be provided with a two-way communication system complying with Section 1007.8.1 and 1007.8.2. **1007.6.3**

Exterior Area for Assisted Rescue

Exterior areas for assisted rescue shall be accessed by an accessible route from the area served. Exterior areas for assisted rescue shall be permitted in accordance with Section 1007.7.1 or 1007.7.2. **1007.7**

Level of Exit Discharge

Where the exit discharge does not include an accessible route from an exit located on a level of exit discharge to a public way, an exterior area of assisted rescue shall be provided on the exterior landing in accordance with Sections 1007.7.3 through 1007.7.6. **1007.7.1**

Outdoor Facilities

Where exit access from the area serving outdoor facilities is essentially open to the outside, an exterior area of assisted rescue is permitted as an alternative to an area of refuge. Every required exterior area of assisted recue shall have direct access to an interior exit stairway, exterior stairway, or elevator serving as an accessible means of egress component. The exterior area of assisted rescue shall comply with Sections 1007.7.3 through 1007.7.6 and shall be provided with a two-way communication system complying with Sections 1007.8.1 and 1007.8.2. **1007.7.2**

Size

Each exterior area for assisted rescue shall be sized to accommodate wheelchair spaces in accordance with Section 1007.6.1. **1007.7.3**

Separation

Exterior walls separating the exterior area of assisted rescue from the interior of the building shall have a minimum fire-resistance rating of 1 hour, rated for exposure to fire from the inside. The fire-resistance-rated exterior wall construction shall extend horizontally 10 feet (3048 mm) beyond the landing on either side of the landing or equivalent fire-resistance-rated construction is permitted to extend out perpendicular to the exterior wall 4 feet (1219 mm) minimum on the side of the landing. The fire-resistance-rated construction shall extend vertically from the ground to a point 10 feet (3048 mm) above the floor level of the area for assisted rescue or to the roof line, whichever is lower. Openings within such fire-resistance-rated exterior walls shall be protected in accordance with Section 716. **1007.7.3**

Openness

The exterior area for assisted rescue shall be open to the outside air. The sides other than the separation walls shall be at least 50 percent open, and the open area shall be distributed so as to minimize the accumulation of smoke or toxic gases. **1007.7.5**

Stairway

Stairways that are part of the means of egress for the exterior area for assisted rescue shall provide a clear width of 48 inches (1219 mm) between handrails. **1007.7.6**

> **EXCEPTION:** The clear width of 48 inches (1219 mm) between handrails is not required at stairways serving buildings equipped throughout with an automatic sprinkler system installed in accordance with Section 903.3.1.1 or 903.3.1.2.

Two-Way Communication

A two-way communication system shall be provided at the elevator landing on each accessible floor that is one or more stories above or below the story of exit discharge complying with Sections 1007.8.1 and 1007.8.2. **1007.8**

EXCEPTIONS:

1. Two-way communication systems are not required at the elevator landing where the two-way communication system is provided within areas of refuge in accordance with Section 1007.6.3.

2. Two-way communication systems are not required on floors provided with ramps conforming to the provisions of Section 1010.

System Requirements

Two-way communication systems shall provide communication between each required location and a central control point location approved by the fire department. Where the central control point is not constantly attended, a two-way communication system shall have a timed automatic telephone dial-out capability to an approved monitoring location. The two-way communication system shall include both audible and visible signals.   **1007.8.1**

Visible Communication Method

[DSA-AC] A button complying with Section 1138A.4 or Sections 11B-205 and 11B-309 in the area of refuge shall activate both a light in the area of refuge indicating that rescue has been requested and a light at the central control point indicating that rescue is being requested. A button at the central control point shall activate both a light at the central control point and a light in the area of refuge indicating that the request has been received. **1007.8.1.1**

Directions

Directions for the use of the two-way communication system, instructions for summoning assistance via the two-way communication system and written identification of the specific story, floor location and building address or other building identifier shall be posted adjacent to the two-way communication system.   **1007.8.2**

Signage

Signage indicating special accessibility provisions shall be provided as shown: **1007.9**

1. Each door providing access to an area of refuge from an adjacent floor area shall be identified by a sign stating: AREA OF REFUGE.

2. Each door providing access to an exterior area for assisted rescue shall be identified by a sign stating: EXTERIOR AREA FOR ASSISTED RESCUE.

   Signage shall comply with Chapter 11A, Section 1143A and Chapter 11B, Section 11B-703.5 as applicable, requirements for visual characters and include the International Symbol of Accessibility complying with Chapter 11B, Section 11B-703.7.2.1. Where exit sign illumination is required by Section 1011.2, the signs shall be illuminated. Additionally, raised character and Braille signage complying with Chapter 11A, Section 1143A and Chapter 11B, Sections 11B-703.1, 11B-703.2, 11B-703.3 and 11B-703.5, and the International Symbol of Accessibility complying with Chapter 11B, Section 11B-703.7.2.1, shall be located at each door to an area of refuge and exterior area for assisted rescue in accordance with Section 1011.4.

Directional Signage

Direction signage complying with Chapter 11B, Section 11B-703.5 indicating the location of the other means of egress and which are accessible means of egress shall be provided at the following: **1007.10**

1. At exits serving a required accessible space but not providing an approved accessible means of egress.

2. At elevator landings.

3. Within areas of refuge.

Instructions

In areas of refuge and exterior areas for assisted rescue, instructions on the use of the area under emergency conditions shall be posted. The instructions shall include all of the following and shall comply with Chapter 11B, Section 11B-703.5:   **1007.11**

1. Persons able to use the exit stairway do so as soon as possible, unless they are assisting others.

2. Information on planned availability of assistance in the use of stairs or supervised operation of elevators and how to summon such assistance.

3. Directions for use of the two-way communications system where provided.

Alarms/Emergency Warning Systems/Accessibility
If emergency warning systems are required, they shall activate a means of warning the hearing impaired. Emergency warning systems as part of the fire-alarm system shall be designed and installed in accordance with NFPA 72 as amended in Chapter 35.   **1007.12**

5. <u>WALKING SURFACES</u>

DEFINITIONS:
"SIDEWALK" is a surfaced pedestrian way contiguous to a street used by the public. (As differentiated from the definition of "WALK".) **202**
"WALK". An exterior prepared surface for pedestrian use, including pedestrian areas such as plazas and courts. (As differentiated from the definition of "Sidewalk"). **202**

<u>Walkways</u>
A walkway shall be provided for pedestrian travel in front of every construction and demolition site unless the applicable governing authority authorizes the sidewalk to be fenced or closed. Walkways shall be of sufficient width to accommodate the pedestrian traffic, but in no case shall they be less than 4 feet (1219 mm) in width. Walkways shall be provided with a durable walking surface. Walkways shall be accessible in accordance with CBC Chapter 11A or 11B as applicable, and shall be designed to support all imposed loads and in no case shall the design live load be less than 150 pounds per square foot (psf) (7.2 kN/m^2). **CBC 3306.2**

<u>General</u>
Walking surfaces that are part of an accessible route shall comply as detailed. **11B-403.1**   *403.1*

<u>Floor or Ground Surface</u>
Floor or ground surfaces shall comply as detailed. **11B-403.2**   *403.2*

_____A.   Floor and ground surfaces comply with Section 7, "FLOOR OR GROUND SURFACES".   **11B-403.2**   *403.2*

<u>Slope</u>

_____B.   The running slope of walking surfaces is not steeper than 1:20 (5%). **11B-403.3**   *403.3*   **Fig. CD-5A**

_____C.   The cross slope of walking surfaces is not steeper than 1:48 (2.08%). **11B-403.3**   *403.3*   **Fig. CD-5A**

> **EXCEPTION:** The running slope of sidewalks shall not exceed the general grade established for the adjacent street or highway.

> ***ADVISORY: Slope.*** *There is an important distinction between a "sidewalk" and a "walk", and they are treated differently under the California Building Code. A sidewalk is contiguous to a street while a walk is not.*
>
> *When the running slope of a walking surface exceeds 5%, it must comply with the accessibility requirements for ramps.*
>
> *A sloping sidewalk with a running slope in excess of 5% is excluded from ramp requirements for landings and handrails but it must comply with cross-slope and width requirements.*

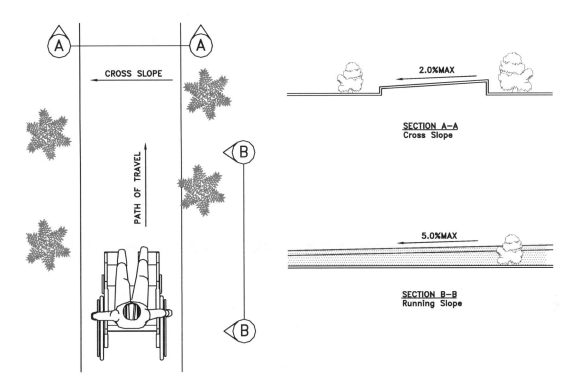

Fig. CD-5A
Allowable Slope and Cross-Slope

Changes in Level

_____D.   Changes in level shall comply with Section 8, "CHANGES IN LEVEL".
11B-403.4   *403.4*

Clearances
Walking surfaces shall provide clearances that comply as detailed.

> **EXCEPTION:** Within employee work areas, clearances on common use circulation paths shall be permitted to be decreased by work area equipment provided that the decrease is essential to the function of the work being performed. **11B-403.5**   *403.5*

Clear Width

_____E.   Except as otherwise specified, the clear width of walking surfaces is 36 inches minimum. **11B-403.5.1**   *403.5.1*   **Fig. CD-5B**

> **EXCEPTIONS:**
>
> 1.   The clear width shall be permitted to be reduced to 32 inches minimum for a length of 24 inches maximum provided that reduced width segment are separated by segments that are 48 inches long minimum and 36 inches wide minimum.   **Fig. CD-5B**
>
> 2.   The clear width for walking surfaces in corridors serving an occupant load of 10 or more shall be 44 inches minimum.

3.   The clear width for sidewalks and walks shall be 48" minimum. When, because of right-of-way restrictions, natural barriers or other existing conditions, the enforcing agency determines that compliance with the 48 inch clear sidewalk would create an unreasonable hardship, the clear width may be reduced to 36 inches.

4.   The clear width for aisles shall be 36 inches minimum if serving elements on only one side, and 44 inches minimum if serving elements on both sides. **Fig. CD-5C**

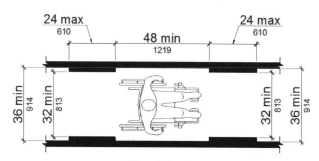

Fig. CD-5B
Clear Width of an Accessible Route
© ICC Reproduced with Permission

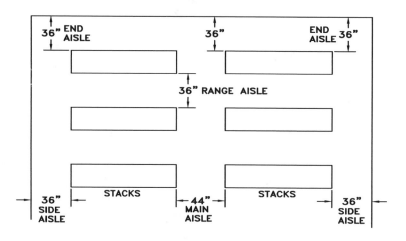

Fig. CD-5C
Clear Width for Aisles

Clear Width at Turn

_____ F.   Where the accessible route makes a 180-degree turn around an element which is less than 48 inches wide, the clear width is 42 inches minimum approaching the turn, 48 inches minimum at the turn and 42 inches minimum leaving the turn. **11B-403.5.2**   *403.5.2*   **Fig. CD-5D**

> **EXCEPTION:** Where the clear width at the turn is 60 inches minimum, compliance with the increased route widths detailed above are not required.

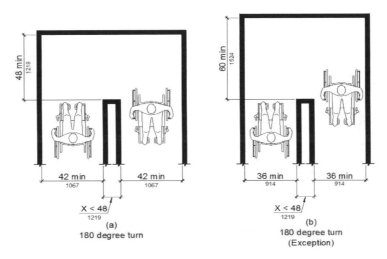

Fig. CD-5D
Clear Width at Turn
© ICC Reproduced with Permission

Passing Spaces

_____G.   An accessible route with a clear width less than 60 inches provides passing spaces at intervals of 200 feet maximum. The passing spaces are either a space 60 inches minimum by 60 inches minimum, or an intersection of two walking surfaces that provide a compliant T-shaped turning space where the base and arms of the T-shaped space extend 48 inches minimum beyond the intersection. **11B-403.5.3**   *403.5.3*   **Fig. CD-5E**

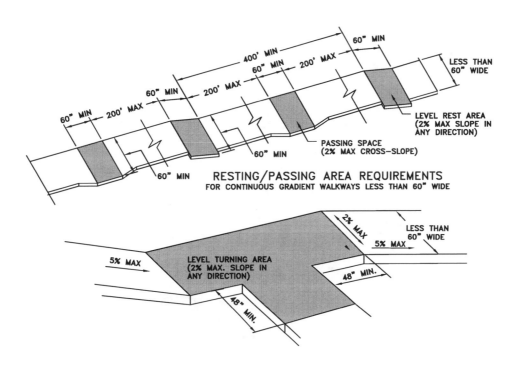

Fig. CD-5E
Resting and Passing Areas

Handrails

_____H.   Where handrails are provided along walking surfaces with running slopes not steeper than 1:20 (5%), they comply with Section 24, "HANDRAILS". **11B-403.6**   *403.6*

Continuous Gradient

_____I.   All walks with continuous gradients have level resting areas, 5 feet in length, at Intervals of 400 feet maximum **11B-403.7**   **Fig, CD-5E**

6. PROTRUDING OBJECTS

General
Protruding objects on circulation paths shall comply as detailed. **11B-204.1** *204.1*

EXCEPTIONS.

1.   Within areas of sport activity, protruding objects on circulation paths shall not be required to comply.

2.   Within play areas, protruding objects on circulation paths shall not be required to comply as detailed provided that ground level accessible routes serving ground level play components provide a vertical clearance of 80 inches high minimum. (**11B-1008.2**, *1008.2*)

Definition of Circulation Path
An exterior or interior way of passage provided for pedestrian travel, including but not limited to, walks, hallways, courtyards, elevators, platform lifts, ramps, stairways and landings. **11B-106.5, 202**   *106.5*

Protrusion Limits

_____A.   Objects with leading edges more than 27 inches and not more than 80 inches above the finished floor or ground protrude a maximum of 4 inches horizontally into the circulation path.   **11B-307.2**   *307.2*   **Fig. CD-6A**

EXCEPTION: Handrails shall be permitted to protrude 4-1/2 inches maximum.

> *ADVISORY. Protrusion Limits. When a cane is used and the element is in the detectable range, it gives a person sufficient time to detect the element with the cane before there is body contact. Elements located on circulation paths, including operable elements, must comply with the requirements for protruding objects. For example, awnings and their supporting structures cannot reduce the minimum required vertical clearance. Similarly, casement windows, when open, cannot encroach more than 4 inches into circulation paths above 27 inches.*

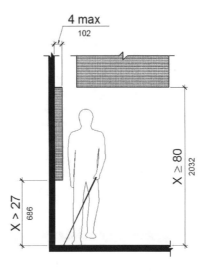

Fig. CD-6A
Limits of Protruding Objects
© ICC Reproduced with Permission

Post-Mounted Objects

_____B.   Free standing objects mounted on posts or pylons overhang circulation paths 12 inches maximum when located 27 inches minimum and 80 inches maximum above the finish floor or ground.   **11B-307.3**   *307.3*   **Fig. CD-6B**

_____C.   Where a sign or other obstruction is mounted between posts or pylons and the clear distance between the posts or pylons is greater than 12 inches, the lowest edge of such sign or obstruction is 27 inches maximum or 80 inches minimum above the finish floor or ground.   **11B-307.3**   *307.*   **Fig. CD-6B**

> **EXCEPTION:** The sloping portions of handrails serving stairs and ramps are not required to comply with these post-mounted object limitations.

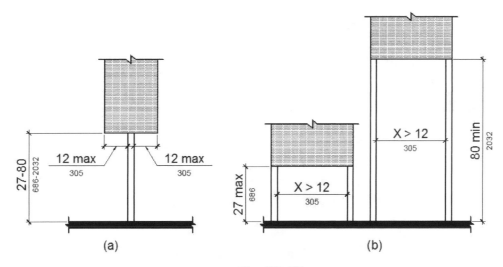

Fig. CD-6B
Post-Mounted Protruding Objects
© ICC Reproduced with Permission

Edges and Corners

_____D.   Where signs or other objects are mounted on posts or pylons, and their bottom edges are less than 80 inches above the floor or ground surface, the edges of such signs shall be rounded or eased and the corners shall have a minimum radius of 1/8 inches.   **11B-307.3.1**

Vertical Clearance

_____E.   Vertical clearance is a minimum of 80 inches high above the finish floor or ground.   **11B-307.4**   *307.4*   **Fig. CD-6C**

_____F.   Guardrails or other barriers are provided where the vertical clearance is less than 80 inches high.   **11B-307.4**   *307.4*   **Fig. CD-6C**

_____G.   The leading edge of such guardrail or barrier is located 27 inches maximum above the finish floor or ground.   **11B-307.4**   *307.4*   **Fig. CD-6C**

_____H.   Where a guy support is parallel to a circulation path including but not limited to sidewalks, a guy brace, sidewalk guy or similar device shall be used to prevent an overhanging obstruction.   **11B-307.4**

>   **EXCEPTION:** Door closers and door stops are permitted to be 78 inches minimum above the finish floor or ground.

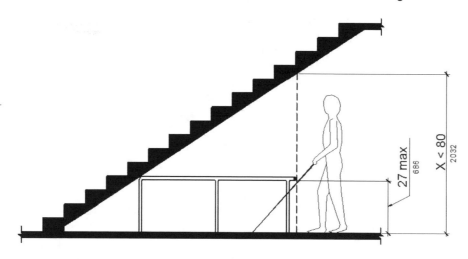

Fig. CD-6C
Vertical Clearance
© ICC Reproduced with Permission

Required Clear Width

_____I.   Protruding objects do not reduce the clear width required for accessible routes.   **11B-307.5**   *307.5*

7. FLOOR OR GROUND SURFACES

General

_____A.   Floor and ground surfaces are stable, firm and slip resistant.
11B-302.1, 11B-303.1 *302.1, 303.1*

> **EXCEPTIONS:**
>
> 1.   Within animal containment areas, floor and ground surfaces shall not be required to be stable, firm, and slip-resistant.
>
> 2.   Areas of sport activity shall not be required to comply.

> ***ADVISORY. General.*** *A stable surface is one that remains unchanged by contaminants or applied force, so that when the contaminant or force is removed, the surface returns to its original condition. A firm surface resists deformation by either indentations or particles moving on its surface. A slip-resistant surface provides sufficient frictional counterforce to the forces exerted in walking to permit safe ambulation.*

> *ADVISORY. General. Outdoor walking surfaces are often constructed of concrete or asphalt. Where permeable surfaces such as compacted decompose granite or similar material are part of the accessible route, stabilizing admixtures or binders can help to provide a firm, stable and slip-resistant surface. Such surfaces should be edged or otherwise contained to provide stability. The maintenance required to keep these surfaces firm and stable should be considered when selecting such materials.*
>
> ***Additional information regarding exterior surfaces is available on the US Access Board website at***
> http://www.access-board.gov/research/completed-research/accessible-exterior-surfaces**.**

Carpet

_____B.   Carpet or carpet tiles are securely attached and have a firm cushion, pad, or backing or no cushion or pad. Carpet or carpet tiles have a level loop, textured loop, level-cut pile or level-cut/uncut pile texture. Pile height is ½ inches maximum.   **11B-302.2** *302.2* **Fig. CD-7A**

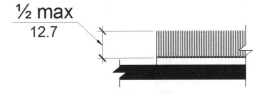

Fig. CD-7A
Carpet Pile Height
© ICC Reproduced with Permission

_____B1.   Exposed edges of carpet are fastened to floor surfaces and have trim on the entire length of the exposed edge. **11B-302.2** *302.2* **Fig. CD-7B**

_____B2.   Carpet edge trim is a maximum of ½ inches in height. **11B-302.2, 11B-303.3**   *302.2, 303.3* **Fig. CD-7B**

_____B3.   Carpet edge trim that is between ¼ inches high minimum and ½ inches high maximum is beveled with a slope not steeper than 1:2. **11B-303.3**   *303.3* **Fig. CD-7B**

_____B4.   Carpet edge trim that is ¼ inches high maximum is permitted to be vertical. **11B-303.2**   *303.2* **Fig. CD-7B**

ADVISORY. Carpet. *Carpets and permanently affixed mats can significantly increase the amount of force (roll resistance) needed to propel a wheelchair over a surface. The firmer the carpeting and backing, the lower the roll resistance. If a backing, cushion or pad is used, it must be firm.*

The accessibility provisions of the California Building Code apply only to the design, alteration, and new construction of buildings and facilities and NOT to moveable floor mats. However, if floor mats are built-in as part of new construction or alterations then they must comply with these requirements and provide a stable, firm, and slip-resistant surface along accessible routes.

The Americans with Disabilities Act however may apply to a moveable mat if it constitutes an architectural "barrier" by not providing a "stable, firm and slip-resistant surface" or by violating any other federal ADA accessibility provision if it is within an accessible route. Moveable floor mats are not required to be "attached" to a floor surface under either the ADA or CBC, just because they are "moveable".

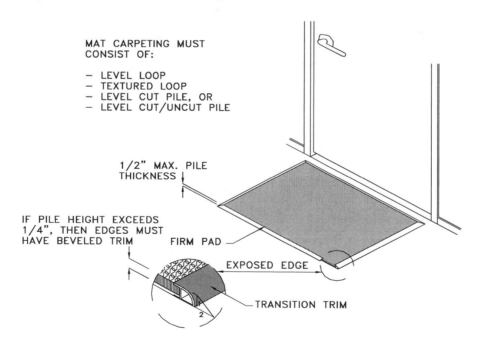

MAT CARPETING MUST
CONSIST OF:

– LEVEL LOOP
– TEXTURED LOOP
– LEVEL CUT PILE, OR
– LEVEL CUT/UNCUT PILE

1/2" MAX. PILE
THICKNESS

IF PILE HEIGHT EXCEEDS
1/4", THEN EDGES MUST
HAVE BEVELED TRIM

FIRM PAD

EXPOSED EDGE

TRANSITION TRIM

**Fig. CD-7B
Carpet Edge**

Openings

_____C.   Openings in floor or ground surfaces do not allow passage of a sphere more than ½ inches in diameter.   **11B-302.3**   *302.3*   **Fig. CD-7C**

_____D.   Elongated openings in floor or ground surfaces are placed so that the long dimension is perpendicular to the dominant direction of travel.   **11B-302.3** *302.3*   **Fig. CD-7C**

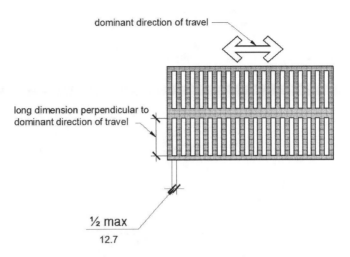

Fig. CD-7C
Elongated Openings in Floor or Ground Surfaces
© ICC Reproduced with Permission

8. CHANGES IN LEVEL

General

Where changes in level are permitted in floor or ground surfaces, they shall comply as detailed. **11B-303.1** *303.1*

EXCEPTIONS:

1. Animal containment areas shall not be required to comply as detailed.

2. Areas of sport activity shall not be required to comply as detailed.

Vertical

_____A. Changes in level of ¼ inches high maximum are permitted to be vertical and without edge treatment. **11B-303.2** *303.2* **Fig. CD-8A**

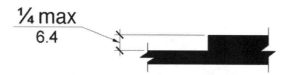

Fig. CD-8A
Vertical Change in Level
© ICC Reproduced with Permission

Beveled

_____B. Changes in level between ¼ inches high minimum and ½ inches high maximum are beveled with a slope not steeper than 1:2. **11B-303.3** *303.3* **Fig. CD-8B**

> **ADVISORY: Beveled.** *A change in level of 1/2 inch (13 mm) is permitted to be 1/4 inch (6.4 mm) vertical plus 1/4 inch (6.4 mm) beveled. However, in no case may the combined change in level exceed 1/2 inch (13 mm). Changes in level exceeding 1/2 inch (13 mm) must comply with the requirements for "Ramps" or "Curb Ramps".*

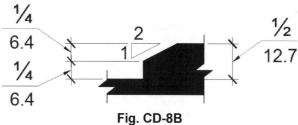

Fig. CD-8B
Beveled Change in Level
© ICC Reproduced with Permission

Ramps

_____ C.   Changes in level greater than ½ inches high are ramped and meet the applicable requirements for an accessible ramp or curb ramp.   **11B-303.4**   *303.4*

Warning Curbs

_____ D.   Abrupt changes in level exceeding 4 inches in a vertical dimension between walks, sidewalks or other pedestrian ways and adjacent surfaces or features shall be identified by warning curbs at least 6 inches in height above the walk or sidewalk surface.   **11B-303.5**   **Fig. CD-8C**

EXCEPTIONS:

1.   A warning curb is not required between a walk or sidewalk and an adjacent street or highway.

2.   A warning curb is not required when a guard or handrail is provided with a guide rail centered 2 inches minimum and 4 inches maximum above the surface of the walk or sidewalk.

ABRUPT CHANGES IN LEVEL, EXCEPT BETWEEN A WALK OR SIDEWALK AND AN ADJACENT STREET OR DRIVEWAY, EXCEEDING FOUR INCHES (4") IN VERTICAL HEIGHT, SUCH AS AT PLANTERS OR FOUNTAINS LOCATED IN OR ADJACENT TO WALKS, SIDEWALKS OR OTHER PEDESTRIAN WAYS, REQUIRE EDGE PROTECTION.

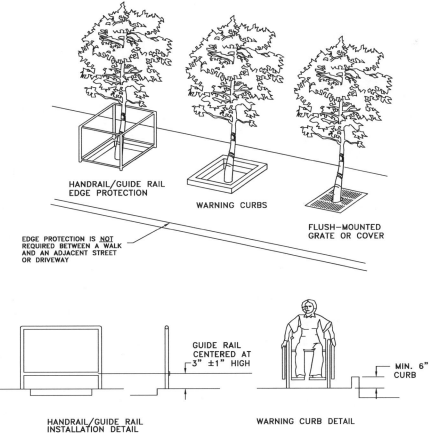

HANDRAIL/GUIDE RAIL
EDGE PROTECTION

WARNING CURBS

FLUSH-MOUNTED
GRATE OR COVER

EDGE PROTECTION IS <u>NOT</u>
REQUIRED BETWEEN A WALK
AND AN ADJACENT STREET
OR DRIVEWAY

GUIDE RAIL
CENTERED AT
3" ±1" HIGH

MIN. 6"
CURB

HANDRAIL/GUIDE RAIL
INSTALLATION DETAIL

WARNING CURB DETAIL

Fig. CD-8C
Warning Curbs

9. **TURNING SPACE**

General
Turning space shall comply as detailed.   **11B-304.1**   *304.1*

Floor or Ground Surfaces
Floor or ground surfaces of a turning space shall comply as detailed. Changes in level are not permitted.   **11B-304.2**   *304.2*

> **EXCEPTION:** Slopes not steeper than 1:48 shall be permitted

_____A.   Floor and ground surfaces comply with Section 7, "FLOOR OR GROUND SURFACES".   **11B-403.2**   *403.2*

> **ADVISORY: Floor or Ground Surface Exception.** *As used in this section, the phrase "changes in level" refers to surfaces with slopes and to surfaces with abrupt rise exceeding that permitted in Section 11B-303.3. Such changes in level are prohibited in required clear floor and ground spaces, turning spaces, and in similar spaces where people using wheelchairs and other mobility devices must park their mobility aids such as in wheelchair spaces, or maneuver to use elements such as at doors, fixtures, and telephones. The exception permits slopes not steeper than 1:48.*

Size

_____B.   The turning space is a compliant Circular Space or T-Shaped Space.
11B-304.3   *304.3*

Circular Space

_____B1.   The circular turning space is a minimum of 60 inches in diameter. The space is permitted to include compliant knee and toe clearance.
11B-304.3.1   *304.3.1*   **11B-304.3.2**
NOTE: (see Section 11 "KNEE AND TOE CLEARANCE" for compliant knee and toe clearance details).

T-Shaped Turning Space

_____B2.   The T-shaped turning space is a T-shaped space within a 60 inches square minimum with arms and base 36 inches wide minimum. Each arm of the T shall be clear of obstructions 12 inches minimum in each direction and the base shall be clear of obstructions 24 inches minimum. The space is permitted to include compliant knee and toe clearance only at the end of either the base or one arm.   **11B-304.3.2**   *304.3.2*
Fig. CD-9A
NOTE: (see Section 11 "KNEE AND TOE CLEARANCE" for compliant knee and toe clearance details).

Door Swing
Doors shall be permitted to swing into turning spaces.   **11B-304.4**   *304.4*

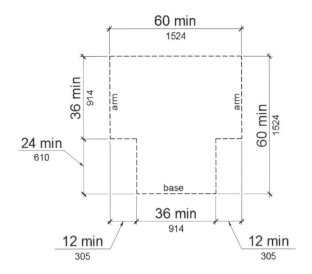

Fig. CD-9A
T-Shaped Turning Space
© ICC Reproduced with Permission

10. <u>CLEAR FLOOR OR GROUND SPACE</u>

<u>General</u>
Clear floor or ground space shall comply as detailed. **11B-305.1** *305.1*

<u>Floor or Ground Surfaces</u>

_____ A. Floor or ground surfaces of a clear space comply with Section 7, "FLOOR OR GROUND SURFACES". Changes in level are NOT permitted. **11B-305.2** *305*

EXCEPTIONS:

1. Slopes not steeper than 1:48 shall be permitted.

<u>Size</u>

_____ B. The clear floor or ground space is 30 inches minimum by 48 inches minimum. **11B-305.3** *305.3* **Fig. CD-10A**

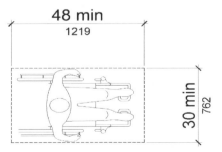

Fig. CD-10A
Clear Floor or Ground Space
© ICC Reproduced with Permission

Knee and Toe Clearance
NOTE: Unless otherwise specified, clear floor or ground space is permitted to include compliant knee and toe clearance (see Checklist Section 11, "KNEE AND TOE CLEARANCE").

Position

_____C.   Unless otherwise specified, clear floor or ground space is positioned for either forward or parallel approach to an element. **11B-305.5** *305.5* **Fig. CD-10B**

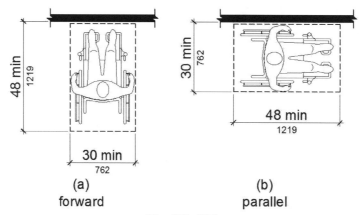

(a)
forward

(b)
parallel

Fig. CD-10B
Position of Clear Floor or Ground Space
© ICC Reproduced with Permission

Approach

_____D.   One full unobstructed side of the clear floor or ground space adjoins an accessible route or adjoins another clear floor or ground space. **11B-305.5** *305.5*

Maneuvering Clearance
Where a clear floor or ground space is located in an alcove or otherwise confined on all or part of three sides, additional maneuvering clearance is provided as follows:

Forward Approach

_____E.   Alcoves are 36 inches wide minimum where the depth exceeds 24 inches **11B-305.7.1** *305.7.1* **Fig. CD-10C**

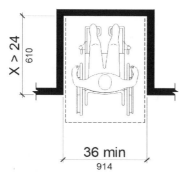

Fig. CD-10C
Maneuvering Clearance in an Alcove, Forward Approach
© ICC Reproduced with Permission

Parallel Approach

_____F.   Alcoves are 60 inches wide minimum where the depth exceeds 15 inches   **11B-305.7.2**   *305.7.2*   **Fig. CD-10D**

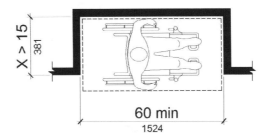

Fig. CD-10D
Maneuvering Clearance in an Alcove, Parallel Approach
© ICC Reproduced with Permission

11. <u>KNEE AND TOE CLEARANCE</u>

General
Where space beneath an element is included as part clear floor or ground space or turning space, the space shall provide compliant knee and toe clearance as detailed. Additional space shall not be prohibited beneath an element but shall not be considered as part of the clear floor or ground spaced or turning space.   **11B-306.1**   *306.1*

> **ADVISORY: General.** *Clearances are measured in relation to the usable clear floor space, not necessarily to the vertical support for an element. When determining clearance under an object for required turning or maneuvering space, care should be taken to ensure the space is clear of any obstructions.*

Toe Clearance

General
Space under an element between the finish floor or ground and 9 inches above the finish floor or ground shall be considered toe clearance and shall comply as detailed. **11B-306.2.1** *306.2.1*

Maximum Depth

_____A.   Toe clearance shall extend 25 inches maximum under an element. **11B-306.2.2**   *306.2.2*   **Fig. CD-11A**

        EXCEPTION: toe clearance shall extend 19 inches maximum Under lavatories that are required to be accessible.

Minimum Required Depth

_____B.   Where toe clearance is required at an element as part of a clear floor space, the toe clearance shall extend 17 inches minimum under the element.   **11B-306.2.3**   *306.2.3*   **Fig. CD-11A**

EXCEPTIONS:

1.   The toe clearance shall extend 19 inches minimum under sinks that are required to be accessible.

2.   The toe clearance shall extend 19 inches minimum under the built-in dining and work surfaces that are required to be accessible.

<u>Additional Clearance</u>
Space extending greater than 6 inches beyond the available knee clearance at 9 inches above the finish floor or ground shall not be considered toe clearance. **11B-306.2.4**   *306.2.4*

<u>Width</u>

_____C.   Toe clearance shall be 30 inches wide minimum. **11B-306.2.5**   *306.2.5* **Fig. CD-11A**

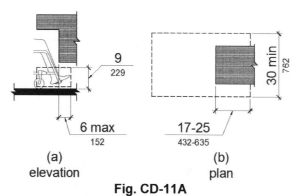

(a)
elevation

(b)
plan

Fig. CD-11A
Toe Clearance
© ICC Reproduced with Permission

<u>Knee Clearance</u>
Space under an element between 9 inches and 27 inches above the finish floor or ground shall be considered knee clearance and shall comply as detailed.   **11B-306.3.1** *306.3.1*

EXCEPTION: At lavatories required to be accessible, space between 9 inches and 29 inches above the finish floor or ground under, shall be considered knee clearance.

<u>Maximum Depth</u>

_____D.   Knee clearance extends 25 inches maximum under an element at 9 inches above the finish floor or ground.   **11B-306.3.2**   *306.3.2* **Fig. CD-11B**

<u>Minimum Required Depth</u>

_____E.   Where knee clearance is required at an element as part of a clear floor space, the knee clearance is 11 inches deep minimum at 9 inches above the finish floor or ground, and 8 inches deep minimum at 27 inches above the finish floor or ground. **11B-306.3.3**   *306.3.3*   **Fig. CD-11B**

EXCEPTIONS:

1. At lavatories required to be accessible, the knee clearance shall be 27 inches high minimum above the finish floor or ground at a depth of 8 inches minimum increasing to 29 inches high minimum above the finish floor or ground at the front edge of a counter with a built-in lavatory or at the front edge of a wall-mounted lavatory fixture.

2. At dining and work surfaces required to be accessible, knee clearance shall extend 19 inches deep minimum at 27 inches above the finish floor or ground.

Clearance Reduction
Between 9 inches and 27 inches above the finish floor or ground, the knee clearance shall be permitted to reduce at a rate of 1 inch in depth for each 6 inches in height. **11B-306.3.4** *306.3.4*

EXCEPTION: The knee clearance shall not be reduced at built-in dining and work surfaces required to be accessible.

Width

_____F. Knee clearance shall be 30 inches wide minimum. **11B-306.3.5** *306.3.5* **Figs. CD-11B**

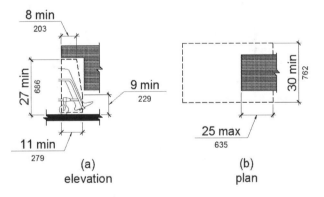

Fig. CD-11B
Knee Clearance
© ICC Reproduced with Permission

12. <u>REACH RANGES</u>

<u>General</u>
Reach ranges shall comply as detailed. **11B-308.1** *308.1*

> **ADVISORY: General.** *The following table provides guidance on reach ranges for children according to age where building elements such as coat hooks, lockers, or operable parts are designed for use primarily by children. These dimensions apply to either forward or side reaches. Accessible elements and operable parts designed for adult use or children over age 12 can be located outside these ranges but must be within the adult reach ranges required by this Section.*

Children's Reach Ranges			
Forward or Side Reach	Ages 3 and 4	Ages 5 through 8	Ages 9 through 12
High (maximum)	36 in (915 mm)	40 in (1015 mm)	44 in (1120 mm)
Low (minimum)	20 in (510 mm)	18 in (455 mm)	16 in (405 mm)

<u>Electrical Switches</u>
Controls and switches intended to be used by the occupant of a room or area to control lighting and receptacle outlets, appliances or cooling, heating and ventilating equipment, shall comply with reach range requirements except the low reach shall be measured to the bottom of the outlet box and the high reach shall be measured to the top of the outlet box. **11B-308.1.1**

<u>Electrical Receptacle Outlets</u>
Electrical receptacle outlets on branch circuits of 30 amperes or less and communication system receptacles shall comply with reach range requirements except the low reach shall be measured to the bottom of the outlet box an the high reach shall be measured to the top of the outlet box. **11B-308.1.2**

<u>FORWARD REACH</u>

<u>Unobstructed</u>

_____A.   Where a forward reach is unobstructed, the high forward reach is 48 inches maximum and the low forward reach is 15 inches minimum above the finish floor or ground. **11B-308.2.1** *308.2.1* **Fig. CD-12A**

_____B.   A clear floor or ground space that is 30 inches wide by 48 inches deep minimum provides for a front approach directly in front of the object.

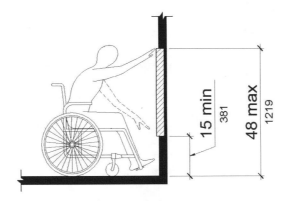

Fig. CD-12A
Unobstructed Forward Reach
© ICC Reproduced with Permission

Obstructed High Reach

_____C.   Where a high forward reach is over an obstruction, a minimum 30 inches by 48 inches clear floor space extends beneath the element for a distance not less than the required reach depth over the obstruction. **11B-308.2.2**   *308.2.2*   **Fig. CD-12B**

_____D.   The high forward reach is 48 inches maximum where the reach depth is 20 inches maximum.   **11B-308.2.2**   *308.2.2*   **Fig. CD-12B**

_____E.   Where the reach depth exceeds 20 inches, the high forward reach is 44 inches maximum and the reach depth is 25 inches maximum. **11B-308.2.2**   *308.2.2*   **Fig. CD-12B**

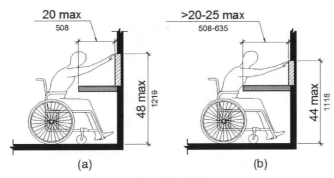

Fig. CD-12B
Obstructed High Forward Reach
© ICC Reproduced with Permission

SIDE REACH

Unobstructed

_____F.   Where a clear floor or ground space allows a parallel approach to an element and the side reach is unobstructed, the high side reach is 48 inches maximum and the low side reach is 15 inches minimum above the finish floor or ground.**11B-308.3.1**   *308.3.1*   **Fig. CD-12C**

EXCEPTIONS:

1. An obstruction shall be permitted between the clear floor or ground space and the element where the depth of the obstruction is 10 inches maximum.

2. Operable parts of fuel dispensers are permitted to be 54 inches maximum measured from the surface of the vehicular way where fuel dispensers are installed on existing curbs.

_____G. A clear floor space that is 30 inches by 48 inches minimum provides a Parallel approach to the element at a distance of 10 inches maximum. **11B-308.3.1** *308.3.1* **Fig. CD-12C**

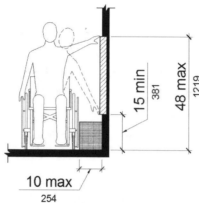

Fig. CD-12C
Unobstructed Side Reach
© ICC Reproduced with Permission

Obstructed High Reach

_____H. Where a clear floor or ground space allows a parallel approach to an element and the high side reach is over an obstruction, the height of the obstruction is 34 inches maximum and the depth of the obstruction is 24 inches maximum. **11B-308.3.2** *308.3.2* **Fig. CD-12D**

_____I. The high side reach is 48 inches maximum for a reach depth of 10 inches maximum. **11B-308.3.2** *308.3.2* **Fig. CD-12D**

_____J. Where the reach depth exceeds 10 inches, the high side reach is 46 inches maximum for a reach depth of 24 inches maximum. **11B-308.3.2** *308.3.2* **Fig. CD-12D**

EXCEPTIONS:

1. The top of washing machines and clothes dryers shall be permitted to be 36 inches maximum above the finish floor.

2. Operable parts of fuel dispensers shall be permitted to be 54 inches maximum measured from the surface of the vehicular way where fuel dispensers are installed on existing curbs.

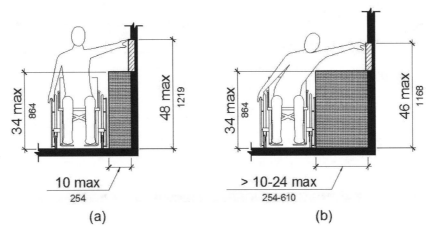

Fig. CD-12D
Obstructed High Side Reach
© ICC Reproduced with Permission

13. <u>OPERABLE PARTS</u>

<u>General</u>

Operable parts on accessible elements, accessible routes, and in accessible rooms and spaces shall comply as detailed.   **11B-205.1, 11B-309.1** *205.1, 309.1*

EXCEPTIONS:

1.   Operable parts that are intended for use only by service or maintenance personnel shall not be required to comply with this section.

2.   Electrical or communication receptacles serving a dedicated use shall not be required to comply with this section.

3.   ***Reserved.***

4.   Floor electrical receptacles shall not be required to comply with this section.

5.   HVAC diffusers shall not be required to comply with this section.

6.   Except for light switches, where redundant controls are provided for a single element, one control in each space shall not be required to comply with this section.

7.   Cleats and other boat securement devices shall not be required to comply with this section.

8.   Exercise machines and exercise equipment shall not be required to comply with this section.

> **ADVISORY:** *General. Controls covered by this section include, but are not limited to, light switches, circuit breakers, duplexes and other convenience receptacles, environmental and appliance controls, plumbing fixture controls, and security and intercom systems.*

Clear Floor Space

_____A.   A clear floor or ground space is provided that complies with Section 10, "CLEAR FLOOR OR GROUND SPACE". **11B-309.2** *309.2*

Height

_____B.   Operable parts are placed within one or more of the reach ranges specified in Section 12, "REACH RANGES". **11B-309.3** *309.3*

Operation

_____C.   Operable parts are operable with one hand and do not require tight grasping, pinching or twisting of the wrist. The force required to activate operable parts is no greater than 5 pounds maximum. **11B-309.4** *309.4*

> **EXCEPTION:** Gas pump nozzles shall (22.2N) maximum. not be required to provide operable parts that have an activating force of 5 pounds.

14. DOORS, DOORWAYS AND GATES

General
Doors, doorways and gates providing user passage shall be provided as detailed herein. **11B-206.5** *206.5*

Entrances
Each entrance to a building or facility required to be accessible shall comply as detailed herein. **11B-206.5.1** *206.5.1*

Rooms and Spaces
Within a building or facility, every door, doorway or gate serving rooms and spaces complying with the California Building Code shall comply as detailed herein. **11B-206.5.2** *206.5.2*

Transient Lodging Facilities
In transient lodging facilities, entrances, doors, and doorways providing user passage into and within guest rooms that are not required to provide mobility features shall provide compliant clear width. **11B-206.5.3** *206.5.3*

> **EXCEPTION:** Shower and sauna doors in guest rooms that are not required to provide mobility features shall not be required to provide compliant clear width.

Residential Dwelling Units
In residential dwelling units required to provide mobility features, all doors and doorways providing user passage shall provide compliant clear width. **11B-206.5.4** *206.5.4*

General

Doors, doorways and gates that are part of an accessible route shall comply as detailed herein.   **11B-404.1**   *404.1*

EXCEPTIONS:

1.   Doors, doorways and gates designed to be operated only by security personnel shall not be required to comply with the accessibility requirements specified for door and gate hardware, closing speed, opening force and maneuvering clearances for manual doors and gates, OR for the automatic and power-assisted door and gate requirements for maneuvering clearance, doors and gates in series, controls, break-out openings, or revolving doors, revolving gates and turnstiles. A sign visible from the approach side that complies with the requirements for accessible visual characters (see Section 30, "SIGNS"), shall be posted stating "Entry restricted and controlled by security personnel".

2.   At detention and correctional facilities, doors, doorways, and gates designed to be operated only by security personnel shall not be required to comply with the requirements for Door and Gate Hardware, Closing Speed, Door and Gate Opening Force, Maneuvering Clearance, Doors in Series and Gates in Series, Controls, Break-out Opening and Revolving Doors, Revolving Gates and Turnstiles.   **11B-404.1**   *404.1*

> ***ADVISORY:*** *Security personnel must have sole control of doors that are eligible for this Exception. It would not be acceptable for security personnel to operate the doors for people with disabilities while allowing others to have independent access.*

Manual Doors, Doorways, and Manual Gates

Manuel doors and doorways and manual gates intended for user passage shall comply as detailed.   **11B-404.2**   *404.2*

Revolving Doors, Gates and Turnstiles

_____A.   Revolving doors, revolving gates and turnstiles are not part of an accessible route.   **11B-404.2.1, 11B-404.3.7**   *404.2.1, 404.3.7*

Double-Leaf Doors and Gates

_____B.   At least one of the active leaves of doorways with two leaves is compliant.   **11B-404.2.2**   *404.2.2*

Clear Width

_____C.   Door openings with swing doors provide a clear width of 32 inches minimum, measured between the face of the door and the stop, with the door open 90-degrees.   **11B-404.2.3**   *404.2.3*   **Fig. CD-14A**

_____D.   Openings more than 24 inches deep provide a clear opening width of 36 inches minimum.   **11B-404.2.3**   *404.2.3*

_____E.　　There are no projections into the required clear opening width lower than 34 inches above the finish floor or ground. **11B-404.2.3** *404.2.3*

_____F.　　Projections into the clear opening width between 34 inches and 80 inches above the finish floor or ground do not exceed 4 inches. **11B-404.2.3** *404.2.3*

EXCEPTIONS:

1.　　In alterations, a projection of 5/8 inches maximum into the required clear width is permitted for the latch side stop.

2.　　Door closers and door stops are permitted to be 78 inches minimum above the finish floor or ground.

3.　　Doors, doorways and gates not providing full user passage shall provide a clear width of 20 inches minimum.

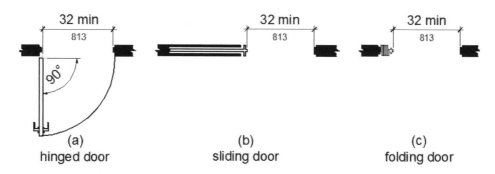

Fig. CD-14A
Clear Width of Doorways
© ICC Reproduced with Permission

Maneuvering Clearances

_____G.　　Minimum maneuvering clearances at doors and gates shall comply with Section 14A, "MANEUVERING CLEARANCES AT DOORS, DOORWAYS AND GATES". Maneuvering clearances shall extend the full width of the doorway and the required latch side or hinge side clearance. **11B-404.2.4** *404.2.4*

Door and Gate Surfaces

_____H.　　Swinging door and gate surfaces within 10 inches of the finish floor or ground measured vertically have a smooth surface on the push side extending the full width of the door or gate. **11B-404.2.10** *404.2.10* **Fig. CD-14B**

_____H1.　　Parts creating horizontal or vertical joints in these surfaces are within 1/16 inch of the same plane as the other and are free of sharp or abrasive edges. **11B-404.2.10** *404.2.10*

_____H2.　　Cavities created by added kick plates are capped. **11B-404.2.10** *404.2.10*

EXCEPTIONS:

1. Sliding doors shall not be required to comply with the 10 inch bottom smooth surface requirement.

2. Tempered glass doors without stiles and having a bottom rail or shoe with the top leading edge tapered at 60 degrees minimum from the horizontal shall not be required to meet the 10 inch bottom smooth surface requirement.

3. Doors and gates that do not extend to within 10 inches of the finish floor or ground are not required to comply with the 10 inches bottom smooth surface requirement.

4. **Reserved**.

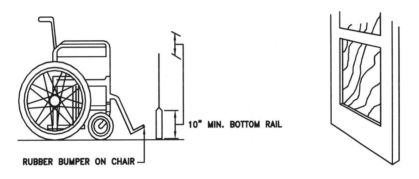

Fig. CD-14B
Door and Gate Surfaces

Vision Lights

_____I. Doors, gates, and side lights adjacent to doors or gates, containing one or more glazing panels that permit viewing through the panels, have the bottom of at least one glazed panel located 43 inches maximum above the finish floor. **11B-404.2.11** *404.2.11* **Fig. CD-14C**

EXCEPTION: Glazing panels with the lowest part more than 66 inches from the finish floor or ground are not required to meet this provision.

VISION LIGHTS
DOORS, GATES AND SIDE LIGHTS ADJACENT TO DOORS AND GATES CONTAINING ONE OR MORE GLAZED PANELS THAT PERMIT VIEWING THROUGH THE PANELS SHALL HAVE THE BOTTOM OF AT LEAST ONE GLAZED PANEL LOCATED A MAXIMUM OF 43" ABOVE THE FINISHED FLOOR.

EXCEPTION: GLAZED PANELS WITH THE LOWEST PART MORE THAN 66" FROM THE FINISHED FLOOR OR GROUND SHALL NOT BE REQUIRED TO COMPLY.

Fig. CD-14C
Vision Lights

Door and Gate Hardware

_____J.   Handles, pulls, latches, locks and other operating parts on doors and gates have a shape that is operable with one hand and does not require tight grasping, tight pinching or twisting of the wrist. **11B-404.2.7, 11B-309.4** *404.2.7, 309.4*

_____K1.   The force required to activate the operable parts on doors and gates is 5 pounds maximum. **11B-404.2.7, 11B-309.4** *404.2.7, 309.4*

_____L.   Operable parts of door and gate hardware are centered between 34 inches Minimum and 44 inches maximum above the finish floor or ground. **11B-404.2.7** *404.2.7* **Fig. CD-14D**

_____M.   Where sliding doors are in the fully open position, operating hardware is exposed and usable from both sides. **11B-404.2.7** *404.2.7*

EXCEPTIONS:

1.   Existing locks shall be permitted in any location at existing glazed doors without stiles, existing overhead rolling doors or grilles, and similar existing doors or grilles that are designed with locks that are activated only at the top or bottom rail.

2.   Access gates in barrier walls and fences protecting pools, spas, and hot tubs shall be permitted to have operable parts of the release of latch on self-latching devices at 54 inches maximum above the finish floor or ground provided the self-latching devices are not also self-locking devices and operated by means of a key, electronic opener, or integral combination lock.

> *ADVISORY: **Door and Gate Hardware.** Door hardware that can be operated with a closed fist or a loose grip accommodates the greatest range of users. Hardware that requires simultaneous hand and finger movements require greater dexterity and coordination, and is not recommended.*
>
> *Designers should also be aware of the Part 12-10-202(f) requirement for lever hardware as adopted by the State Fire Marshal. The lever of lever-actuated latches or locks must be curved with a return to within ½ inch of the door to prevent catching on the clothing of persons during egress.*

Closing Speed

_____N.   **Door Closers and Gate Closers**. Door closers and gate closers shall be adjusted so that from an open position of 90 degrees, the time required to move the door to a position of 12 degrees from the latch is 5 seconds minimum. **11B-404.2.8.1** *404.2.8.1*

_____O.   **Spring Hinges**. Door and gate spring hinges shall be adjusted so that from the open position of 70 degrees, the door or gate shall move to the closed position in 1.5 seconds minimum. **11B-404.2.8.2** *404.2.8.2*

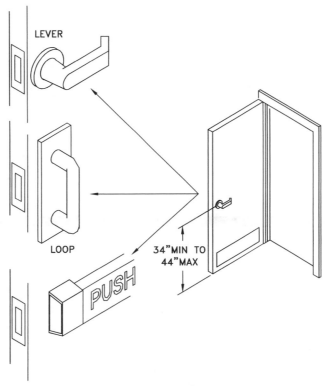

Fig. CD-14D
Door and Gate Hardware

Door and Gate Opening Force
The force for pushing or pulling open a door or gate other than fire doors is as follows:

> **ADVISORY: Door and Gate Opening Force.** *The maximum force pertains to the continuous application of force necessary to fully open a door, not the initial force needed to overcome the inertia of the door. These forces do not apply to the force required to retract latch bolts or disengage other devices that hold the door or gate in a closed position.*

____P.   The force for pushing or pulling open <u>interior hinged doors and gates</u> is 5 pounds maximum. **11B-404.2.9**   *404.2.9*

____Q.   The force for pushing or pulling open <u>sliding or folding doors</u> is 5 pounds maximum. **11B-404.2.9**   *404.2.9*

____R.   The force for pushing or pulling open <u>required fire doors</u> the minimum opening force allowable by the appropriate administrative authority, not to exceed 15 pounds. **11B-404.2.9**

____S.   The force for pushing or pulling open <u>exterior hinged doors</u> is 5 pounds maximum. **11B-404.2.9**

EXCEPTIONS:

1.   Exterior doors to machinery spaces including, but not limited to, elevator pits or elevator penthouses; mechanical, electrical or communications

---

equipment rooms; piping or equipment catwalks; electric substations and transformer vaults; and highway and tunnel utility facilities.

2.   When, at a single location, one of every 8 exterior door leafs, or fraction of 8 is a powered door, other exterior doors at the same location, serving the same interior space, may have a maximum opening force of 8.5 pounds (37.8 N). The powered leaf(s) shall be located closest to the accessible route.

   a.   Powered doors shall comply as specified herein.   Powered doors shall be fully automatic doors complying with Builders Hardware Manufacturer's Association (BHMA) A156.10 or low energy operated doors complying with BHMA A156.19.

   b.   Powered doors serving a building or facility with an occupancy of 150 or more shall be provided with a back-up battery or back-up generator. The back-up power source shall be able to cycle the door a minimum of 100 cycles.

   c.   Powered doors shall be controlled on both the interior and exterior sides of the doors by sensing devices, push plates, vertical actuation bars or other similar operating devices that comply with Section 13, "OPERABLE PARTS."

   At each location where push plates are provided there shall be two push plates; the centerline of one push plate shall be 7 inches (7") minimum and eight inches (8") maximum above the floor or ground surface and the second push plate shall be 30 inches (30") minimum and 44 inches (44") maximum above the floor or ground surface. Each push plate shall be a minimum of 4 inches (4") diameter or a minimum of four inches (4") by four inches (4") square and shall display the International Symbol of Accessibility in a compliant manner.

   At each location where vertical actuation bars are provided the operable portion shall be located so the bottom is 5 inches (5") maximum above the floor or ground surface and the top is 35 inches (35") maximum above the floor or ground surface. The operable portion of each vertical actuation bar shall be a minimum of 2 inches (2") wide and shall display the International Symbol of Accessibility in a compliant manner.

   Where push plates, vertical actuation bars or other similar operating devices are provided, they shall be placed in a conspicuous location. A level and clear floor or ground space for an accessible forward or parallel approach shall be provided, centered on the operating device. Doors shall not swing into the required clear floor or ground space.

   d.   Signage identifying the accessible entrance (see Section 3, "ENTRANCES") shall be placed on, or immediately adjacent to, each powered door. Signage shall be provided in compliance with BHMA A156.10 or BHMA 156.19, as applicable.

   e.   In addition to the requirements of Item d, were a powered door is provided in buildings or facilities containing assembly occupancies of 300 or more, a sign displaying the International Symbol of Accessibility measuring 6 inches by 6 inches (6" x 6") in a compliant manner shall be provided above the door on both the interior and exterior sides of each powered door.

Automatic and Power-Assisted Doors and Gates

Full-powered automatic doors shall comply with ANSI/BHMA A156.10. Low-energy and power-assisted doors shall comply with ANSI/BHMA A156.19. **11B-404.3** *404.3*

> ***ADVISORY: Automatic and Power-Assisted Doors and Gates.*** *Automatic and power-assisted doors are often used by designers to provide accessibility when door closer pressure would exceed the allowable opening force for interior or exterior doors. Heavy doors are difficult to open for persons using wheelchairs because holding the door open with one hand requires the user to let go of one wheel – not enough control remains to prevent the chair from twisting out of the intended direction. An automatic door may be used as equivalent facilitation when it is technically infeasible to provide sufficient strike-side clearance for code compliance, but a power-assisted door may not.*

Clear Width

_____T.   Doorways provide a clear opening of 32 inches minimum in power-on and power-off mode. The minimum clear width for automatic door systems in a doorway provide a clear, unobstructed opening of 32 inches with one leaf positioned at an angle of 90 degrees from its closed position. **11B-404.3.1** *404.3.1*

Break Out Opening

_____U.   Where doors and gates without standby power are a part of a means of egress, the clear break out opening at swinging or sliding doors and gates is 32 inches minimum when operated in emergency mode. **11B-404.3.6** *404.3.6*

> **EXCEPTION:** Where manual swinging doors and gates already comply with the required disabled access provisions, and also serve the same means of egress, the break-out opening capability is not required.

Maneuvering Clearance

_____V.   The clearances at power-assisted doors and gates, and the clearances at automatic doors and gates without standby power that serve an accessible means of egress, comply with Section 14A, "MANEUVERING CLEARANCES AT DOORS AND GATES". **11B-404.3.2** *404.3.2*

> **EXCEPTION:** Where automatic doors and gates remain open in the power-off condition, compliance with the following Section 14A, "MANEUVERING CLEARANCES AT DOORS AND GATES", is not required.

Controls

_____W.   Manually operated controls comply with Section #13 "OPERABLE PARTS". The clear floor space adjacent to the control is located beyond the arc of the door swing. **11B-404.3.5** *404.3.5*

> ***ADVISORY: Controls.*** *Where push plates, vertical actuation bars or other similar operating mechanisms are provided, they must be placed in a conspicuous location. A compliant level and clear floor or ground space for forward or parallel approach should be provided that is centered on the operating device. Doors should not swing into the required clear floor or ground space at the operating device.*

Revolving Doors, Gates and Turnstiles

_____X.   Revolving doors, revolving gates and turnstiles are not part of an accessible route.   **11B-404.3.7**   *404.3.7*

14A.   <u>MANEUVERING CLEARANCES AT DOORS AND GATES</u>

<u>Maneuvering Clearances</u>. The minimum maneuvering clearances listed shall extend the full width of the doorway and the required latch side or hinge side clearance. **11B-404.2.4**   *404.2.4*

Floor or Ground Surfaces

_____A.   Floor or ground surfaces within the required maneuvering clearances comply with Section 7, "FLOOR OR GROUND SURFACES". Changes in level are NOT permitted. **11B-404.2.4.4, 11B-302.1** *404.2.4.4, 302.1*

EXCEPTIONS:

    1.   Slopes not steeper than 1:48 shall be permitted.

    2.   Changes in level at compliant thresholds is permitted.

Changes in Level, Thresholds

_____B.   Thresholds, if provided at doorways, are ½ inches high maximum. Raised thresholds and changes in level at doorways shall comply with Section 8, "CHANGES IN LEVEL" **11B-404.2.5**   *404.2.5*

Doors in Series and Gates in Series

_____L.   The distance between two hinged or pivoted doors in series and gates in series is 48 inches minimum plus the width of the doors or gates swinging into the space. **11B-404.2.6** *404.2.6* **Fig CD-14E**

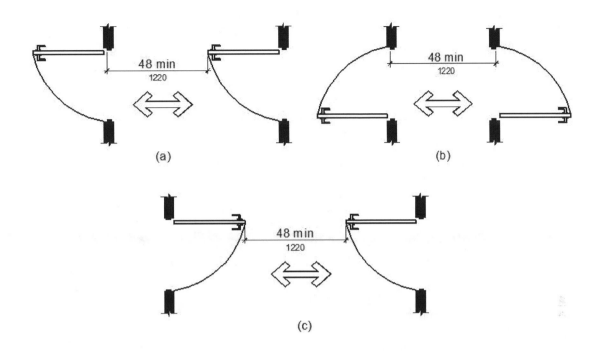

Fig. CD-14E
Doors in Series and Gates in Series
© ICC Reproduced with Permission

Swinging Doors and Gates. Swinging doors and gates shall have maneuvering clearances that comply with the dimensions listed below in Table 14A. **11B-404.2.4.1** *404.2.4.1*

Table 14A – Maneuvering Clearances at Manual Swinging Doors and Gates

Type of Use		Minimum Maneuvering Clearance	
Approach Direction	**Door or Gate Side**	**Perpendicular to Doorway**	**Parallel to Doorway (beyond latch side unless noted)**
From front	Pull	60 inches (*1524 mm*)	18 inches (*457* mm)[5]
From front	Push	48 inches (*1219 mm*)	0 inches (*0 mm*)[1]
From hinge side	Pull	60 inches (*1524 mm*)	36 inches (*914 mm*)
From hinge side	Push	*44 inches (1118 mm)*[2]	22 inches (*559 mm*)[3]
From latch side	Pull	*60 inches (1524 mm)*	24 inches (610 mm)
From latch side	Push	*44 inches (1118 mm)*[4]	24 inches (610 mm)

1. Add 12 inches (305 mm) if closer and latch are provided.
2. Add *4 inches (102 mm)* if closer and latch are provided.
3. Beyond hinge side.
4. Add *4 inches (102 mm)* if closer is provided.
5. Add *6 inches (152 mm) at exterior side of exterior doors.*

MANEUVERING CLEARANCES AT MANUAL SWINGING DOORS AND GATES

FRONT APPROACH, PULL SIDE

_____M.   The floor or ground area is the width of the door and a minimum of 60 inches deep, perpendicular to the doorway.   **11B-404.2.4**   *404.2.4*   **Fig. CD-14F(a)**

_____N.   The floor or ground area beyond the latch side of the door and parallel to the doorway is 18 inches minimum at <u>interior</u> doors.   **11B-404.2.4**   *404.2.4*   **Fig.CD-14F(a)**

_____O.   The floor or ground area beyond the latch side of the door and parallel to the doorway is 24 inches minimum at <u>exterior</u> doors.   **11B-404.2.4**   *404.2.4*   **Fig. CD-14F(c)**

HINGE SIDE APPROACH, PULL SIDE

_____R.   The floor or ground area is the width of the door and a minimum of 60 inches deep, perpendicular to the doorway.   **11B-404.2.4**   *404.2.4*   **Fig. CD-14F(d)**

_____S.   The floor or ground area beyond the latch side of the door and parallel to the doorway is 36 inches minimum.   **11B-404.2.4**   *404.2.4*   **Fig. CD-14F(d)**

HINGE SIDE APPROACH, PUSH SIDE

_____T.   The floor or ground area is the width of the door and a minimum of 44 inches deep, perpendicular to the doorway.   **11B-404.2.4**   *404.2.4*   **Fig. CD-14F(f)**

_____U.   The floor or ground area is the width of the door and a minimum of 48 inches deep, perpendicular to the doorway, if <u>BOTH</u> A CLOSER <u>AND</u> A LATCH ARE PROVIDED.   **11B-404.2.4**   *404.2.4*   **Fig. CD-14G(g)**

_____V.   The floor or ground area beyond the hinge side of the door and parallel to the doorway is 22 inches minimum.   **11B-404.2.4**   *404.2.4*   **Figs. CD-14G(f) & (g)**

LATCH SIDE APPROACH, PULL SIDE

_____W.   The floor or ground area is the width of the door and a minimum of 60 inches deep, perpendicular to the doorway.   **11B-404.2.4**   *404.2.4*   **Fig. CD-14G(h)**

_____X.   The floor or ground area beyond the latch side of the door and parallel to the doorway is 24 inches minimum.   **11B-404.2.4**   *404.2.4*   **Fig. CD-14G(h)**

LATCH SIDE APPROACH, PUSH SIDE

_____Y.   The floor or ground area is the width of the door and a minimum of 44 inches deep, perpendicular to the doorway.   **11B-404.2.4**   *404.2.4*   **Fig. CD-14G(j)**

_____Z.   The floor or ground area is the width of the door and a minimum of 48 inches deep, perpendicular to the doorway, if A CLOSER IS PROVIDED.   **11B-404.2.4**   *404.2.4*   **Fig. CD-14G(k)**

_____AA.   The floor or ground area beyond the latch side of the door and parallel to the doorway is 24 inches minimum.   **11B-404.2.4**   *404.2.4*   **Figs. CD-14G(j) & (k)**

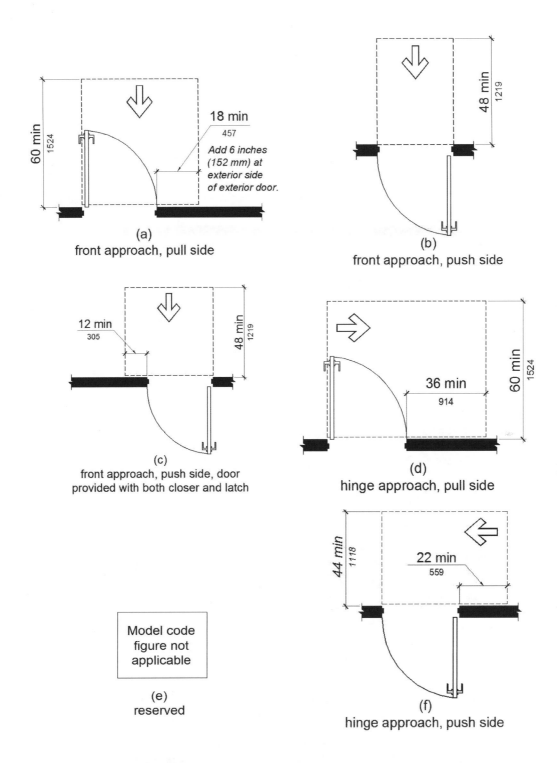

(a)
front approach, pull side

18 min
457

Add 6 inches (152 mm) at exterior side of exterior door.

60 min
1524

(b)
front approach, push side

48 min
1219

(c)
front approach, push side, door provided with both closer and latch

12 min
305

48 min
1219

(d)
hinge approach, pull side

36 min
914

60 min
1524

Model code figure not applicable

(e)
reserved

(f)
hinge approach, push side

44 min
1118

22 min
559

Fig. CD-14F
Maneuvering Clearances at Manual Swinging Doors and Gates
© ICC Reproduced with Permission

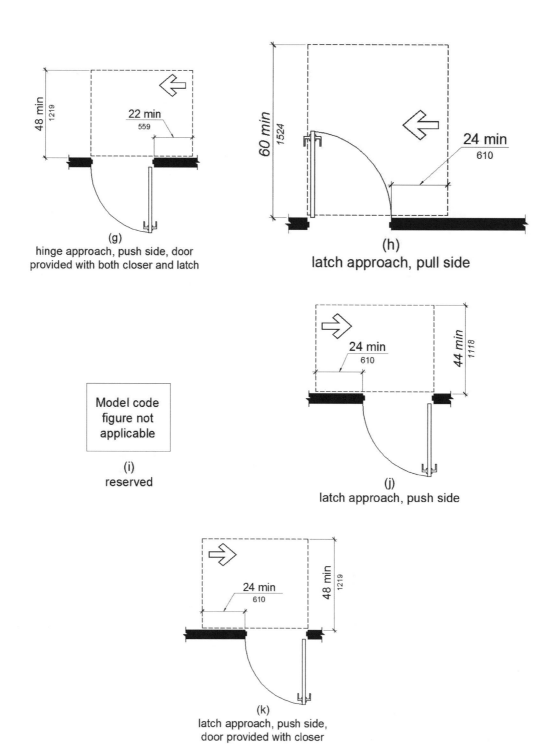

Fig. CD-14G
Maneuvering Clearances at Manual Swinging Doors and Gates
© ICC Reproduced with Permission

Doorways Without Doors or Gates, Sliding Doors, and Folding Doors
Doorways less than 36 inches wide without doors or gates, sliding doors, or folding doors shall have maneuvering clearances that comply with the dimensions below in Table 14B.
11B-404.2.4.2   *404.2.4.2*

Table 14B
Maneuvering Clearances at Doorways Without Doors or Gates,
Manual Sliding Doors, and Manual Folding Doors

Approach Direction	Minimum Maneuvering Clearance	
	Perpendicular to Doorway	**Parallel to Doorway (beyond stop/latch side unless noted)**
From front	48 inches (*1219* mm)	0 inches (0 mm)
From side[1]	42 inches (*1067* mm)	0 inches (0 mm)
From pocket/hinge side	42 inches (*1067* mm)	22 inches (*559* mm)[2]
From stop/latch side	42 inches (*1067* mm)	24 inches (610 mm)

1. Doorway with no door only.
2. Beyond pocket/hinge side.

MANEUVERING CLEARANCES AT DOORWAYS WITHOUT DOORS OR GATES, MANUAL SLIDING DOORS, AND MANUAL FOLDING DOORS

FRONT APPROACH

_____BB.   The floor or ground area is the width of the door and a minimum of 48 inches deep, perpendicular to the doorway.   **11B-404.2.4.2**   *404.2.4.2*   **Fig. CD-14H(a)**

SIDE APPROACH (doorway with no door only)

_____CC.   The floor or ground area is the width of the door and a minimum of 42 inches deep, perpendicular to the doorway.   **11B-404.2.4.2**   *404.2.4.2*   **Fig. CD-14I(b)**

POCKET/HINGE SIDE APPROACH

_____DD.   The floor or ground area is the width of the door and a minimum of 42 inches deep, perpendicular to the doorway.   **11B-404.2.4.2**   *404.2.4.2*   **Fig. CD-14H(c)**

_____EE.   The floor or ground area beyond the stop/latch side of the door and parallel to the doorway is 22 inches minimum beyond the pocket/hinge side   **11B-404.2.4.2**   *404.2.4.2*   **Fig. CD-14H(c)**

STOP/LATCH SIDE APPROACH

_____FF.   The floor or ground area is the width of the door and a minimum of 42 inches deep, perpendicular to the doorway.   **11B-404.2.4.2**   *404.2.4.2*   **Fig. CD-14H(d)**

_____GG.   The floor or ground area beyond the stop/latch side of the door and parallel to the doorway is 24 inches minimum.   **11B-404.2.4.2**   *404.2.4.2*   **Fig. CD-14H(d)**

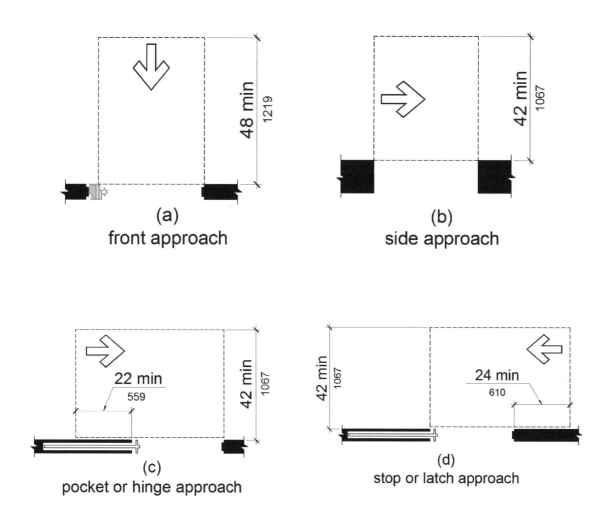

Fig. CD-14H
Maneuvering Clearances at Doorways without Doors, Sliding Doors,
Gates, and folding Doors
© ICC Reproduced with Permission

MANEUVERING CLEARANCES AT RECESSED DOORS AND GATES

Recessed Doors and Gates

Maneuvering clearances for forward approach shall be provided when any obstruction within 18 inches of the latch side of a doorway projects more than 8 inches beyond the face of the door, measured perpendicular to the face of the door or gate. **11B-404.2.4.3** *404.2.4.3*

FRONT APPROACH, PULL SIDE

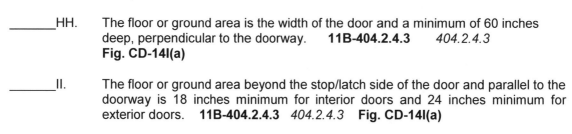

_____HH.   The floor or ground area is the width of the door and a minimum of 60 inches deep, perpendicular to the doorway. **11B-404.2.4.3** *404.2.4.3* **Fig. CD-14I(a)**

_____II.   The floor or ground area beyond the stop/latch side of the door and parallel to the doorway is 18 inches minimum for interior doors and 24 inches minimum for exterior doors. **11B-404.2.4.3** *404.2.4.3* **Fig. CD-14I(a)**

FRONT APPROACH, PUSH SIDE

_____JJ.   The floor or ground area is the width of the door and a minimum of 48 inches deep, perpendicular to the doorway.   **11B-404.2.4.3**   *404.2.4.3*   **Fig. CD-14I(b)**

FRONT APPROACH, PUSH SIDE WITH BOTH LATCH AND CLOSER

_____KK.   The floor or ground area is the width of the door and a minimum of 48 inches deep, perpendicular to the doorway.   **11B-404.2.4.3**   *404.2.4.3*   **Fig. CD-14I(c)**

_____LL.   The door is provided with both a latch and a closer and the floor or ground area beyond the stop/latch side of the door and parallel to the doorway is 12 inches minimum.   **11B-404.2.4.3**   *404.2.4.3*   **Fig. CD-14I(c)**

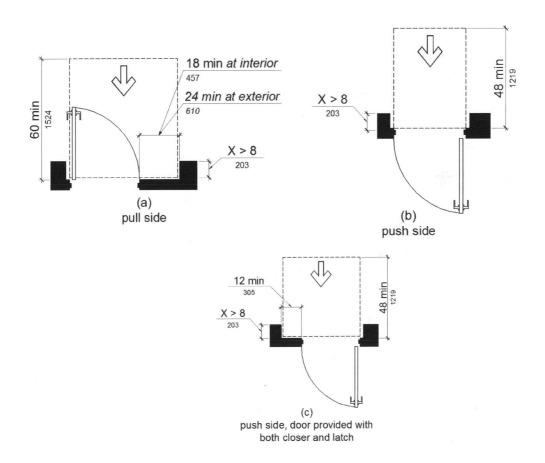

Fig. CD-14I
Maneuvering Clearances at Recessed Doors and Gates
© ICC Reproduced with Permission

15. <u>RAMPS</u>

<u>DEFINITION</u>
RAMP is a walking surface which has a running slope steeper than one unit vertical in 20 units horizontal (5% slope).   **202**

<u>General</u>
Ramps on accessible routes shall comply as detailed. **11B-405.1**   *405.1*

> **EXCEPTION:** In assembly areas, aisle ramps adjacent to seating and not serving elements required to be on an accessible route shall not be required to comply.

<u>Slope</u>

_____A.   Ramp runs have a running slope not steeper than 1:12 (8.33%). **11B-405.2** *405.2*

> ***ADVISORY: Slope.** To accommodate the widest range of users, consideration should be given to providing ramps with the least possible running slope and, wherever possible, accompany ramps with stairs for use by those individuals for whom distance presents a greater barrier than steps, e.g., people with heart disease or limited stamina.*

<u>Cross Slope</u>

_____B.   Cross slope of ramp runs are not steeper than 1:48 (2.08%).   **11B-405.3** *405.3*

> ***ADVISORY: Cross Slope.** Cross slope is the slope of the surface perpendicular to the direction of travel. Cross slope is measured the same way as slope is measured (i.e., the rise over the run).*
>
> *Curved ramps are not prohibited, however, requirements for maximum slope and cross-slope cannot be exceeded.*

<u>Floor or Ground Surfaces</u>

_____C.   Floor or ground surfaces of ramp runs comply with Section 7 "FLOOR OR GROUND SURFACES".   **11B-405.4**   *405.4*

<u>Clear Width</u>

_____D.   The clear width of a ramp run shall be 48 inches minimum. **11B-405.2**   *405.5*

> **EXCEPTIONS:**
>
> 1.   Within employee work areas, the required clear width of ramps that are a part of common use circulation paths shall be permitted to be decreased by work area equipment provided that the decrease is essential to the function of the work being performed.

2.   Handrails may project into the required clear width of the ramp at each side 3-1/2 inches maximum at the handrail height.

3.   The clear width of ramps in residential uses serving an occupant load of fifty or less shall be 36" minimum between handrails.

Rise

_____E.   The rise for any ramp run is 30 inches maximum. **11B-405.6**   *405.6*

Landings

_____F.   Ramps have landings at the top and the bottom if each ramp run.
11B-405.7   *405.7*   **Fig. CD-15A**

> *ADVISORY: Landings. Ramps that do not have level landings at changes in direction can create a compound slope that will not meet the requirements of this document. Circular or curved ramps continually change direction. Curvilinear ramps with small radii also can create compound cross slopes and cannot, by their nature, meet the requirements for accessible routes. A level landing is needed at the accessible door to permit maneuvering and simultaneously door operation.*

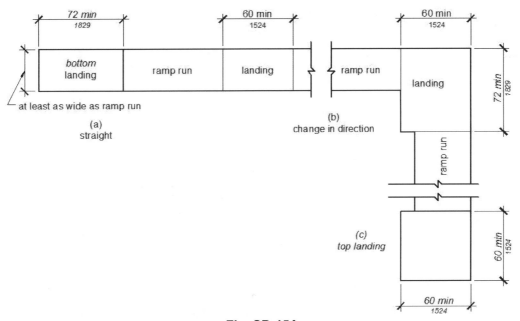

Fig. CD-15A
Ramp Landings
© ICC Reproduced with Permission

Slope

_____G.   Changes in level at ramp landings are not permitted. **11B-405.7.1**
405.7.1

EXCEPTION: slopes not steeper than 1:48 (2.08%) shall be permitted.

Floor or Ground Surfaces

_____H.   Floor or ground surfaces of ramp landings comply with Section 7, "FLOOR OR GROUND SURFACES". **11B-405.7.1** *405.7.1*

Width

_____I.   The landing clear width is at least as wide as the widest ramp run leading to the landing. **11B-405.7.2** *405.7.2* **Fig. CD-15A(a)**

Length

_____J.   The landing clear length is 60 inches long minimum. **11B-405.7.3** *405.7.3*

_____K.   Bottom landings extend 72 inches minimum in the direction of ramp run.
11B-405.7.3.1 *405.7.3.1* **Fig. CD-15A(a)**

Change in Direction

_____L.   Ramps that change direction between runs at landings have a clear landing 60 inches minimum by 72 inches minimum in the direction of downward travel from the upper ramp run. **11B-405.7.4** *405.7.4* **Fig. CD-15A(b)**

Doorways
Where doorways are located adjacent to a ramp landing, the required doorway maneuvering clearances are permitted to overlap the required landing area. **11B-405.7.5** *405.7.5*

_____M.   Doors, when fully open, do not reduce the required ramp landing width by more than 3 inches. Doors, in any position, do not reduce the minimum dimension of the ramp landing to less than 42 inches. **11B-405.7.5** **Fig. CD-15B**

_____N.   Door width plus 42 inches minimum landing dimension is provided. **11B-405.7.5** **Fig. CD-15B**

_____O.   Door does not reduce the required ramp landing width by more than 3 inches when fully open. **11B-405.7.5** **Fig. CD-15B**

> ***ADVISORY: Doorways.*** *The placement of a pedestrian ramp at a doorway requires a good deal of consideration. Since doors cannot reduce the required width of a ramp (this <u>includes</u> the landing) by more than 3" when fully open, a door that encroaches by more than 3" into the required landing width will require the provision of a full landing <u>prior</u> to the door. The length of this landing would be dependent on whether it was at the top or bottom of the ramp slope; 60" minimum at the top of a ramp and 72" minimum at the bottom of a ramp. This design requirement makes considerable sense; an individual in a wheelchair may not be able to avoid hitting a fully open door that extended into a landing with only 42" of clearance between the terminus of the ramp and the door.*

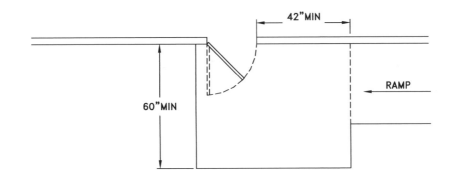

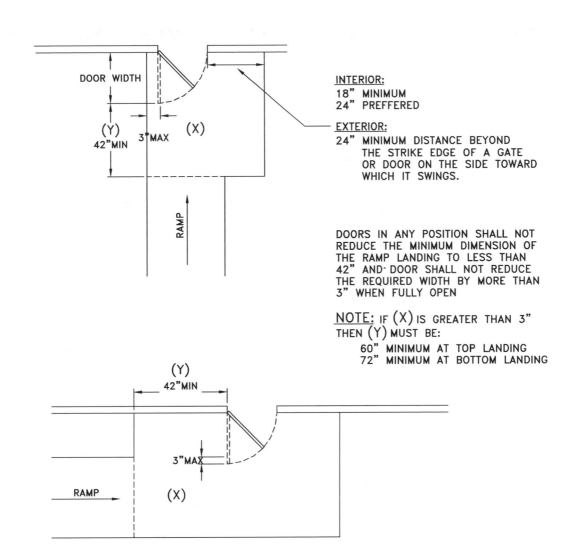

DOOR WIDTH

(Y)
42"MIN

3"MAX

(X)

RAMP

INTERIOR:
18" MINIMUM
24" PREFFERED

EXTERIOR:
24" MINIMUM DISTANCE BEYOND
THE STRIKE EDGE OF A GATE
OR DOOR ON THE SIDE TOWARD
WHICH IT SWINGS.

DOORS IN ANY POSITION SHALL NOT
REDUCE THE MINIMUM DIMENSION OF
THE RAMP LANDING TO LESS THAN
42" AND DOOR SHALL NOT REDUCE
THE REQUIRED WIDTH BY MORE THAN
3" WHEN FULLY OPEN

NOTE: IF (X) IS GREATER THAN 3"
THEN (Y) MUST BE:
60" MINIMUM AT TOP LANDING
72" MINIMUM AT BOTTOM LANDING

(Y)
42"MIN

3"MAX

RAMP

(X)

Fig. CD-15B
Ramps at Doorways

Wet Conditions

_____P.   Landings subject to wet conditions are designed to prevent the accumulation of water.   **11B-405.10**   *405.10*

Handrails

_____Q.   Ramp runs have handrails that comply with Section #24 "HANDRAILS". **11B-405.8**   405.8

EXCEPTIONS:

1. **Reserved.**

2. Handrails are not required at ramps immediately adjacent to fixed seating in assembly areas.

3. Curb ramps do not require handrails.

4. At door landings, handrails are not required on ramp runs less than 6 inches in rise or 72 inches in length.

Edge Protection

_____R.   Edge protection consisting of a curb or barrier that complies as detailed below is provided on each side of ramp runs and at each side of ramp landings. **11B-405.9**   *405.9*   **Figs. CD-15C & D**

EXCEPTIONS:

1. Edge protection shall not be required on ramps that are not required to have handrails and have flared sides consistent to a parallel curb ramp and are not steeper than 1:10 (10%).

2. Edge protection shall not be required on the sides of ramp landings serving an adjoining ramp run or stairway.

3. Edge protection shall not be required on the sides of ramp landings having a vertical drop-off of ½ inches maximum within 10 inches horizontally of the minimum required landing area.

Curb or Barrier

_____S.   A curb, 2 inches high minimum, or barrier is provided that prevents the Passage of a 4 inch diameter sphere, where any portion of the sphere is within 4 inches of the finish floor or ground surface.   **11B-405.9.2** *405.9.2* **Fig. CD-15C & D**

_____T.   To prevent wheel entrapment, the curb or barrier provides a continuous and uninterrupted barrier along the length of the ramp.   **11B-405.9.2 Fig. CD-15C & D**

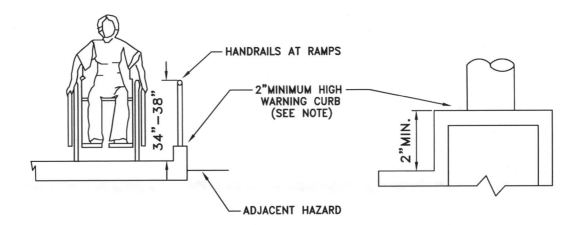

HANDRAILS AT RAMPS

2"MINIMUM HIGH WARNING CURB (SEE NOTE)

34" – 38"

2" MIN.

ADJACENT HAZARD

NOTE: IF A DROP–OFF OF MORE THAN 4 INCHES EXISTS BETWEEN THE RAMP SURFACE AND THE ADJACENT GRADE, A 6 INCH WARNING CURB MUST BE UTILIZED.

Fig. CD-15C
Curb Edge Protection

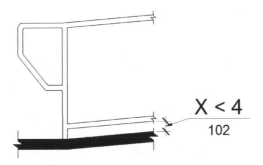

X < 4
102

Fig. CD-15D
Barrier Edge Protection
© ICC Reproduced with Permission

16. <u>CURB RAMPS, BLENDED TRANSITIONS AND ISLANDS</u>

DEFINITION
CURB RAMP is a sloping pedestrian way, intended for pedestrian traffic, which provides access between a walk or sidewalk to a surface located above or below an adjacent curb face. **202**

DEFINITION
BLENDED TRANSITION. A raised pedestrian street crossing, depressed corner, or similar connection between the pedestrian access route at the level of the sidewalk and the level of the pedestrian street crossing that has a grade of 5 percent (5%) or less. **202**

General
Curb ramps, blended transitions and islands on accessible routes shall comply as detailed. Curb ramps may be perpendicular, parallel, or a combination of perpendicular and parallel. **11B-406.1** *406.1*

PERPENDICULAR CURB RAMPS

Slope

_____A.   Ramp runs have a running slope not steeper than 1:12 (8.33%). **11B-406.2.1** *405.2* **Fig. CD-16A**

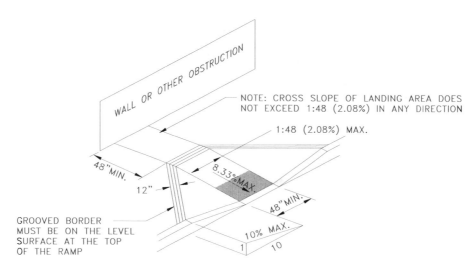

NOTE: CROSS SLOPE OF LANDING AREA DOES NOT EXCEED 1:48 (2.08%) IN ANY DIRECTION

WALL OR OTHER OBSTRUCTION

1:48 (2.08%) MAX.

48" MIN.

12"

8.33% MAX.

48" MIN.

GROOVED BORDER MUST BE ON THE LEVEL SURFACE AT THE TOP OF THE RAMP

10% MAX.

1 / 10

NOTE: TRUNCATED DOME DETECTABLE WARNING SURFACE IS REQUIRED ON SHADED PORTION OF THE RAMP.

Fig. CD-16A
Perpendicular Curb Ramp

Sides of Curb Ramps

_____B.   Where provided, curb ramp flares are not steeper than 1:10 (10%). **11B.406-2.2** *406.3* **Figs. CD-16B**

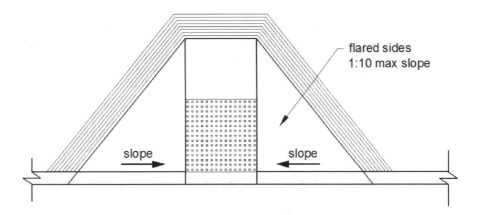

Fig. 16B
Sides of Curb Ramps
© ICC Reproduced with Permission

Common Requirements

_____C.　　Perpendicular curb ramps comply with COMMON REQUIREMENTS sections I through DD, below.　**11B-406.2, 11B-406.5**

PARALLEL CURB RAMPS

Slope

_____D.　　The running slope of the curb ramp segments are in-line with the direction of sidewalk travel.　Ramp runs have a running slope not steeper than 1:12 (8.33%).　**11B-406.3.1**　*405.2*　**Figs. CD-16C**

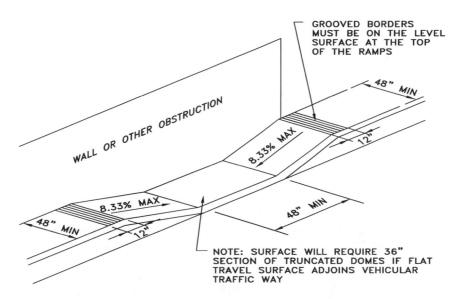

Fig. CD-16C
Parallel Curb Ramp

Turning Space

_____E. A turning space 48 inches minimum by 48 inches minimum is provided at the bottom of the curb ramp. The slope of the turning space in all directions is 1:48 maximum (2.08%). **11B.406-3.2   Fig. CD-16D**

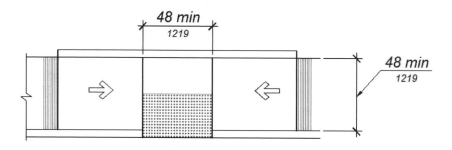

Fig. CD-16D
Parallel Curb Ramp
© ICC Reproduced with Permission

Common Requirements

_____F. Parallel curb ramps comply with comply with COMMON REQUIREMENTS sections I through DD, below.   **11B-406.2, 11B-406.5**

BLENDED TRANSITIONS

Slope

_____G. Blended transitions have a running slope not steeper than 1:20 (5%). **11B-406.4.1**

Common Requirements

_____H. Blended transitions comply with COMMON REQUIREMENTS sections I through DD, below.   **11B-406.4, 11B-406.5**

COMMON REQUIREMENTS

Location

_____I. Curb ramps and the flared sides of curb ramps are located so they do not project into vehicular traffic lanes, parking spaces, or parking access aisles. **11B-406.5.1**   *406.5*

_____J. Curb ramps at marked crossings are wholly contained within the markings, excluding any flared sides. **11B-406.5.1**   *406.5*   **Fig. CD-16E**

> **EXCEPTION:** Diagonal curb ramps shall comply as detailed below.

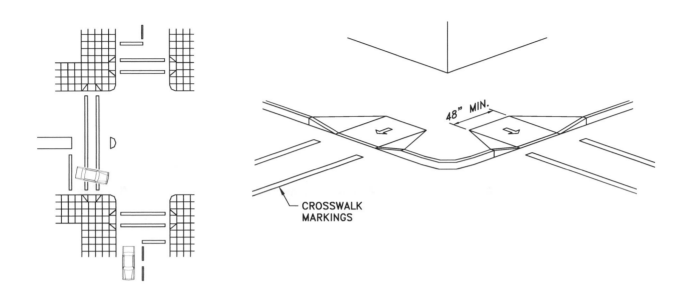

Fig. CD-16E
Curb Ramps at Marked Crossings

Width

_____K. The clear width of curb ramp runs (excluding any flared sides), blended transitions, and turning spaces is 48" minimum. **11B-406.5.2** *405.5*

Landings

_____L. Landings are provided at the tops of curb ramps and blended transitions. **11B-406.5.3** *406.4*

_____M. The landing clear length is 48 inches minimum. **11B-406.5.3 Fig. CD-16F**

_____N. The landing clear width is at least as wide as the curb ramp, excluding any flared sides, or the blended transition leading to the landing. **11B-406.5.3** *406.4* **Fig. CD-16F**

_____O. The slope of the landing in all directions is 1:48" maximum (2.08%). **11B-406.5.3** *406.1, 405.7.1* **Fig. CD-16F**

> **EXCEPTION:** Parallel curb ramps shall not be required to Comply with items L through O, above.

> **ADVISORY: Landings.** *A level landing, 48 inches deep and at least as wide as the curb ramp, is required at the top of a curb ramp. The landing provides an area for users to exit the curb ramp and proceed along the walking surface at the top of the curb ramp.*

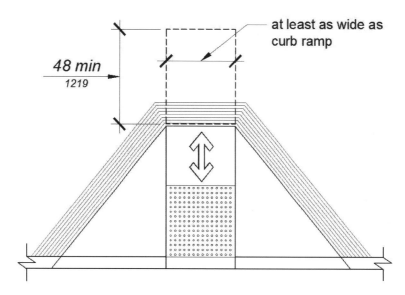

Fig. CD-16F
Landings at the Top of Curb Ramps
© ICC Reproduced with Permission

Floor or Ground Surfaces

_____P.   Floor or ground surfaces of ramp runs comply with Section 7, "FLOOR OR GROUND SURFACES". **11B-406.5.4**   *406.1, 405.4*

Wet Conditions

_____Q.   Curb ramps and blended transitions subject to wet conditions are designed to prevent the accumulation of water.   **11B-406.5.5**
406.1, 405.10

Grade Breaks

_____R.   Grade breaks at the top and bottom of curb ramp runs are perpendicular to the direction of the ramp run.   **11B-406.5.6**

_____S.   There are no grade breaks on the surface of ramp runs or turning spaces. **1B-406.5.6**

_____T.   Surface slopes that meet at grade breaks are flush.   **11B-406.5.6**

Cross Slope

_____U.   The cross slope of curb ramps and blended transitions is 1:48 maximum (2.08%). **11B-406.5.7**   *406.1, 405.3*

Counter Slope

_____V.   Counter slopes of adjoining gutters and road surfaces immediately adjacent to and within 24 inches of the curb ramp are not steeper than 1:20 (5%). **11B-406.5.8**   *406.2*   **Fig. CD-16G**

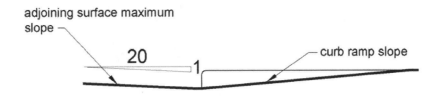

Fig. CD-16G
Counter Slope of Surfaces Adjacent to Curb Ramps
© ICC Reproduced with Permission

_____W.   The adjacent surfaces at transitions at curb ramps to walks, gutters and street are at the same level.   **11B-406.5.8**   *406.2*

Clear Space at Diagonal Curb Ramps

_____X.   The bottom of diagonal curb ramps shall have a clear space 48 inches minimum outside active traffic lanes of the roadway.   **11B-406.5.9   Fig. CD-16H**

_____Y.   Diagonal curb ramps provided at marked crossings shall provide the 48 inches minimum clear space within the markings.   **11B-406.5.9**   *406.5*   **Fig. CD-16H**

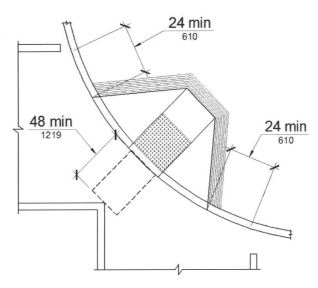

Fig. CD-16H
Diagonal or Corner Type Curb Ramps
© ICC Reproduced with Permission

Diagonal Curb Ramps

_____Z.   Diagonal or corner type curb ramps with returned curbs or other well-defined edges have the edges parallel to the direction of pedestrian flow.   **11B-406.5.10**   *406.6*   **Fig. CD-16I**

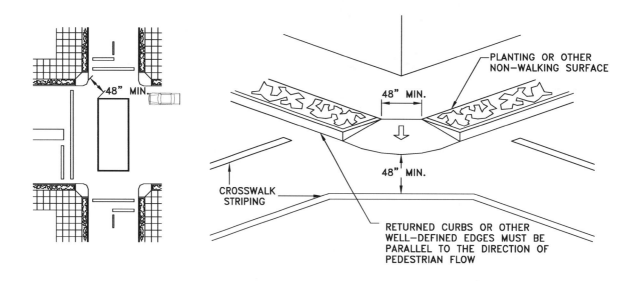

Fig. CD-16I
Curb Ramps with Returned Curbs or Other Well Defined Edges

_____AA.   Diagonal curb ramps with flared sides have a segment of curb 24 inches long minimum located on each side of the curb ramp and within the marked crossing. **11B-406.5.10** *406.6* **Fig. CD-16H**

Grooved Border

_____BB.   Curb ramps have a grooved border 12 inches wide along the top of the curb ramp at the level surface of the top landing and at the outside edges of the flared sides. **11B-406.5.11**

_____CC.   The grooved border consists of a series of grooves ¼ inches wide by ¼ inches deep, at ¾ inches on center. **11B-406.5.11**

EXCEPTIONS:

1.   At parallel curb ramps, the grooved border is on the level surface of each top landing across the full width of the curb ramp.

2.   A grooved border shall not be required at blended transitions.

Detectable Warnings

_____DD.   Curb ramps and blended transitions have detectable warnings that comply with Section 32, "DETECTABLE WARNINGS AND DETECTABLE DIRECTIONAL TEXTURE". **11B-406.5.12**

ISLANDS

_____EE.   Raised islands in crossings are cut through level with the street or have curb ramps at both sides.   **11B-406.6**   *406.7*   **Fig. CD-16J**

_____FF.   The clear width of the accessible route at islands is 60 inches wide minimum. **11B-406.6**   *406.7*   **Fig. CD-16J**

_____GG.   Where curb ramps are provided, they shall comply.   **11B-406.6**

_____HH.   Compliant landings and the accessible route shall be allowed to overlap. **11B.406.6**

_____II.   Islands have detectable warnings that comply with Section 32, "DETECTABLE WARNINGS AND DETECTABLE DIRECTIONAL TEXTURE".   **11B-406.6** **Fig. CD-16J**

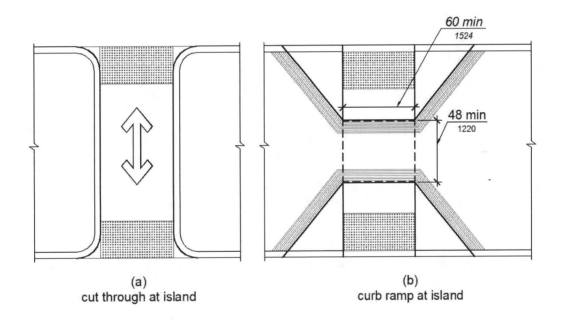

(a)
cut through at island

(b)
curb ramp at island

Fig. 11B-16J
Islands in Crossings
© ICC Reproduced with Permission

17. <u>ELEVATORS</u>

<u>Elevators</u>
Elevators provided for passengers shall comply as detailed. Where multiple elevators are provided, each elevator shall comply as detailed. **11B-206.6** *206.6*

EXCEPTIONS:

1. In a building or facility permitted to use the exceptions for Multi-Story Buildings and Facilities or permitted to use a platform lift, elevators complying with the requirements for accessible Limited-Use/Limited-Application Elevators shall be permitted.

2. Elevators complying with the requirements for accessible Limited-Use/Limited-Activation Elevators or Private Residence Elevators shall be permitted in multi-story residential dwelling units. Elevators provided as a means of access within a private residence shall be installed so that they are not accessible to the general public or to the other occupants of the building.

<u>Existing Elevators</u>
Where elements of existing elevators are altered, the same element shall also be altered in all elevators that are programmed to respond to the same hall call control as the altered elevator and shall comply with the requirements for Elevators for the altered element. **11B-206.6.1** *206.6.1*

<u>Signs at Elevators When Not All Are Accessible</u>
Where existing elevators do not comply with Section 17, "ELEVATORS", elevators that do comply shall be clearly identified with the International Symbol Accessibility. Existing buildings that have been remodeled to provide specific elevators for public use that comply with these buildings standards shall have the location of and the directions to these elevators posted in the building lobby on a sign complying with accessible Visual Characters, including the International Symbol of Accessibility. **11B-216.7** *216.7*

<u>General</u>
Elevators shall comply with these provisions and with ASME A17.1. They shall be passenger elevators as classified by ASME A17.1. Elevator operation shall be automatic. **11B-407.1** *407.1*

> *ADVISORY: The ADA and other Federal civil rights laws require that accessible features be maintained in working order so that they are accessible to and usable by those people they are intended to benefit. Building owners should note that the ASME Safety Code for Elevators and Escalators requires routine maintenance and inspections. Isolated or temporary interruptions in service due to maintenance or repairs may be unavoidable; however, failure to take prompt action to effect repairs could constitute a violation of Federal laws and these requirements.*

<u>Combined Passenger and Freight Elevators</u>
Where the only elevators provided for use by the public and employees are combination passenger and freight elevators, they shall comply with these provisions and ASME A17.1. **11B-407.1.1**

17A. <u>ELEVATOR LANDING REQUIREMENTS</u>

<u>Elevator Landing requirements</u>
Elevator landings shall comply as detailed.   **11B-407.2**   *407.2*

<u>Call Controls</u>

_____A.   Where elevator call buttons or keypads are provided, they shall be operable with one hand and not require tight grasping, pinching or twisting of the wrist. The force required to operate controls is 5 pounds maximum.   **11B-407.2.1, 11B-309.4**   *407.2.1, 309.4*

<u>Height</u>

_____B.   Call buttons and keypads are located within one of the accessible reach ranges, measured to the centerline of the highest operable part.   **11B-407.2.1.1, 11B-308**   *407.2.1.1, 308*

<u>Size and Shape</u>

_____C.   Call buttons have square shoulders, are ¾ inches minimum in the smallest dimension and are raised 1/8 inches plus or minus 1/32 inches above the surrounding surface. The buttons shall be activated by a mechanical motion that is detectable. **11B-407.2.1.2**   *407.2.1.2*

<u>Clear Floor or Ground Space at Call Controls</u>

_____D.   Floor or ground surfaces within the clear floor or ground space at car controls comply with Section 7, "FLOOR OR GROUND SURFACES". Changes in level are NOT permitted. **11B-407.2.1.3, 11B-305, 11B-302.1**   *407.2.1.3, 305, 302.1*

> **EXCEPTION:**   Slopes not steeper than 1:48 shall be permitted.

> **ADVISORY. Clear Floor or Ground Space.** *The clear floor or ground space required at elevator call buttons must remain free of obstructions including ashtrays, plants, and other decorative elements that prevent wheelchair users and others from reaching the call buttons. The height of the clear floor or ground space is considered to be a volume from the floor to 80 inches above the floor. Recessed ashtrays should not be placed near elevator call buttons so that persons who are blind or visually impaired do not inadvertently contact them or their contents as they reach for the call buttons.*

<u>Location</u>

_____E.   The call button that designates the up direction is located above the call button that designates the down direction.   **11B-407.2.1.4**   *407.2.1.4*

> **EXCEPTION:**   Destination-oriented elevators are not required to comply with this provision.

> **ADVISORY: Location Exception.** *A destination-oriented elevator system provides lobby controls enabling passengers to select floor stops, lobby indicators designating which elevator to use, and a car indicator designating the floors at which the car will stop. Responding cars are programmed for maximum efficiency by reducing the number of stops any passenger experiences.*

Signals

_____F.   Call buttons have visual signals that will activate when each call is registered and will distinguish when each call is answered. Call buttons are internally illuminated with a white light over the entire surface of the button.   **11B-407.2.1.5**   *407.2.1.5*

> **EXCEPTION:** Destination-oriented elevators shall not be required to comply with this provision provided that visible and audible signals that comply with Section 17B, "ELEVATOR HALL SIGNALS", indicating which elevator car to enter are provided.

Keypads

_____G.   Where keypads are provided, keypads are in a standard telephone keypad arrangement. **11B-407.2.1.6**   *407.2.1.6*

_____G1.   Keypads are identified by accessible visual characters that shall be centered on the corresponding keypad button (see Section 30, "SIGNS").   **11B-407.4.7.2, 11B-703.5** *407.4.7.2, 703.5*

_____G2.   The number five key shall have a single raised dot. The dot shall be 0.118 inch to 0.120 inch base diameter and in other aspects comply with Table 30A (see Section 30, "SIGNS"). **11B-407.7.2** *407.7.2*

17B. ELEVATOR HALL SIGNALS

Hall Signals
Hall signals, including in-car signals, shall comply as detailed.   **11B-407.2.2**   *407.2.2*

Visible and Audible Signals

_____A.   A visual and audible signal shall be provided at each hoistway entrance to indicate which car is answering a call and the car's direction of travel. Where in-car signals are provided, they shall be visible from the floor area adjacent to the hall call buttons.   **11B-407.2.2.1**   *407.2.2.1* **Fig. CD-17BA**

> **EXCEPTION:** Visible and audible signals shall not be required at each destination-oriented elevator where a visible and audible signal is provided indicating the elevator car designation information.

Visible Signals

_____B.   Visible signal fixtures shall be centered at 72 inches minimum above the finish floor or ground. The visible signal elements shall be a minimum 2-1/2 inches high by 2-1/2 inches wide. Signals shall be visible from the floor area adjacent to the hall call button.   **11B-407.2.2.2** *407.2.2.2*   **Fig. CD-17BA**

> **EXCEPTION:** Destination-oriented elevators shall be permitted to have signals visible from the floor area adjacent to the hoistway entrance.

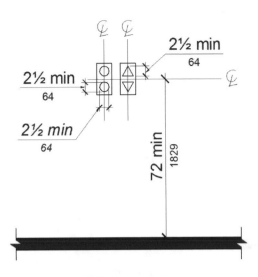

Fig. CD-17BA
Visible Hall Signals
© ICC Reproduced with Permission

Audible Signals

____C. Audible signals shall sound once for the up direction and twice for the down direction or shall have verbal annunciators that indicate the direction of elevator car travel. Audible signals shall have a frequency of 1500 Hz maximum. Verbal annunciators shall have a frequency of 300 Hz minimum and 3000 Hz maximum. The audible signal and verbal annunciator shall be 10 dB minimum above ambient, but shall not exceed 80 dB, measured at the hall call button. **11B-407.2.2.3** *407.2.2.3*

> **EXCEPTION:** Destination-oriented elevators shall not be required to comply with this provision provided that the audible tone and verbal announcement is the same as those given at the call button or call button keypad.

Differentiation

____D. Each destination-oriented elevator in a bank of elevators shall have audible and visual means for differentiation. **11B-407.2.2.4** *407.2.2.4*

17C. <u>ELEVATOR HOISTWAY SIGNS</u>

Hoistway Signs
Signs at elevator hoistways shall comply as detailed. **11B-407.2.3** *407.2.3*

Floor Designation

____A. Floor designations shall be provided on both jambs of elevator hoistway entrances in both tactile characters and Braille. **11B-407.2.3.1** *407.2.3.1* **Fig. CD-17CA**

____B. Raised characters shall be 2 inches high minimum. **11B-407.2.3.1** *407.2.3.1* **Fig. CD-17CA**

____C. A raised star, placed to the left of the floor designation, shall be provided on both jambs at the main entry level. **11B-407.2.3.1** *407.2.3.1* **Fig. CD-17CA**

_____C1.   The outside diameter of the star shall be 2 inches high and all points shall be of equal length.   **11B-407.2.3.1**   *407.2.3.1*   **Fig. CD-17CA**

_____C2.   Contracted (Grade 2) Braille shall be placed below the corresponding raised characters and the star. The Braille translation for the star shall be "MAIN".   **11B-407.2.3.1** *407.2.3.1*   **Fig. CD-17CA**

           NOTE: Applied plates are acceptable if they are permanently fixed to the jamb.

_____D.   Raised characters, including the star, shall be white on a black background. **11B-407.2.3.1**   **Fig. CD-17CA**

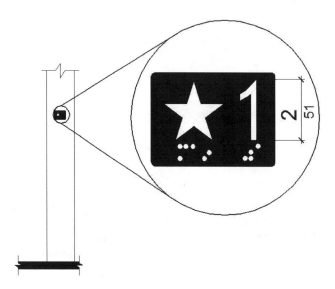

Fig. CD-17CA
Floor Designations on Jambs of Elevator Hoistway Entrances
© ICC Reproduced with Permission

Height Above Finish Floor or Ground

_____E.   Tactile characters on signs shall be located 48 inches minimum above the finish floor or ground surface, measured from the baseline of the lowest Braille cells and 60 inches maximum above the finish floor or ground surface, measured from the baseline of the highest line of raised characters.

           EXCEPTION: Tactile characters for elevator car controls shall not be required to comply with this requirement. **11B-703.4.1**   **Fig. CD-17CB**

Depth

_____F.   Raised characters shall be 1/32 inches minimum above their background. **11B-703.2.1**   *703.2.1*

Case

_____G.   Characters shall be uppercase. **11B-703.2.2**   *703.2.2*   **Fig. CD-17CB**

Style

_____H.   Characters shall be sans serif. Characters shall not be italic , oblique, script, highly decorative, or of other unusual forms. **11B-703.2.3**   *703.2.3*

Character Proportions

_____I.   Characters shall be selected from fonts where the width of the uppercase letter "O" is 60 percent minimum and 110 percent maximum of the height of the uppercase letter "I". **11B-703.2.4**   *703.2.4*

Character Height

_____J.   Character height measured vertically from the baseline of the character shall be 5/8 inches minimum and 2 inches maximum based on the height of the uppercase letter "I". **11B-703.2.5**   *703.2.5*

> **EXCEPTION:** Where separate raised and visual characters with the same information are provided, raised character height shall be permitted to be ½ inches minimum.

Stroke Thickness

_____K.   Stroke thickness of the uppercase letter "I" shall be 15 percent maximum of the height of the character. **11B-703.2.6**   *703.2.6*

Character Spacing

_____L.   Character spacing shall be measured between the two closest points of adjacent raised characters within a message, excluding word spaces. Where characters have rectangular cross sections, spacing between individual raised characters shall be 1/8 inches minimum and 4 times the raised character stroke width maximum. Where characters have other cross sections, spacing between individual raised characters shall be 1/16 inches minimum and 4 times the raised character stroke width maximum at the base of the cross sections, and 1/8 inches minimum and 4 times the raised character stroke width maximum at the top of the cross sections. Characters shall be separated from raised borders and decorative elements 3/8 inches minimum.   **11B-703.2.7**   *703.2.7*

Line Spacing

_____M.   Spacing between the baselines of separate lines of raised characters within a message shall be 135 percent minimum and 170 percent maximum of the raised character height. **11B-703.2.8**   *703.2.8*

Format

_____N.   Text shall be in a horizontal format. **11B-703.2.9**

Braille

_____O.   Raised characters are duplicated in Braille. Braille shall be contracted (Grade 2) and comply as detailed. **11B-703.2, 11B-703.3**
703.2, 703.3   **Fig. CD-17CB**

Dimensions and Capitalization

_____P. Braille dots shall have a domed or rounded shape and shall comply with Table 30A (see Section 30, "SIGNS"). The indication of an uppercase letter or letters shall only be used before the first word of sentences, proper nouns and names, individual letters of the alphabet, initials, and acronyms. **11B-703.3.1** *703.3.1*

Position

_____Q. Braille shall be positioned below the corresponding text in a horizontal format, flush left or centered. If text is multi-lined, braille shall be placed below the entire text. Braille shall be separated 3/8 inches minimum and ½ inches maximum from any other tactile characters and 3/8 inches minimum from raised borders and decorative elements. **11B-703.3.2** *703.3.2*

Car Designations

_____R. Destination-oriented elevators shall provide tactile car identification on both jambs of the hoistway immediately below the floor designation. The identification complies with items B, D and F through Q, above. **11B-407.2.3.2** *407.2.3.2* **Fig. CD-17CB**

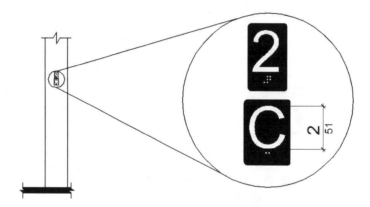

Fig. CD-17CB
Car Designations on Jambs of Destination-Oriented Elevator Hoistway Entrances
© ICC Reproduced with Permission

17D. <u>ELEVATOR DOOR REQUIREMENTS</u>

<u>Elevator Door Requirements</u>
Hoistway and car doors shall comply as detailed. **11B-407.3** *407.3*

<u>Type</u>

_____A. Elevator doors shall be the horizontal sliding type. Car gates shall be prohibited. **11B-407.3.1** *407.3.1*

<u>Operation</u>

_____B. Elevator hoistway and car doors shall open and close automatically. **11B-407.3.2** *407.3.2*

> **EXCEPTION:** Existing manually operated hoistway swing doors shall be permitted provided that they comply with the requirements for "Clear Width" and "Door and Gate Opening Force", contained in Section 14, "DOORS, DOORWAYS AND GATES". Car door closing shall not be initiated until the hoistway door is closed.

<u>Reopening Device</u>

_____C. Elevator doors shall be provided with a reopening device that complies as detailed that shall stop and reopen a car door and hoistway door automatically if the door becomes obstructed by an object or person. **11B-407.3.3** *407.3.3*

> **EXCEPTION:** Existing elevators with manually operated doors shall not be required to have door reopening devices that remain effective for 20 seconds minimum.

<u>Height</u>

_____C1. The device shall be activate by sensing an obstruction passing through the opening at 5 inches nominal and 29 inches nominal above the finish floor. **11B-407.3.3.1** *407.3.3.1*

<u>Contact</u>

_____C2. The device shall not require physical contact to be activated, although contact is permitted to occur before the door reverses. **11B-407.3.3.2** *407.3.3.2*

<u>Duration</u>

_____C3. Door reopening devices shall remain effective for 20 seconds minimum. **11B-407.3.3.3** *407.3.3.3*

<u>Door and Signal Timing</u>

_____D. The minimal acceptable time from notification that a car is answering a call or notification of the car assigned at the means for the entry of destination information until the doors of that car start to close shall be calculated from the following equation:

> $T = D/(1.5 \text{ ft/s})$ or $T = D/(455\text{mm/s}) = 5$ seconds minimum where T equals the total time in seconds and D equals the distance (in feet or millimeters) from the point in the lobby or corridor 60" directly in front of the farthest call button controlling that car to the centerline of its hoistway door. **11B-407.3.4** *407.3.4*

EXCEPTIONS:

1.   For cars with in-car lanterns, T shall be permitted to begin when the signal is visible from the point 60 inches directly in front of the farthest hall call button and the audible signal is sounded.

2.   Destination-oriented elevators shall not be required to comply with door and signal timing requirements.

Door Delay

_____E.   Elevator doors shall remain fully open in response to a car call for 5 seconds minimum. **11B-407.3.5**   *407.3.5*

Width

_____F.   The width of elevator doors shall comply with Table 17EA.   **11B-407.3.6**   *407.3.6*

EXCEPTION: In existing elevators, a power-operated car door that complies with the requirement for "Clear Width" in the Checklist section for "Doors, Doorways and Gates" shall be permitted.

17E. <u>ELEVATOR CAR REQUIREMENTS</u>

Elevator Car Requirements
Elevator cars shall comply as detailed.   **11B-407.4**   *407.4*

Car Dimensions

_____A.   Inside dimensions of elevator cars and clear width of elevator doors shall comply with Table 17EA.   **11B-407.4.1**   *407.4.1*   **Figs. CD-17EA & 17EB**

EXCEPTION: In existing buildings, where existing shaft configuration or technical infeasibility prohibits strict compliance with these provisions, existing elevator car configurations that provide a clear floor area of 18 square feet minimum and also provide an inside clear depth 54 inches minimum and a clear width 48 inches minimum shall be permitted.

Table 17EA
Elevator Car Dimensions

Door Location	Minimum Dimensions			
	Door Clear Width	Inside Car, Side to Side	Inside Car, Back Wall to Front Return	Inside Car, Back Wall to Inside Face of Door
Centered	42 inches (1067 mm)	80 inches (2032 mm)	51 inches (1295 mm)	54 inches (1372 mm)
Side (off-centered)	36 inches (914 mm)[1]	68 inches (1727 mm)	51 inches (1295 mm)	54 inches (1372 mm)
Any	36 inches (914 mm)[1]	54 inches (1372 mm)	80 inches (2032 mm)	80 inches (2032 mm)
Any	36 inches (914 mm)[2]	60 inches (1524 mm)[2]	60 inches (1524 mm)[2]	60 inches (1524 mm)[2]

1. A tolerance of minus ⅝ inch (15.9 mm) is permitted.
2. Other car configurations that provide a turning space complying with *Section 11B*-304 with the door closed shall be permitted.

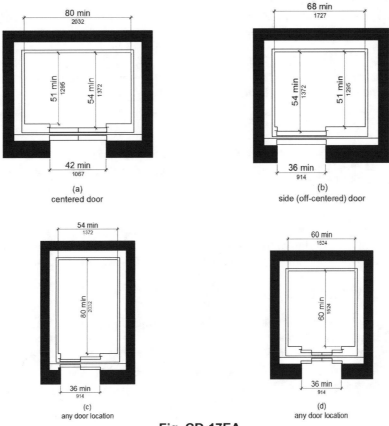

Fig. CD-17EA
Elevator Car Dimensions
© ICC Reproduced with Permission

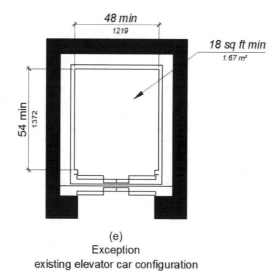

(e)
Exception
existing elevator car configuration

Fig. CD-17EB
Elevator Car Dimensions
© ICC Reproduced with Permission

Floor Surfaces

_____B.   Floor surfaces in elevator cars comply with Section 7, "FLOOR OR GROUND SURFACES, and Section 8, CHANGES IN LEVEL".
11B-407.4.2, 11B-303, 11B-302.1   *407.4.2, 303, 302.1*

Platform to Hoistway Clearance

_____C.   The clearance between the car platform sill and the edge of any hoistway landing shall be 1-1/4 inches maximum.   **11B-407.4.3**   *407.4.3*

Leveling

_____D.   Each car shall be equipped with a self-leveling feature that will automatically bring and maintain the car at floor landings within a tolerance of ½ inches under rated loading to zero loading conditions.   **11B-407.4.4**   *407.4.4*

Illumination

_____E.   The level of illumination at the car controls, platform, car threshold and landing sill shall be 5 footcandles (54 lux) minimum.   **11B-407.4.5**   *407.4.5*

Emergency Communication

_____F.   Emergency two-way communication systems shall comply with accessible reach ranges. Raised symbols and braille shall be provided adjacent to the device and shall comply with the requirements for accessible raised characters. Emergency two-way communication systems between the elevator and a point outside the hoistway shall comply with ASME A17.1
11B-407.4.9   *407.4.9*

Support Rail

_____G.   Support rails shall be provided on one wall of the car.   **11B-407.4.10**

ADVISORY: **Support Rail.** *Support rails in elevator cabs are used in a different way than handrails along a walking surface such as a ramp or stairway. Support rails in elevator cabs are required to comply with the requirements of this section, not the requirements for handrails on walking surfaces.*

Location

_____G1. Clearance between support rails and adjacent surfaces shall be 1-1/2 Inches minimum. Top of support rails shall be 31 inches minimum to 33 inches maximum above the floor of the car. The ends of the support rails shall be 6 inches maximum from adjacent walls. **11B-407.4.10.1**

Surfaces

_____G2. Support rails shall be smooth and any surface adjacent to them shall be free of sharp or abrasive elements. **11B-407.4.10.2**

Structural Strength

_____G3. Allowable stresses shall not be exceeded for materials used when a vertical or horizontal force of 250 pounds is applied at any point on the support rail, fastener, mounting device, or supportive structure. **11B- 407.4.10.3**

17F. <u>ELEVATOR CAR CONTROLS</u>

Elevator Car Controls

_____A. Where provided, elevator car controls shall comply as detailed. Operable parts shall be operable with one hand and shall not require tight grasping, pinching or twisting of the wrist. The force required to activate operable parts shall be 5 pounds maximum. **11B-407.4.6, 11B-309.4** *407.4.6, 309.4*

> **EXCEPTION:** in existing elevators, where a new car operating panel complying with these requirements is provided, existing car operating panels may remain operational and shall not be required to comply with these requirements.

Location

_____B. Controls shall be located within one of the accessible reach ranges (see Section 12, "REACH RANGES". **11B-407.4.6.1** *407.4.6.1*

> **EXCEPTIONS:**
>
> 1. Where the elevator panel serves more than 16 openings and a parallel approach is provided, buttons with floor designations shall be permitted to be 54" maximum above the finish floor.
>
> 2. In existing elevators, car control buttons with floor designations shall be permitted to be 54 inches maximum above the finish floor where a parallel approach is provided.

Buttons

_____C.   Car control buttons with floor designations shall comply as detailed below. **11B-407.4.6.2** *407.4.6.2*

Size and Shape

_____C1.   Buttons shall have square shoulders, be ¾ inches minimum in their smallest dimension and be raised a minimum 1/8 inches plus or minus 1/32 inches above the surrounding surface. **11B-407.4.6.2.1** *407.4.6.2.1*

Arrangement

_____C2.   Buttons shall be arranged with numbers in ascending order. When two or more columns of buttons are provided they shall be read from left to right. **11B-407.4.6.2.2** *407.4.6.2.2*

Illumination

_____C3.   Car control buttons shall be illuminated. **11B-407.4.6.2.3**

Operation

_____C4.   Car control buttons shall be activated by a mechanical motion that is detectable. **11B-407.4.6.2.4**

Keypads

_____D.   Car control keypads shall be in a standard telephone keypad arrangement and shall comply as detailed. **11B-407.4.6.3** *407.4.6.3*

_____D1.   Keypads are identified by accessible visual characters that shall be centered on the corresponding keypad button (see Section 30, "SIGNS"). **11B-407.4.7.2, 11B-703.5** *407.4.7.2, 703.5*

_____D2.   The number five key shall have a single raised dot. The dot shall be 0.118 inch to 0.120 inch base diameter and in other aspects comply with Table 30A (see Section 30, "SIGNS"). **11B-407.7.2** *407.7.2*

Emergency Controls

_____E.   Emergency controls shall comply as detailed. **11B-407.4.6.4** *407.4.6.4*

Height

_____E1.   Emergency control buttons shall have their centerlines 35 inches minimum above the finish floor. **11B-407.4.6.4.1** *407.4.6.4.1*

Location

_____E2.   Emergency controls, including the emergency alarm, shall be grouped at the bottom of the panel.

17G. <u>ELEVATOR DESIGNATIONS AND INDICATORS OF CAR CONTROLS</u>

<u>Elevator Designations and Indicators of Car Controls</u>

_____A.   Designations and indicators of car controls shall comply as detailed.
11B-407.4.7   *407.4.7*

> **EXCEPTION:** In existing elevators, where a new car operating panel that complies as detailed is provided, existing car operating panels may remain operational and shall not be required to comply as detailed.

<u>Buttons</u>

_____B.   Car control buttons shall comply as detailed. **11B-407.4.7.1**   *407.4.7.1*

<u>Type</u>

_____B1.   Control buttons shall be identified by accessible raised characters or symbols, white on a black background, and compliant Braille.
11B-407.4.7.1.1   *407.4.7.1.1*   **Table 17G**

<u>Location</u>

_____B2.   Raised characters and Braille designations shall be placed immediately to the left of the control button to which the designations apply (see Section 30, "SIGNS)".
11B-407.4.7.1.2   *407.4.7.1.2*

<u>Symbols</u>

_____B3.   The control button for the emergency stop, alarm, door open, door close, main entry floor, and phone, shall be identified with raised symbols and Braille as shown in Table 17G. **11B-407.4.7.1.4**   *407.4.7.1.4*

<u>Visible Indicators</u>

_____B4.   Buttons with floor designations shall be provided with visible indicators to show that a call has been registered. The visible indication shall extinguish when the car arrives at the designated floor. **11B-407.4.7.1.4**   *407.4.7.1.4*

<u>Button Spacing</u>

_____B5.   A minimum clear space of 3/8 inch or other suitable means of separation shall be provided between rows of control buttons. **11B-407.4.7.1.5**

<u>Keypads</u>

_____C.   Keypads shall be identified by compliant visible characters and shall be centered on the corresponding keypad button. The number 5 key shall have a single raised dot. The dot shall be 0.118 inches to 0.120 inches base diameter and in other aspects comply with Table 30A (see Section 30, "SIGNS"). **11B-407.4.7.2**   *407.4.7.2*

Table 17G
Elevator Control Button Identification

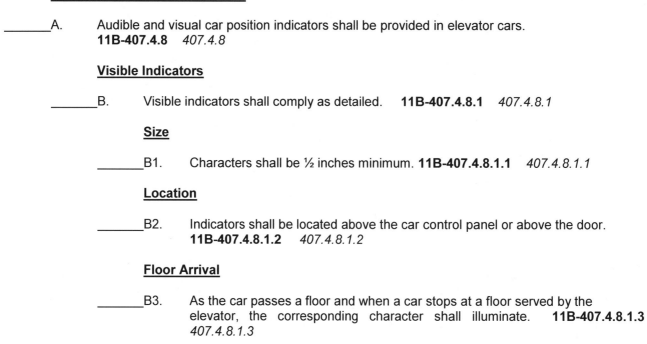

Control Button	Raised Symbol	Braille Message
Emergency Stop	⊗	"ST"OP Three Cells
Alarm	🔔	AL"AR"M Four Cells
Door Open	◀‖▶	OP"EN" Three Cells
Door Close	▶◀	CLOSE Five Cells
Main Entry Floor	★	MA"IN" Three Cells
Phone	☎	PH"ONE" Four Cells

17H. ELEVATOR CAR POSITION INDICATORS

Elevator Car Position Indicators

_____A. Audible and visual car position indicators shall be provided in elevator cars. **11B-407.4.8** *407.4.8*

Visible Indicators

_____B. Visible indicators shall comply as detailed. **11B-407.4.8.1** *407.4.8.1*

Size

_____B1. Characters shall be ½ inches minimum. **11B-407.4.8.1.1** *407.4.8.1.1*

Location

_____B2. Indicators shall be located above the car control panel or above the door. **11B-407.4.8.1.2** *407.4.8.1.2*

Floor Arrival

_____B3. As the car passes a floor and when a car stops at a floor served by the elevator, the corresponding character shall illuminate. **11B-407.4.8.1.3** *407.4.8.1.3*

> **EXCEPTION:** Destination-oriented elevators shall not be required to comply provided that the visible characters extinguish when the call ` has been answered.

Destination Indicator

_____B4.   In destination-oriented elevators, a display shall be provided in the car with the visible indicators to show car designations. **11B-407.4.8.1.4** *407.4.8.1.4*

Audible Indicators

_____C.   Audible indicators shall comply as detailed. **11B-407.4.8.2** *407.4.8.2*

Signal Type

_____C1.   The signal shall be an automatic verbal annunciator which announces the floor at which the car is about to stop. **11B-407.4.8.2.1** *407.4.8.2.1*

> **EXCEPTION:** For elevators other than destination-oriented elevators that have a rated speed of 200 feet per minute (1 m/s) or less, a non-verbal audible signal with a frequency oif 1500 Hz maximum which sounds as the car passes or is about to stop at a floor served by the elevator shall be permitted.

Signal Level

_____C2.   The verbal annunciator shall be 10 dB minimum above ambient, but shall not exceed 80 dB, measured at the annunciator. **11B-407.4.8.2.2** *407.4.8.2.2*

Frequency

_____C3.   The verbal annunciator shall have a frequency of 300 Hz minimum to 3000 Hz maximum. **11B-407.4.8.2.3** *407.4.8.2.3*

18. <u>LIMITED USE/LIMITED APPLICATION ELEVATORS</u>

<u>General</u>
Limited-use/limited application elevators shall comply as detailed below and with ASME A17.1. They shall be passenger elevators as classified by SME A17.1.

_____A.   Elevator operation is automatic.   **11B-408.1**   *408.1*

<u>Elevator Landings</u>
Landings serving limited-use/limited-application elevators shall comply as detailed.
11B-408.2   *408.2*

<u>Call Buttons</u>

_____B.   Elevator call buttons and keypads shall comply with Section 17A, "ELEVATOR LANDING REQUIREMENTS". **11B-408.2.1**   *408.2.1*

<u>Hall Signals</u>

_____C.   Hall signals shall comply with Section 17B, "ELEVATOR HALL SIGNALS". **11B-408.2.2**   *408.2.2*

<u>Hoistway Signs</u>

_____D.   Signs at elevator hoistways shall comply with Section 17C, "ELEVATOR HOISTWAY SIGNS. **11B-408.2.3**   *408.2.3*

<u>Elevator Doors</u>
Elevator hoistway doors shall comply as detailed. **11B-408.3**   *408.3*

<u>Sliding Doors</u>

_____E.   Sliding hoistway and car doors shall comply with Section 17D, "ELEVATOR DOOR REQUIREMENTS", Items "A" through "C3". **11B-408.3.1**   *408.3.1*

<u>Swinging Doors</u>

_____F.   Swinging hoistway doors shall open and close automatically and shall comply with Section 14, "DOORS, DOORWAYS AND GATES, Section 14A, "MANEUVERING CLEARANCES AT DOORS AND GATES", and the requirements below for power-operation and duration. **11B-408.3.2**   *408.3.2*

<u>Power Operation</u>

_____G.   Swinging doors shall be power-operated and shall comply with ANSI/BHMA A156.19 (1997 or 2002 edition). **11B-408.3.2.1**   *407.3.2.1*

<u>Duration</u>

_____H.   Power-operated swinging doors shall remain open for 20 Seconds minimum when activated. **11B-407.3.2.2**   *407.3.2.2*

Elevator Cars
Elevator cars shall comply as detailed. **11B-408.3**   *408.3*

Car Dimensions and Doors

_____I.   Elevator cars shall provide a clear width 42 inches minimum and a clear depth 54 inches minimum. Car doors shall be positioned at the narrow ends of cars and shall provide 32 inches minimum clear width. **11B-408.4.1**   *408.4.1*   **Fig. CD-18A**

> **EXCEPTIONS:** Cars that provide a clear width 51 inches minimum and a clear depth 51 inches minimum provided that the car does provide a clear opening 36 inches wide minimum.

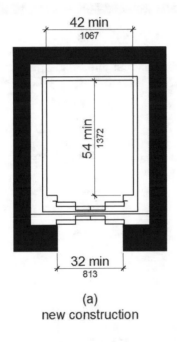

(a)
new construction

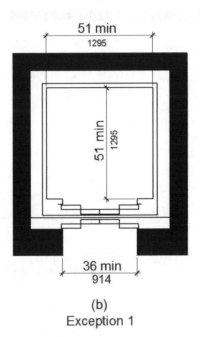

(b)
Exception 1

Model code
figure not
applicable

(c)
reserved

Fig. CD-18A
Limited Use-Limited Application (LULA) Elevator Car Dimensions
© ICC Reproduced with Permission

Floor Surfaces

_____J.   Floor surfaces in elevator cars complies with Section 7, "FLOOR OR GROUND SURFACES", and Section 8, "CHANGES IN LEVEL". **11B-408.4.2**   *408.4.2*

Platform to Hoistway Clearance

_____K.   The clearance between the car platform sill and the edge of any Hoistway landing shall be 1-1/4 inch maximum. **11B-408.4.3, 11B-407.4.3**   *408.4.3, 407.4.3*

Leveling

_____L.   Each car shall be equipped with a self-leveling feature that will automatically bring and maintain the car at floor landings within a tolerance of ½ inch under rated loading to zero loading conditions. **11B-408.4.4, 11B-407.4.4**   *409.4.4, 407.4.4*

Illumination

_____M.   The level of illumination at the car controls, platform, car threshold and car landing sill shall be 5 footcandles minimum. **11B-408.4.5, 11B-407.4.5**   *408.4.5, 407.4.5*

Car Controls

_____N.   Elevator Car Controls shall comply with Section 17F, "ELEVATOR CAR CONTROLS". Control panels shall be centered on a side wall. **11B-408.4.6, 11B-407.4.6**   *408.4.6, 407.4.6*

Designations and Indicators of Car Controls

_____O.   Designators and indicators of car controls shall comp[ly with Section 17G, "DESIGNATORS AND INDICATORS OF CAR CONTROLS". **11B-408.4.7, 11B-407.4.7**   *408.4.7, 407.4.7*

Emergency Communication

_____P.   Car emergency signaling devices comply with Section 17I, "ELEVATOR CAR EMERGENCY COMMUNICATION". **11B-408.4.8, 11B-407.4.9**   *408.4.8, 407.4.9*

19. <u>PRIVATE RESIDENCE ELEVATORS</u>

<u>General</u>

Private residence elevators that are provided within a residential dwelling unit required to provide mobility features shall comply as detailed and with ASME A17.1. They shall be passenger elevators as classified by ASME A17.1. Elevator operation shall be automatic. **11B-409.1**   *409.1*

<u>Call Buttons</u>

_____A.   Call buttons are ¾ inch minimum in the smallest dimension and comply with Section 13, "OPERABLE PARTS".   **11B-409.2**   *409.2*

<u>Elevator Doors</u>

_____B.   Hoistway Doors, car doors and car gates shall comply as detailed and with Section 14, "DOORS, DOORWAYS AND GATES".   **11B-409.3**   *409.3*

> **EXCEPTION:** Doors shall not be required to comply with the maneuvering clearance requirements for approaches to the push side of swinging doors.

<u>Power Operation</u>

_____C.   Elevator car and hoistway doors and gates shall be power operated and shall comply with ANSI/BHMA A156.19 (1997 or 2002 edition). Power operated doors and gates shall remain open for 20 seconds minimum when activated.   **11B-409.3.1**   *409.3.1*

> **EXCEPTION:** In elevator cars with more than one opening, hoistway doors and gates shall be permitted to be of the manual-open, self-close type.

<u>Location</u>

_____D.   Elevator car doors or gates shall be positioned at the narrow end of the clear floor spaces detailed below.   **11B-409.3.2**   *409.3.2*

<u>Elevator Cars</u>

<u>Inside Dimension of Elevator Cars</u>

_____E.   Elevator cars shall provide a clear floor space of 36 inches minimum by 48 inches minimum and shall comply with Section 10, "CLEAR FLOOR OR GROUND SPACE".   **11B-409.4.1**   *409.4.1*   **Fig. CD-18A**

<u>Floor Surfaces</u>

_____F.   Floor surfaces in elevator cars complies with Section 6, "FLOOR OR GROUND SURFACES", and Section 7, "CHANGES IN LEVEL".   **11B-409.4.2**   *409.4.2*

<u>Platform to Hoistway Clearance</u>

_____G.   The clearance between the car platform sill and the edge of any hoistway landing shall be 1-1/2 inch maximum.   **11B-409.4.3**   *409.4.3*

Leveling

_____H.   Each car shall automatically stop at a floor landing within a tolerance of ½ inch under rated loading to zero loading conditions. **11B-409.4.4**   *409.4.4*

Illumination Levels

_____I.   The level of illumination at the car controls, platform, car threshold and car landing sill shall be 5 footcandles minimum. **11B-409.4.5, 11B-407.4.5**   *409.4.5, 407.4.5*

Car Controls

_____J.   Elevator car control buttons shall be raised or flush and shall be placed within one of the reach ranges specified in Section 12, "REACH RANGES". **11B-409.4.6** *409.4.6*

Operation

_____K.   Operable parts are operable with one hand and do not require tight grasping, pinching or twisting of the wrist. The force required to activate operable parts is no greater than 5 pounds maximum. **11B-309.4**   *309.4*

Size

_____L.   Control buttons shall be ¾ inch minimum in their smallest dimension. **11B-409.4.61** *409.4.6.1*

Location

_____M.   Control panels shall be on a side wall, 12 inches minimum from any adjacent wall **11B-409.4.62** *409.4.6.2* **Fig. CD-19A**

Emergency Communication
Emergency two-way communication systems shall comply as detailed.

Type

_____L.   A telephone and emergency signal device shall be provided in the car. **11B-409.4.7.1** *409.4.7.1*

Operable Parts

_____M.   The telephone and emergency signaling device shall be within one of the reach ranges specified in Section 12, "REACH RANGES". **11B-409.4.7.2** *409.4.7.2*

Compartment

_____N.   If the telephone or device is in a closed compartment the compartment door hardware shall comply with Section 13, "OPERABLE PARTS". **11B-409.4.7.3**   *409.4.7.3*

Cord

_____L.   The telephone cord shall be 29 inches long minimum. **11B-409.4.7.4**   *409.4.7.4*

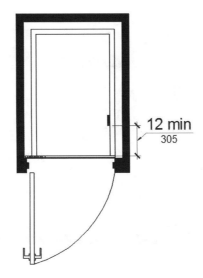

Fig. CD-19A
Location of Private Residence Elevator Control Panel
© ICC Reproduced with Permission

20. <u>PLATFORM LIFTS</u>

Platform Lifts
Platform lifts shall comply as required. Platform lifts shall be permitted as a component of an accessible route in new construction as detailed. Platform lifts shall be permitted as a component of an accessible route in an existing building or facility. **11B-206.7** *206.7*

Performance Areas and Speakers' Platforms
Platform lifts shall be permitted to provide accessible routes to performance areas and speakers' platforms. **11B-206.7.1** *206.7.1*

Wheelchair Spaces
Platform lifts shall be permitted to provide an accessible route to comply with the wheelchair space dispersion and line-of-sight requirements for Assembly Areas, Wheelchair Spaces, Companion Seats, Designated Aisle Seats and Semi-Ambulant Seats. **11B-206.7.2** *206.7.2*

Incidental Spaces
Platform lifts shall be permitted to provide an accessible route to incidental spaces which are not public use spaces and which are occupied by five persons maximum.
11B-206.7.3 *206.7.3*

Judicial Spaces
Platform lifts shall be permitted to provide an accessible route to: jury boxes and witness stands; raised courtroom stations including, judges' benches, clerks' stations, bailiffs' stations, deputy clerks' stations, and court reporters' stations; and to depressed areas such as the well of the court. **11B-206.7.4** *206.7.4*

Existing Site Constraints
Platform lifts shall be permitted where existing exterior site constraints make use of a ramp or elevator infeasible. **11B-206.7.5** *206.7.5*

Platform Lifts in Guest Rooms and Residential Dwelling Units

Platform lifts shall be permitted to connect levels within transient lodging guest rooms required to provide mobility features complying with Section 39, "TRANSIENT LODGING GUEST ROOMS", or residential dwelling units required to provide mobility features complying with Section 43, "RESIDENTIAL FACILITIES" and adaptable features complying with Chapter 11A, Division IV.   **11B-206.7.6**   *206.7.6*

Amusement Rides

Platform lifts shall be permitted to provide accessible routes to load and unload areas serving amusement rides.   **11B-206.7.7**   *206.7.7*

Play Areas

Platform lifts shall be permitted to provide accessible routes to play components or soft contained play structures.   **11B-206.7.8**   *206.7.8*

Team or Player Seating

Platform lifts shall be permitted to provide accessible routes to team or player seating areas serving areas of sport activity.   **11B-206.7.9**   *206.7.9*

Recreational Boating Facilities and Fishing Piers and Platforms

Platform lifts shall be permitted to be used instead of gangways that are part of accessible routes serving recreational boating facilities and fishing piers and platforms.
11B-206.7.10   *206.7.10*

General

Platform lifts shall comply with ASME A18.1. Platform lifts shall not be attendant-operated and shall provide unassisted entry and exit from the lift.
11B-410.1   *410.1*

Floor Surfaces

_____A.   Floor surfaces in platform lifts shall comply with Section 7, "FLOOR OR GROUND SURFACES", and Section 8, "CHANGES IN LEVEL".   **11B-410.2**   *410.2*

Clear Floor Space

_____B.   Clear floor space in platform lifts shall comply with Section 10, "CLEAR FLOOR OR GROUND SPACE". **11B-410.2**   *410.2*

Platform to Runway Clearance

_____C.   The clearance between the platform sill and the edge of any runway landing shall be 1-1/4 inch maximum.   **11B-410.4**   *410.4*

Operable Parts

_____D.   Controls for platform lifts shall comply with Section 13, "OPERABLE PARTS". **11B-410.2**   *410.2*

Doors and Gates

_____E.   Platform lifts shall have low-energy power-operated doors or gates that comply with the requirements for "Automatic and Power-Assisted Doors and Gates", in Section 14, "DOORS, DOORWAYS AND GATES".   **11B-410.6**   *410.6*   **Fig. CD-20A**

_____E1.   Doors shall remain open for 20 seconds minimum.

_____E2.   End doors and gates shall provide a clear width 32 inches minimum.

_____E3.   Side doors and gates provide a clear width 42 inches minimum.

> **EXCEPTION:** Platform lifts serving two landings maximum and having doors or gates on opposite sides shall be permitted to have self-closing manual doors or gates.

Landing Size

_____F.   The minimum size of landings at platform lifts shall be 60 inches by 60 inches.   **11B-410.7**

Restriction Sign

_____G.   A sign complying with the requirements for accessible visual characters (see Section 30, "SIGNS", shall be posted in a conspicuous place at each landing and within the platform enclosure stating "No Freight" and include the International Symbol of Accessibility. **11B-410.8**

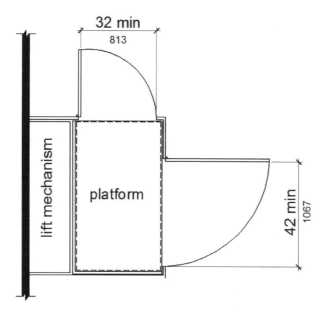

Fig. CD-20A
Platform Lift Doors and Gates
© ICC Reproduced with Permission

ADVISORY:

<u>For lifts with a signed installation contract BEFORE May 1, 2008</u>
CCR, Title 8, Section 3094 provides safety regulations for Vertical Platform (Wheelchair) Lifts. This section includes requirements which address platform size and gate configurations for lifts with 90-degree egress, and additionally requires signs posted at the landings as follows:

(1)   International Symbol of Accessibility
(2)   lift capacity
(3)   the telephone number to call in case of emergency, and
(4)   the lift shall not be used to transport materials or equipment.

<u>For lifts with a signed installation contract ON OR AFTER May 1, 2008</u>
In addition to the accessibility requirements of the building code, Vertical Platform (Wheelchair) Lifts are required to comply with the applicable provisions of the Elevator Code, CCR, Title 8, Section 3142.1, Which provides safety regulations for lifts. This section incorporated ASME A18.1-2003, sections 2 and 5 by reference, and requires compliance with CCR, Title 8, Sections 3094.2(d). 3094.2(e), 2094.2(g), and 3094.2 (p).

These regulations address technical requirements for lifts, including platform size, gate configurations for lifts with 90-degree egress, and required signs as follows::

(1)   lift capacity, and
(2)   "No Freight" prohibiting the transport of materials or equipment.

21. <u>PARKING SPACES</u>

<u>Parking</u>
Parking spaces that comply shall be identified by signs that comply as detailed.
11B-216.5 *216.5*

> **EXCEPTIONS:**
>
> > 1.   Reserved.
> >
> > 2.   In residential facilities, where parking spaces are assigned to specific residential dwelling units, identification of accessible parking spaces shall not be required.

<u>Valet Parking</u>
Parking facilities that provide valet parking services shall provide at least one accessible passenger loading zone. The accessible parking requirements apply to facilities with valet parking. **11B-209.4** *209.4*

<u>Mechanical Access Parking Garages</u>
Mechanical access parking garages shall provide at least one accessible passenger loading zone at vehicle drop-off and vehicle pick-up areas. **11B-209.5** *209.5*

<u>General</u>
Where parking spaces are provided, parking spaces shall be provided as detailed herein. **11B-208.1** *208.1*

> **EXCEPTION:** Parking spaces used exclusively for buses, trucks, other delivery vehicles, or vehicular impound shall not be required to comply as detailed provided that lots accessed by the public are provided with an accessible passenger loading zone.

<u>Minimum Number</u>
Accessible parking spaces shall be provided in accordance with Table 21A except as required for Hospital Outpatient Facilities, Rehabilitation Facilities, Outpatient Physical Therapy Facilities and Residential Facilities. Where more than one parking facility is provided on a site, the number of accessible spaces provided on the site shall be calculated according to the number of spaces required for each parking facility.
11B-208.2 *208.2*

> *ADVISORY. **Minimum Number.** The term "parking facility" is used instead of the term "parking lot" so that it is clear that both parking lots and parking structures are required to comply with this section. The number of parking spaces required to be accessible is to be calculated separately for each parking facility: the required number is not based on the total number of parking spaces provided in all of the parking facilities provided on the site.*
>
> *Accessible spaces can be provided in other facilities or locations, or, in the case of parking structures, on one level only when equal or greater access is provided in terms of proximity to an accessible entrance, cost, and convenience. For example, accessible spaces required for outlying parking facilities may be located in a parking facility closer to an accessible entrance. The minimum number of spaces must still be determined separately for each facility even if the spaces are to be provided in other facilities or locations. Accessible spaces may be grouped on one level of a parking structure in order to achieve greater access. However, where parking levels serves different building entrances, accessible spaces should be dispersed so that access is provided to each entrance.*

Table 21A Parking Spaces

Total Number of Parking Spaces Provided in Parking Facility	Minimum Number of Required Accessible Parking Spaces
1 to 25	1
26 to 50	2
51 to 75	3
76 to 100	4
101 to 150	5
151 to 200	6
201 to 300	7
301 to 400	8
401 to 500	9
501 to 1000	2 percent of total
1001 and over	20, plus 1 for each 100, or fraction thereof, over 1000

Hospital Outpatient Facilities

Ten percent of patient and visitor parking spaces provided to serve hospital outpatient facilities, and free-standing buildings providing outpatient clinical services of a hospital shall comply.   **11B-208.2.1** *208.2.1*

ADVISORY. Hospital Outpatient Facilities. The term "outpatient facility" is not defined but is intended to cover facilities or units that are located in hospitals and that provide regular and continuing medical treatment without an overnight stay. Doctor's offices, independent clinics, or other facilities not located in hospitals are not considered hospital outpatient facilities for purposes of these requirements.

The higher percentages required for hospital outpatient facilities or rehabilitation facilities specializing in treating conditions that affect mobility and outpatient physical therapy facilities are intended primarily for visitor and patient parking.   If there are separate lots for visitors or patients and employees, the 10% or 20% requirement shall be applied to the visitor/patient lot while accessible parking could be provided in the employee parking lot according to the general scoping requirements in Table 21A, above. If a lot serves both visitors or patients and employees, 10% or 20% of the spaces intended for use by visitors or patients must be accessible.

At hospitals or other facilities where parking does not specifically serve an outpatient unit, only a portion of the lot would need to comply with the 10% scoping requirement. A local zoning code that requires a minimum number of parking spaces according to occupancy type and square footage may be an appropriate guide in assessing the number of spaces in the lot that "belong" to the outpatient unit. These spaces would be held to the 10% requirement while the rest of the lot would be subject to general scoping requirement in the Table. Those accessible spaces required for the outpatient unit should be located at the accessible entrance serving the unit. This method may also be used in applying the 20% requirement to hospitals or other facilities where only a portion of the unit provides specialized rehabilitation or physical therapy treatment or services for persons with mobility impairments.

Rehabilitation Facilities and Outpatient Physical Therapy Facilities

Twenty percent of patient and visitor parking spaces provided to serve rehabilitation facilities specializing in treating conditions that affect mobility and outpatient physical therapy facilities shall comply.   **11B-208.2.2**   *208.2.2*

ADVISORY. Rehabilitation Facilities and Outpatient Physical Therapy Facilities. Conditions that affect mobility include conditions requiring the use of a brace, cane, crutch, prosthetic device, wheelchair, or powered mobility aid; arthritic, neurological, or orthopedic conditions that severely limit one's ability to walk; respiratory diseases and other conditions which may require the use of portable oxygen; and cardiac conditions that impose significant functional limitations.

Residential Facilities
Parking spaces provided to serve residential facilities shall comply as follows: **11B-208.2.3**   *208.2.3*

Parking for Residents
Where at least one parking space is provided for each residential dwelling unit, at least one parking space that complies shall be provided for each residential dwelling unit required to provide mobility features (see Section 43 "RESIDENTIAL FACILITIES").   **11B-208.2.3.1** *208.2.3.1*

Additional Parking Spaces for Residents
Where the total number of parking spaces provided for each residential dwelling unit exceeds one parking space per residential dwelling unit, 2 percent, but no fewer than one space, of all the parking spaces not covered by "Parking for Residents", above, shall comply. **11B-208.2.3.2**   *208.2.3.2*

Parking for Guests, Employees, and Other Non-Residents
Where parking spaces are provided for persons other than residents, parking shall be provided in accordance with Table 21A.   **11B-208.2.3.3**   *208.2.3.3*

Requests for Accessible Parking Spaces
When assigned parking is provided, designated accessible parking for the adaptable residential dwelling units shall be provided on requests of residents with disabilities one the same terms and with the full range of choices (e.g. off-street parking, carport or garage) that are available to other residents.
11B-208.2.3.4

Van Parking Spaces
For every six or fraction of six parking spaces required by Table 21A to comply, at least one shall be a van parking space that complies.   **11B-208.2.4**   *208.2.4*

Location
Parking spaces shall comply as detailed.   **11B-208.3**   *208.3*

General
Accessible parking spaces that serve a particular building or facility shall be located on the shortest accessible route from parking to an accessible entrance. Where parking serves more than one accessible entrance, accessible parking spaces shall be dispersed and located on the shortest accessible route to the accessible entrances. In parking facilities that do not serve a particular building or facility, accessible parking spaces shall be located on the shortest accessible route to an accessible pedestrian entrance of the parking facility.   **11B-208.3.1**   *208.3.1*

> **EXCEPTIONS:**
>
> 1.   All van parking spaces shall be permitted to be grouped on one level within a multi-story parking facility.
>
> 2.   Parking spaces shall be permitted to be located in different parking facilities if substantially equivalent or greater accessibility is provided in terms of distance from an accessible entrance or entrances, parking fee, and user convenience.

ADVISORY. **General Exception 2.** *Factors that could affect "user convenience" include, but are not limited to, protection from the weather, security, lighting, and comparative maintenance of the alternative parking site.*

Residential Facilities

In residential facilities containing residential units required to provide mobility features (see Section 43 "RESIDENTIAL FACILITIES"), and adaptable features complying with CBC Chapter 11A, Division IV Parking spaces provided for Residential Facilities shall be located on the shortest accessible route to the residential dwelling unit entrance they serve. Spaces provided in accordance with Residential Facilities shall be dispersed throughout all types of parking provided for the residential dwelling units. **11B-208.3.2** *208.3.2*

> **EXCEPTION:** Parking spaces provided in accordance with "Residential Facilities" shall not be required to be dispersed throughout all types of parking if substantially equivalent or greater accessibility is provided in terms of distance from an accessible entrance, parking fee, and user convenience.

ADVISORY. **Residential Facilities Exception.** *Factors that could affect "user convenience" include, but are not limited to, protection from the weather, security, lighting, and comparative maintenance of the alternative parking site.*

Private Garages Accessory to Residential Dwelling Units

Private garages accessory to residential dwelling units shall comply. Private garages include individual garages and multiple individual garages grouped together.
11B-208.3.3

> Detached private garages accessory to residential dwelling units, shall be accessible.
> **11B-208.3.3.1**

> Attached private garages directly serving a single residential dwelling unit shall provide at least one of the following options: **11B-208.3.3.2**

> > 1. A door leading directly from the residential dwelling unit which immediately enters the garage.

> > 2. An accessible route from the residential dwelling unit to an exterior door entering the garage.

> > 3. An accessible route from the residential dwelling unit's primary entry door to the vehicular entrance of the garage.

_____ A. The correct number of standard accessible and van-accessible parking stalls is provided on the site. **11B-208.2, 11B-208.2.4** *208.2, 208.2.4* **Table 21A**

_____ B. Parking space(s) that serve a particular building or facility are located on the shortest route from parking to an accessible entrance. **11B-208.3.1** *208.3.1*

_____ C. Where parking serves more than one accessible entrance, accessible parking spaces are dispersed and located on the shortest accessible route to the accessible entrances. **11B-208.3.1** *208.3.1*

EXCEPTION: Parking spaces provided in accordance with "Residential Facilities" shall not be required to be dispersed throughout all types of parking if substantially equivalent or greater accessibility is provided in terms of distance from an accessible entrance, parking fee, and user convenience.

_____D. In parking facilities that do not serve a particular building or facility, accessible parking spaces shall be located on the shortest accessible route to an accessible pedestrian entrance of the parking facility. **11B-208.3.1** *208.3.1*

_____E. Where parking spaces are marked with lines, width measurements of parking spaces and access aisles are made from the centerline of the markings. **11B-502.1** *502.1*

EXCEPTION: Where parking spaces or access aisles are not adjacent to another parking space or access aisle, measurements shall be permitted to include the full width of the line defining the parking space or access aisle.

Vehicle Spaces

_____F. Car and van parking spaces are 216 inches long (18.0 feet) minimum.
11B-502.2 *502.2* **Fig. CD-21A**

_____G. Car parking spaces are 108 inches wide (9.0 feet) minimum.
11B-502.2 *502.2* **Fig. CD-21A**

_____H. Van parking spaces are 144 inches wide (12.0 feet) minimum.
11B-502.2 *502.2* **Fig. CD-21A**

_____I. The parking spaces are marked to define the width, and have an adjacent access aisle.
11B-502.2 *502.2* **Fig. CD-21A**

EXCEPTION: Van parking spaces are permitted to be 108 inches wide (9.0 feet) minimum where the access aisle is 96 inches wide (8.0 feet) minimum. **Fig. CD-21B**

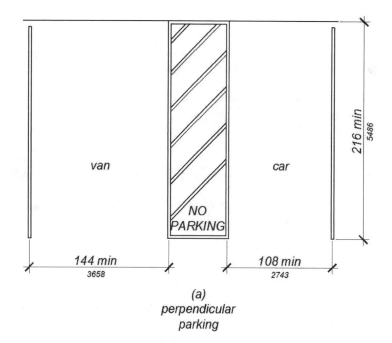

(a)
perpendicular
parking

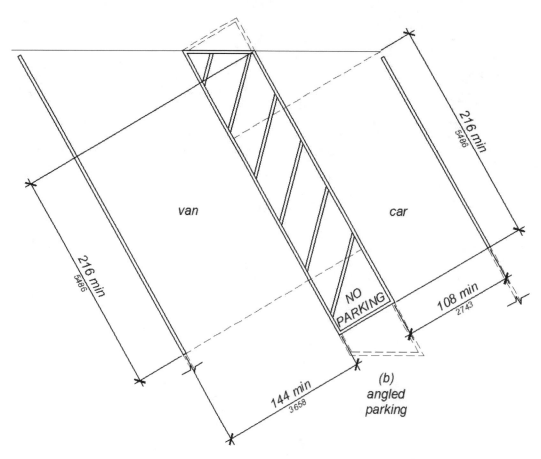

(b)
angled
parking

Fig. CD-21A
Vehicle Parking Spaces
© ICC Reproduced with Permission

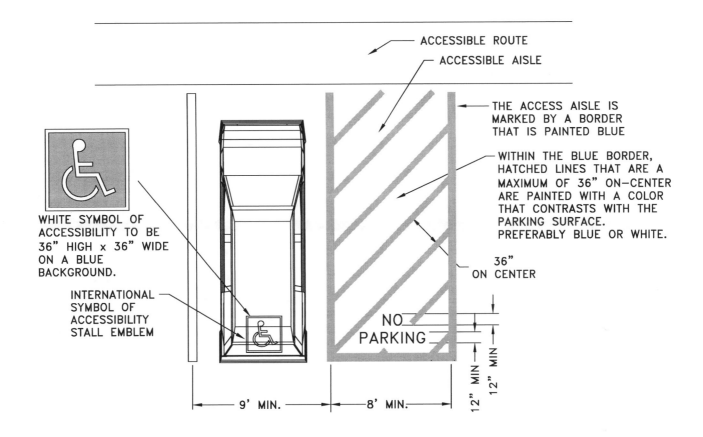

ACCESSIBLE ROUTE

ACCESSIBLE AISLE

THE ACCESS AISLE IS MARKED BY A BORDER THAT IS PAINTED BLUE

WITHIN THE BLUE BORDER, HATCHED LINES THAT ARE A MAXIMUM OF 36" ON-CENTER ARE PAINTED WITH A COLOR THAT CONTRASTS WITH THE PARKING SURFACE. PREFERABLY BLUE OR WHITE.

36" ON CENTER

12" MIN.

12" MIN.

NO PARKING

9' MIN.

8' MIN.

WHITE SYMBOL OF ACCESSIBILITY TO BE 36" HIGH x 36" WIDE ON A BLUE BACKGROUND.

INTERNATIONAL SYMBOL OF ACCESSIBILITY STALL EMBLEM

Fig. CD-21B
Van Accessible Parking Space

Access Aisle

Width

_____ J.　Access aisles serving car and van parking spaces are 60 inches wide (5.0 feet) minimum.　**11B-502.3.1**　*502.3.1*　**Fig. CD-21C**

Length

_____ K.　Access aisles extend the full required length of the parking space they serve.
11B-502.3.2　*502.3.*　**Fig. CD-21C**

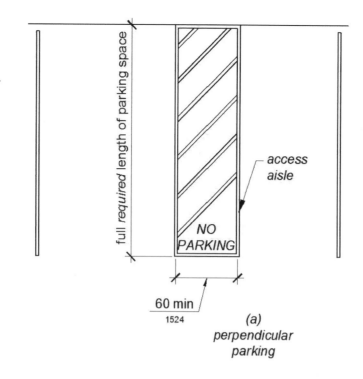

access aisle

60 min
1524

(a)
perpendicular
parking

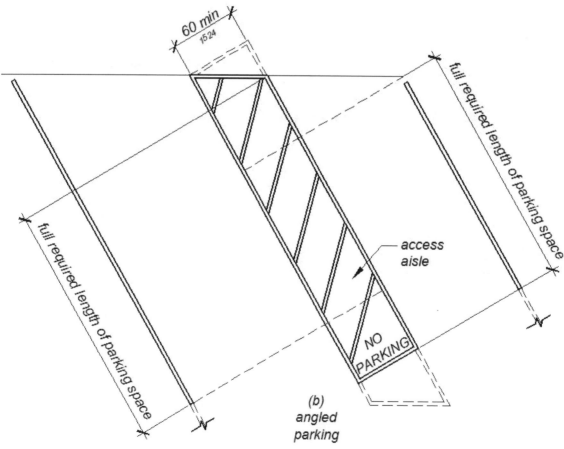

Fig. CD-21C
Parking Space Access Aisle
© ICC Reproduced with Permission

Marking

_____L.   Access aisles are marked with a blue painted borderline around their perimeter.
11B-502.3.3   **Fig. 21B & D**

_____M.   The area within the blue borderlines is marked with hatched lines a maximum of 36 inches (3.0 feet) on center in a color contrasting with that of the aisle surface, preferably blue or white. **11B-502.3.3**   **Fig. CD-21B & D**

_____N.   The words "NO PARKING" are painted on the surface within each access aisle in white letters a minimum of 12 inches (1.0 foot) in height and located to be visible from the adjacent vehicular way.   **11B-502.3.3**   **Fig. CD-21B & D**

> **EXCEPTION:** Access aisle markings may extend beyond the minimum required length.

ADVISORY. Marking. The requirement that the hatching at the loading and unloading access aisle is a suitable contrasting color to the parking space is intended to ensure that the hatching is visually distinct from the background to which it is applied, and thus can be more easily seen. As hatching is generally recognized as a no parking area, this difference in contrast assists drivers by providing a conspicuous visual deterrent to parking in the loading and unloading access aisle.

Asphalt is often the parking surface material used at accessible parking spaces. Asphalt is generally considered to be fairly dark in appearance. In order to provide a suitable contrasting color at the hatched area of the loading and unloading access aisle, a light color hatching should be used at locations where asphalt is the parking surface material. Although white paint is preferred (and traditionally the color most often used), its use is not mandatory under the California Building Code (CBC).

In order to provide a suitable contrast at the hatched area of the loading and unloading access aisle in locations where light concrete is used as the parking surface material (such as at concrete parking garages), a dark color hatching should be used. Although blue paint is preferred, its use is not mandatory under the California Building Code (CBC).

Location

_____O.   Access aisles do not overlap the vehicular way. Access aisles are permitted to be placed on either side of the parking space except for van parking spaces shall have the access aisles located on the passenger side of the parking space.   **11B-502.3.4**   *502.3.4* **Fig. CD-21B & C**

ADVISORY. Location. Wheelchair lifts typically are installed on the passenger side of vans. Many drivers, especially those who operate vans, find it more difficult to back into parking spaces than to back out into comparatively unrestricted vehicle lanes. For this reason, where a van and car share an access aisle, or where a separate van parking stall is placed, the van space is required to be located on the passenger side of the vehicle.

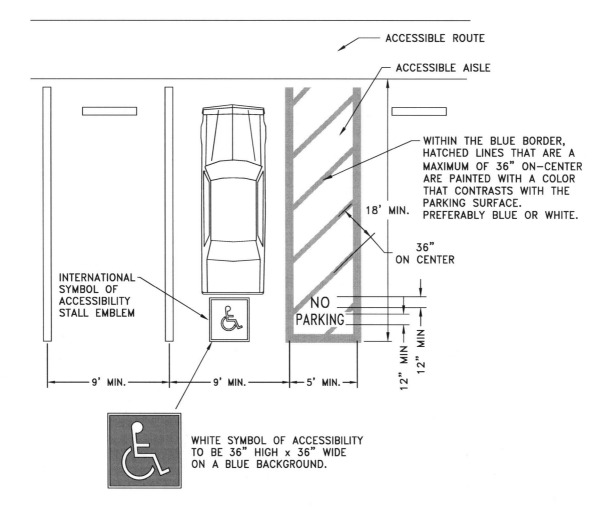

ACCESSIBLE ROUTE

ACCESSIBLE AISLE

WITHIN THE BLUE BORDER, HATCHED LINES THAT ARE A MAXIMUM OF 36" ON-CENTER ARE PAINTED WITH A COLOR THAT CONTRASTS WITH THE PARKING SURFACE. PREFERABLY BLUE OR WHITE.

18' MIN.

36" ON CENTER

INTERNATIONAL SYMBOL OF ACCESSIBILITY STALL EMBLEM

NO PARKING

12" MIN   12" MIN

9' MIN.   9' MIN.   5' MIN.

WHITE SYMBOL OF ACCESSIBILITY TO BE 36" HIGH x 36" WIDE ON A BLUE BACKGROUND.

**Fig. CD-21D
Parking Space Marking**

Floor or Ground Surfaces

_____P.   Parking spaces and access aisles serving them shall comply with Section 7, "FLOOR OR GROUND SURFACES". Access aisles shall be at the same level as the parking spaces they serve. Changes in level are NOT permitted.
11B-502.4, 11B-305, 11B-302.1 *502.4, 305, 302.1*

> **EXCEPTION:**   Slopes not steeper than 1:48 shall be permitted.

> ***ADVISORY. Floor or Ground Surfaces.*** *Accessible parking spaces and access aisles are required to be level (maximum 1:48 slope) in all directions to provide a surface for wheelchair transfer to and from vehicles. Built-up curb ramps are not permitted to project into access aisles and parking spaces because they would create slopes greater than 1:48. The access aisle needs to be adjacent to, and the same required length as the accessible space it serves. Where two parking stalls share the same access aisle, a full depth aisle is required; in diagonal parking, the geometric layout will require additional access aisle depth to satisfy the loading area for both of the adjacent parking stalls as indicated in Fig. CD-21A.*

Vertical Clearance

_____Q.　Parking spaces, access aisles and vehicular routes serving them shall provide a vertical clearance of 98 inches (8.2 feet) minimum.　**11B-502.5**　*502.5*

> **ADVISORY.　*Vertical Clearance.*** *Signs provided at entrances to parking facilities informing drivers of clearances and the location of van accessible parking spaces can provide useful customer assistance.*

Identification

_____R.　Parking space identification signs shall include the International Symbol of Accessibility.　**11B-502.6**　*502.6*　**Fig. CD-21E & F**

_____U.　Signs identifying van parking spaces contain additional language or an additional sign with the designation "van accessible".
11B-502.6　*502.6*　**Fig. CD-21E**

_____V.　Signs shall be 60 inches (5.0 feet) minimum above the finish floor or ground Surface measured to the bottom of the sign. **11B-502.6** *502.6* **Fig. CD-21E & F**

> **EXCEPTION:**　Signs located within an accessible route shall be a minimum of 80 inches above the finish floor or ground surface to the bottom of the sign.

> **ADVISORY.　*Identification.*** *The required "van-accessible" designation is intended to be informative, not restrictive, in identifying those spaces that are better suited for van use.　Enforcement of motor vehicle laws, including parking privileges, is a local matter.*
>
> *This section requires that an additional "van accessible" sign or additional language be posted below the International Symbol of Accessibility when designating an accessible parking space that will accommodate a van.　A single parking sign incorporating both the International Symbol of Accessibility and the language "van accessible" may be used.*

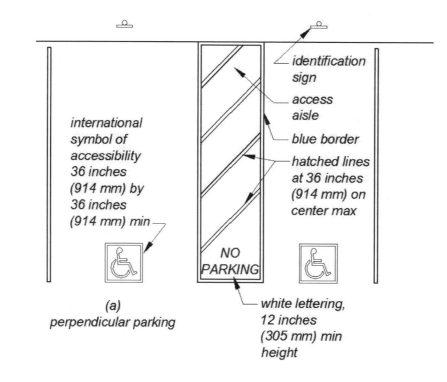

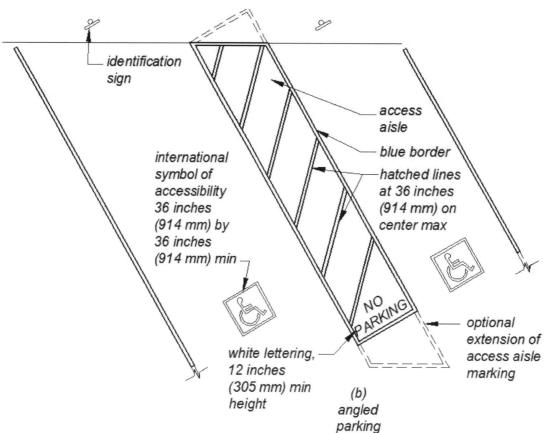

Fig. CD-21E
Angled and Perpendicular Parking Identification
© ICC Reproduced with Permission

Finish and Size

_____W.   Parking identification signs shall be reflectorized with a minimum area of 70 square inches.
11B-502.6.1   Fig. CD-21F

Minimum Fine

_____X.   Additional language or an additional sign below the International Symbol of Accessibility shall
state "Minimum Fine $250."   **11B-502.6.2   fig. CD-21F**

Location

_____Y.   A parking space identification sign shall be visible from each parking space, shall be
permanently posted either immediately adjacent to the parking space or within the projected
parking space width at the head end of the parking space. Signs may also be permanently
posted on a wall at the interior end of the parking space. **11B-502.6.3   Fig. CD-21E & F**

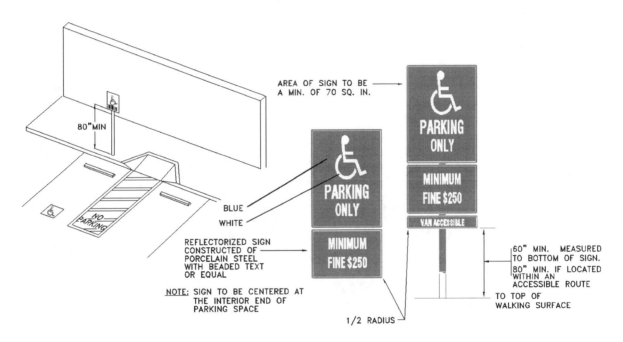

Fig. CD-21F
Parking Space Identification Signs

Marking

_____Z.   Each accessible car and van space shall have surface identification that complies with either of
the following options: **11B-502.6.4**

_____AA.   The parking space shall be marked with an International Symbol of Accessibility in
white on a blue background a minimum 36 inches wide by 36 inches high.
11B-502.6.4.1   Fig. CD-21B, D & E

_____BB.   The centerline of the International Symbol of Accessibility is a maximum of 6
inches from the centerline of the parking space, its sides parallel to the length
of the parking space and its lower corner at, or lower side aligned with the end
of the parking space length.   **11B-502.6.4.1   Fig. CD-21B, D & E**

OR

_____CC. The parking space shall be outlined or painted blue and shall be marked with an International Symbol of Accessibility in white on a blue background a minimum 36 inches wide by 36 inches high in white or a suitable contrasting color. The centerline of the International Symbol of Accessibility is a maximum of 6" from the centerline of the parking space, its sides parallel to the length of the parking space and its lower corner at, or lower side aligned with the end of the parking space length.. **11B-502.6.4.2   Fig. CD-21B, D & E**

Relationship to Accessible Route

_____DD. Parking spaces and access aisles are designed so that cars and vans, when parked, cannot obstruct the required clear width of adjacent accessible routes. **11B-502.7**   *502.7*

_____EE. Access aisles shall adjoin an accessible route. Two parking spaces shall be permitted to share a common access aisle. **11B-502.3**   *502.3*

> ***ADVISORY. Relationship to Accessible Routes.** Accessible routes must connect parking spaces to accessible entrances. Where possible, it is required that the accessible route not pass behind parked vehicles.*

Arrangement

_____FF. Parking spaces and access aisles shall be designed so that persons using them are not required to travel behind parking spaces other than to pass behind the parking space in which they parked. **11B-502.7.1**

> ***ADVISORY.   Arrangement.** Accessible parking spaces located so that the accessible route passes behind parked vehicles create a safety hazard, especially for wheelchair users.*
>
> *Wheelchair users traveling behind parked vehicles may be obscured from the view of drivers backing out of parking spaces, especially when passing behind high profile vehicles. This section requires that persons with disabilities not be required to travel behind parking spaces other than the one in which they have parked.*

Wheel Stops

_____GG. A curb or wheel stop shall be provided if required to prevent encroachment of vehicles over the required clear width of adjacent accessible routes.
11B-502.7.2

Additional Signage

_____HH. An additional sign shall be posted either in a conspicuous place at either each entrance to an off-street parking facility or immediately adjacent to on-site accessible parking and visible from each parking space. **11B-502.8   Fig. CD-21G**

Size

_____II. The additional sign shall not be less than 17 inches wide by 22 inches high. **11B-502.8.1   Fig. CD-21G**

Lettering

_____JJ.　The additional sign shall clearly state in letters with a minimum height of 1 inches the following:　**11B-502.8.2　Fig. CD-21G**

"Unauthorized vehicles parked in designated accessible spaces not displaying distinguishing placards or special license plates issued for persons with disabilities will be towed away at the owner's expense.　Towed vehicles may be reclaimed at ___ _____ or by telephoning _____."　**11B-502.8.2　Fig. CD-21G**

_____LL.　Blank spaces shall be filled in with appropriate information as a permanent part of the sign.　**11B-502.8.2　Fig. CD-21G**

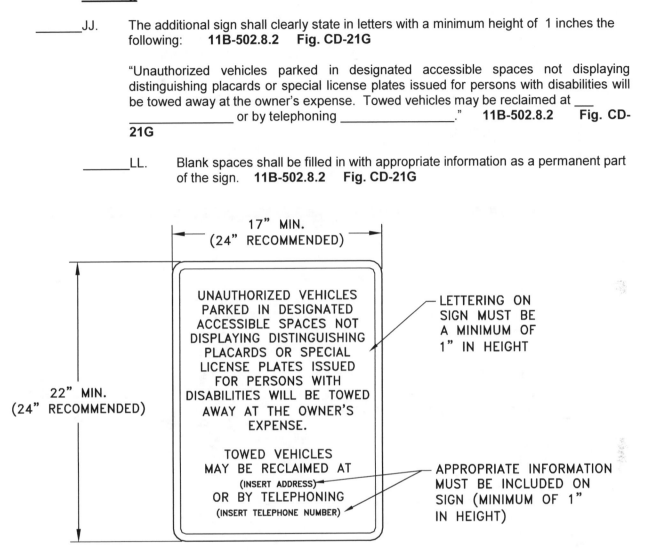

Fig. CD-21G
Additional Signage Warning Unauthorized Vehicles
Not to Park in Accessible Parking Spaces

22. PASSENGER LOADING ZONES

General
Passenger loading zones shall be provided as detailed. **11B-209.1** *209.1*

Type
Where provided, passenger loading zones shall comply as detailed. **11B-209.2** *209.2*

Passenger Loading Zones
Passenger loading zones, except those in Bus Loading Zones and On-street Bus Stops, shall provide at least one accessible passenger loading zone in every continuous 100 linear feet (30480 mm) of loading zone space, or fraction thereof. **11B-209.2.1** *209.2.1*

Medical Care and Long Term Care Facilities
At least one compliant passenger loading zone shall be provided at an accessible entrance to licensed medical care and licensed long-term care facilities where the period of stay may exceed twenty-four hours. **11B-209.3** *209.3*

Valet Parking
Parking facilities that provide valet parking services shall provide at least one compliant passenger loading zone. The requirements for accessible parking still apply to facilities with valet parking. **11B-209.4** *209.4*

> *ADVISORY. Valet Parking. Valet parking does not eliminate the requirements to provide accessible parking spaces. Some vehicles may be adapted with hand controls or lack a driver's seat, and may not be operable by a valet parking attendant. The accessible parking space requirements apply to valet parking, including the requirement for an accessible route of travel to the entrance of the facility In addition, when valet parking is provided, a compliant passenger loading zone shall be located on an accessible route to the entrance of the facility.*

Mechanical Access Parking Garages
Mechanical access parking garages shall provide at least one compliant passenger loading zone at vehicle drop-off and vehicle pick-up areas. **11B-209.5** *209.5*

General
Passenger drop-off and loading zones shall comply as detailed. **11B-503.1** *503.1*

Vehicle Pull-Up Space

_____A.   Passenger drop-off and loading zones shall provide a vehicle pull-up space 96 inches wide minimum and 20 feet long minimum. **11B-503.2** *503.2* **Fig. CD-22A**

Access Aisle

_____B.   Passenger drop-off and loading zones shall provide compliant access aisles adjacent and parallel to the vehicle pull-up space. **11B-503.3** *503.3*
Fig. CD-22A

_____C.   Access aisles shall adjoin an accessible route and shall not overlap the vehicular way. **11B-503.3** *503.3*

Width

_____D.   Access aisles serving vehicle pull-up spaces shall be 60 inches wide minimum
11B-503.3.1   *503.3.1*   **Fig. CD-22A**

Length

_____E.   Access aisles shall extend the full length of the vehicle pull-up spaces they
serve.   **11B-503.3.2**   *503.3.2*   **Fig. CD-22A**

Marking

_____F.   Access aisles shall be marked with a painted borderline around their perimeter.
The area within the borderlines shall be marked with hatched lines a maximum
of 36 inches on center in a color contrasting with that of the aisle surface.
11B-503.3.2   *503.3.2*   **Fig. CD-22A**

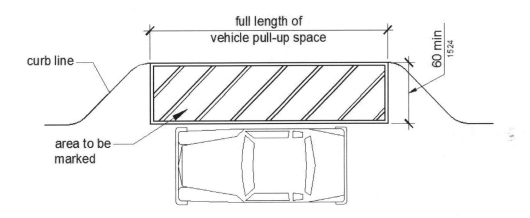

Fig. CD-22A
Passenger Drop-Off and Loading Zone Access Aisle
© ICC Reproduced with Permission

Floor and Ground Surfaces

_____G.   Vehicle pull-up spaces and access aisles serving comply with Section 7,
"FLOOR OR GROUND SURFACES. Access aisles shall be at the same level as the
vehicle pull-up space they serve. Changes in level are NOT permitted.   **11B-503.4,
11B-305, 11B-302.1**   *503.4, 305, 302.1*

EXCEPTION: Slopes not steeper than 1:48 shall be permitted.

Vertical Clearance

_____H.   Parking spaces, access aisles serving them, and a vehicular route from
an entrance to the passenger loading zone and from the passenger loading zone to a
vehicular exit shall provide a vertical clearance of 114 inches minimum.   **11B-503.5**
503.5

ADVISORY: Vertical Clearance. *The minimum vertical clearance at designated passenger drop-off and loading zones, and along at least one vehicle access route to the area from site entrances and exits, is 114 inches. This differs from the 98 inch vertical clearance requirement for accessible parking spaces. Commercial vans used for accessible transit may be taller than passenger vans.*

Identification

_____I.   Each passenger loading zone designated for persons with disabilities shall be identified with a compliant reflectorized sign. It shall be permanently posted immediately adjacent to and visible from the passenger loading zone stating "Passenger Loading Zone Only" and including the International Symbol of Accessibility in white on a dark blue background.   **11B-503.6   Fig. CD-22B**

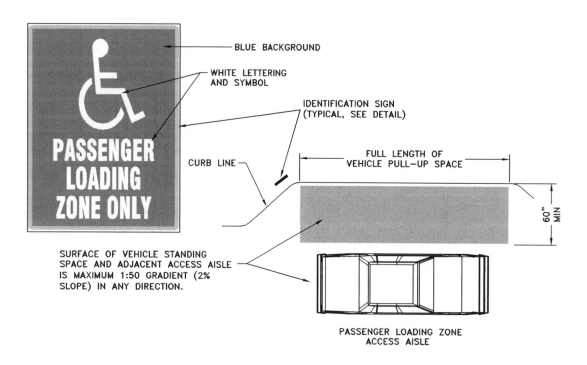

Fig. CD-22B
Identification for Passenger Drop-Off and Loading Zones

23. <u>STAIRWAYS</u>

<u>General</u>
Interior and exterior stairs shall comply as detailed. **11B-210.1, 11B-504.1** *210.1, 504.1*

EXCEPTIONS:

1. In detention and correctional facilities, stairs that are not located in public use areas shall not be required to comply.

2. In alterations, stairs between levels that are connected by an accessible route shall not be required to comply, except that compliant tread striping and compliant handrails shall be provided when the stairs are altered.

3. In assembly areas, aisle stairs shall not be required to comply with these requirements except that compliant tread striping shall be provided.

4. Stairs that connect play components shall not be required to comply, except that compliant tread striping shall be required.

<u>Stairs and Escalators in Existing Buildings</u>
In alterations and additions, where an escalator or stair is provided where none existed previously and major structural modifications are necessary for the installation, an accessible route shall be provided between the levels served by the escalator or stair unless exempted by the Multi-Story Buildings and Facilities Exceptions 1 through 7. **11B-206.2.3.1** *206.2.3.1*

<u>Treads and Risers</u>

_____A.	All steps on a flight of stairs shall have uniform riser heights and uniform tread depths. **11B-504.2** *504.2*

_____B.	Risers shall be 4 inches high minimum and 7 inches high maximum. **11B-504.2** *504.2*

_____C.	Treads shall be 11 inches deep minimum. **11B-504.2** *504.2*

<u>Open Risers</u>

_____D.	Open risers are not permitted. **11B-504.3** *504.3*

EXCEPTIONS:

1. On exterior stairways, an opening of not more than ½ inch may be permitted between the base of the riser and the tread.

2. On exterior stairways, risers constructed of grating containing openings of not more than ½ inch may be permitted.

<u>Tread Surface</u>

_____E.	Stair treads shall comply wioth Section 7, "GROUND OR FLOOR SURFACES". Changes in level are not permitted. **11B-504.4** *504.4*

EXCEPTION: Treads shall be permitted to have a slope not greater than 1:48.

<u>Contrasting Stripe</u>

_____F.	Interior stairs shall have the upper approach and lower tread marked by a stripe providing clear visual contrast. **11B-504.4.1**	**Fig. CD-23A**

_____G.   Exterior stairs shall have the upper approach and all treads marked by a stripe providing clear visual contrast.   **11B-504.4.1   Fig. CD-23A**

_____H.   The stripe shall be a minimum of 2 inches wide to a maximum of 4 inches wide placed parallel to and not more than 1 inch from the nose of the step or upper approach. **11B-504.4.1   Fig. CD-23A**

_____I.   The stripe shall extend the full width of the step or upper approach and shall be of a material that is at least as slip-resistant as the other treads of the stair.   A painted stripe shall be acceptable. Grooves shall not be used to satisfy this requirement. **11B-504.4.1**

> **ADVISORY: Contrasting Stripe.**   *Some designers propose to provide a group of tooled grooves in lieu of the contrasting color. While grooves do provide shadow lines in some lighting conditions, these shadows cannot be relied upon under all lighting conditions. In addition, the spaces between the grooves do not provide any contrast.*

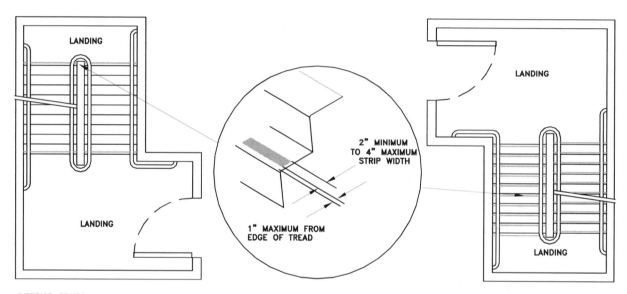

INTERIOR STAIRS
UPPER APPROACH AND LOWER
TREADS MUST BE MARKED

WARNING STRIPS MUST BE CLEARLY CONTRASTING COLOR FROM ADJOINING
SURFACES. THE STRIP MUST BE MADE OF A MATERIAL THAT IS AT LEAST
AS SLIP RESISTANT AS THE OTHER TREADS OF THE STAIR. (PAINTED STRIPES
ARE ACCEPTABLE)

EXTERIOR STAIRS
UPPER APPROACH AND ALL
TREADS MUST BE MARKED

**Fig. CD-23A
Contrasting Striping**

Nosings

_____J.   The radius of curvature at the leading edge of the tread shall be ½ inch maximum. **11B-504.5**   *504.5*   **Fig. CD-23B**

_____K.   Nosings that project beyond risers shall have the underside of the leading edge curved or beveled. Risers shall be permitted to slope under the tread at an angle of 30 degrees maximum from vertical.   **11B-504.5**   *504.5*

_____L.   The permitted projection of the nosing shall extend 1-1/4 inches maximum over the tread below **11B-504.5**   *504.5*   **Fig. CD-23B**

EXCEPTION: In existing buildings there is no requirement to retroactively alter existing nosing projections of 1-1/2 inches which were constructed in compliance with the building code in effect at the time of original construction.

ADVISORY. Nosings. While Chapter 10 may allow larger radius of curvature at the leading edge (nosing) of the tread, this radius must be no larger than ½ inch to comply with both Chapter 10 and Chapter 11B requirements.

Additionally, because it is possible to catch the top of one's shoe on the underside of stair nosings, access compliance regulations require the underside of nosings to be free of abrupt changes – a beveled slope or non-square underside.

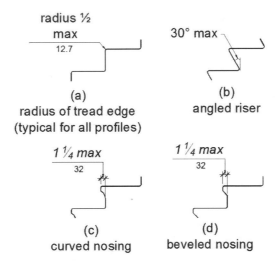

(a)
radius of tread edge
(typical for all profiles)

(b)
angled riser

(c)
curved nosing

(d)
beveled nosing

Fig. CD-23B
Stair Nosings
© ICC Reproduced with Permission

Handrails

_____ M.   Stairs shall have handrails that comply with Section 24, "HANDRAILS". **11B-504.6**   *504.6*

Wet Conditions

_____ N.   Stair treads and landings subject to wet conditions shall be designed to prevent the accumulation of water. **11B-504.7**   *504.7*

Floor Identification

_____ O.   Floor identification signs required by CBC Chapter 10, section 1022.9 and complying with the requirements for accessible "Signs", "Raised Characters", "Braille" and "Visual Characters" shall be located at the landing of each floor level, placed adjacent to the door on the latch side, in all enclosed stairways in buildings two or more stories in height to identify the floor level. At the exit discharge level, the sign shall include a raised five pointed star located to the left of the identifying floor level. The outside diameter of the star shall be the same as the height if the raised characters. **11B-504.8**

24. <u>HANDRAILS</u>

<u>General</u>

Handrails provided along accessible walking surfaces, at accessible ramps that are required to have handrails, and at accessible stairs shall comply as detailed. **11B.505.1** *505.1*

> ***ADVISORY. Handrails.** Handrails are not required on walking surfaces with running slopes less than 1:20. However, handrails are required to be accessible and comply with this section when they are provided on walking surfaces with running slopes less than 1:20.*

<u>Where Required</u>

_____A.   Handrails are provided on both sides of stairs and ramps. **11B-505.2** *505.2*

        EXCEPTIONS:

           1.   In assembly areas, handrails shall not be required on both sides of aisle ramps where a handrail is provided at either side or within the aisle width.

           2.   Curb ramps do not require handrails.

           3.   At door landings, handrails are not required when the ramp run is less than 6 inches (152 mm) in rise or 72 inches (1829 mm) in length.

<u>Continuity</u>

_____B.   Handrails are continuous within the full length of each stair flight or ramp run. **11B-505.3** *505.3*

_____C.   Inside handrails on switchback or dogleg stairs and ramps are continuous between flights or runs. **11B-505.3** *505.3*

        EXCEPTION: In assembly areas, ramp handrails adjacent to seating or within the aisle width shall not be required to be continuous in aisles serving seating.

<u>Height</u>

_____D.   Top of gripping surfaces of handrails are 34 inches minimum and 38 inches maximum vertically above walking surfaces, stair nosings, and ramp surfaces. **11B-505.4** *505.4*
        Fig. CD-24A

> ***ADVISORY: Height.** The requirements for stair and ramp handrails are for adults. When children are the principal users in a building or facility (e.g., elementary schools), a second set of handrails at an appropriate height can assist them and aid in preventing accidents. A maximum height of 28 inches measured to the top of the gripping surface from the ramp surface or stair nosing is recommended for handrails designed for children. Sufficient vertical clearance between upper and lower handrails, 9 inches minimum, should be provided to help prevent entrapment.*

_____E.   Handrails are at a consistent height above walking surfaces, stair nosings, and ramp surfaces. **11B-505.4** *505.4*

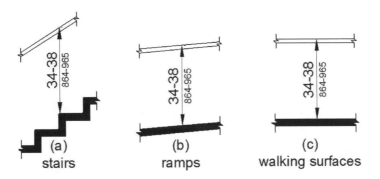

Fig. CD-24A
Handrail Height
© ICC Reproduced with Permission

Clearance

_____F.   Clearance between handrail gripping surfaces and adjacent surfaces is 1-1/2 inches minimum. Handrails may be located in a recess if the recess is 3 inches maximum deep and 18 inches minimum clear above the top of the handrail. **11B-505.5**   *505.5*   **Fig. CD-24B**

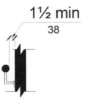

Fig. CD-24B
Handrail Clearance
© ICC Reproduced with Permission

Gripping Surfaces

_____G.   Handrail gripping surfaces are continuous along their length and shall not be obstructed along their tops or sides.   **11B-505.5**   *505.5*

_____H.   The bottoms of handrail gripping surfaces are not obstructed for more than 20% of their length   **11B-505.6**   *505.6*

_____I.   Where provided, horizontal projections occur 1-1/2 inches minimum below the bottom of the handrails gripping surface.   **11B-505.6**   *505.6*   **Fig. CD-24C**

EXCEPTIONS:

1.   Where handrails are provided along walking surfaces with slopes not steeper than 1:20 (5%), the bottoms of handrail gripping surfaces shall be permitted to be obstructed along their entire length where they are integral to crash rails or bumper guards.

2.   The distance between horizontal projections and the bottom of the gripping surface shall be permitted to be reduced by 1/8 inch for each ½ inch of additional handrail perimeter dimension that exceeds 4 inches.

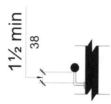

Fig. CD-24C
Horizontal Projections Below Gripping Surface
© ICC Reproduced with Permission

Cross Section

Handrail gripping surfaces have either a circular or non-circular cross section that meets the applicable specifications: **11B-505.7** *505.7*

_____J.   **Circular Cross Section.** Handrail gripping surfaces with a circular cross section have an outside diameter of 1-1/4 inches minimum and 2 inches maximum. **11B-505.7.1** *505.7.1*

_____K.   **Non-Circular Cross Section.** Handrail gripping surfaces with a non- circular cross section have a perimeter dimension of 4 inches minimum and 6-1/4 inches maximum, and a cross-section dimension of 2-1/4 inches maximum. **11B-505.7.2** *505.7.2* **Fig. CD-24D**

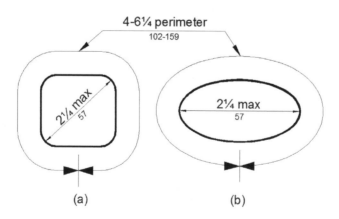

(a)   (b)

Fig. CD-24D
Handrail Non-Circular Cross Section
© ICC Reproduced with Permission

Surfaces

_____L.   Handrail gripping surfaces and any surfaces adjacent to them are free of sharp or abrasive elements and have rounded edges. **11B-505.8** *505.8*

Fittings

_____M. Handrails do not rotate within their fittings. **11B-505.9** *505.9*

Handrail Extensions
Handrail gripping surfaces shall extend beyond and in the same direction of stair flights and ramp runs as specified. **11B-505.10** *505.10*

EXCEPTIONS

1. Extensions shall not be required for continuous handrails at the inside turn of switchback or dogleg stairs and ramps.

2. In assembly areas, extensions shall not be required for ramp handrails in aisles serving seating where the handrails are discontinuous to provide access to seating and to permit crossovers within aisles.

3. In alterations, where the extension of the handrail in the direction of the ramp run would create a hazard, the extension of the handrail may be turned 90 degrees from the ramp run.

Top and Bottom Extensions at Ramps

_____N. Ramp handrails extend horizontally above the landing for 12 inches minimum beyond the top and bottom of ramp runs. **11B-505.10.1** *505.10.1* **Fig. CD-24E**

_____O. Extensions return to a wall, guard, or the landing surface, or shall be continuous to the handrail of an adjacent ramp run. **11B-505.10.1** *505.10.1* **Fig. CD-24E**

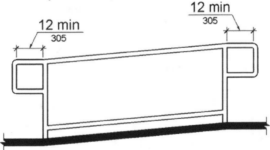

Fig. CD-24E
Top and Bottom Handrail Extension at Ramps
© ICC Reproduced with Permission

Top Extensions at Stairs

_____P. At the top of a stair flight, handrails shall extend horizontally above the landing for 12 inches minimum beginning directly above the first riser nosing. **11B-505.10.2** *505.10.2* **Fig. CD-24F**

_____Q. Extensions return to a wall, guard, or the landing surface, or shall be continuous to the handrail of an adjacent stair flight. **11B-505.10.2** *505.10.2* **Fig. CD-24F**

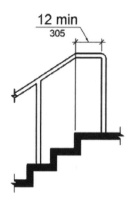

Fig. CD-24F
Top Handrail Extension at Stairs
© ICC Reproduced with Permission

Bottom Extension at Stairs

_____R.   At the bottom of a stair flight, handrails extend at the slope of the stair flight for a horizontal distance equal to one tread depth beyond the last riser nosing. **11B-505.10.3**   *505.10.3*   **Fig. CD-24F**

_____S.   Such extension shall continue with a horizontal extension or shall be continuous to the handrail of an adjacent stair flight or shall return to a wall, guard, or the walking surface.   **11B-505.10.3**   *505.10.3*   **Fig. CD-24F**

_____T.   At the bottom of a stair flight, a horizontal extension of a handrail shall be 12 inches long minimum and a height equal to that of the sloping portion of the handrail as measured above the stair nosings.   **11B-505.10.3**   *505.10.3*   **Fig. CD-24F**

_____U.   The extension shall return to a wall, guard, or the landing surface, or is continuous to the handrail of an adjacent stair flight.   **11B-505.10.3**   *505.10.3*

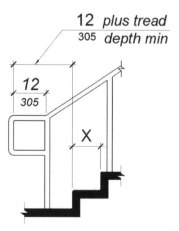

Note: X = tread depth

Fig. CD-24F
Bottom Handrail Extension at Stairs
© ICC Reproduced with Permission

25. <u>DRINKING FOUNTAINS</u>

<u>General</u>
Where drinking fountains are provided on an exterior site, on a floor, or within a secured area they shall be provided in accordance with these requirements. **11B-211.1** *211.1*

> **EXCEPTION:** In detention or correctional facilities, drinking fountains only serving holding or housing cells not required to provide mobility features shall not be required to be accessible.

<u>Minimum Number</u>
No fewer than two drinking fountains shall be provided. One drinking fountain shall be wheelchair accessible and one drinking fountain shall be for standing persons.
11B-211.2 *211.2*

> **EXCEPTION:** Where a single drinking fountain complies with the requirements for a wheelchair accessible fountain and a drinking fountain for standing persons, it shall be permitted to be substituted for two separate drinking fountains.

> *ADVISORY: **Minimum number.** The 2013 California Plumbing Code (CPC), Table 4-1 provides minimum plumbing fixture requirements for new buildings, and changes of occupancy or use in an existing building resulting in increased occupant load. CPC Table 4-1 should be consulted in conjunction with the requirements of this section.*

<u>More than Minimum Number</u>
Where more than the minimum number of drinking fountains are provided, 50 percent of the total number of drinking fountains provided shall be wheelchair accessible, and 50 percent of the total number of drinking fountains provided shall be for standing persons. **11B-211.3** *211.3*

> **EXCEPTION:** Where 50 percent of the drinking fountains yields a fraction, 50 percent shall be permitted to be rounded up or down provided that the total number of compliant drinking fountains equals 100 percent of drinking fountains.

<u>General</u>
Drinking fountains shall comply as detailed. **11B-602.1** *602.1*

<u>Clear Floor Space</u>

_____ A. Units shall have a clear floor or ground space that complies with Section 10, "CLEAR FLOOR OR GROUND SPACE", positioned for a forward approach and centered on the unit. **11B-602.2** *602.2*

> **EXCEPTION:** A compliant parallel approach shall be permitted at units for children's use where the spout is 30 inches maximum above the finish floor or ground and is 3-1/2 inches maximum from the front edge of the unit, including bumpers.

<u>Knee and Toe Clearance</u>

_____ B. Knee and toe clearance that complies with Section 11, "KNEE AND TOE CLEARANCE", shall be provided. **11B-602.2, 11B-306** *602.2, 306*

<u>Operable Parts</u>

_____ C. Operable parts comply with Section 13, "OPERABLE PARTS". The flow of water shall be activated by a manually operated system that is front mounted or side mounted and located within 6 inches of the front edge of the fountain or electronically controlled device. **11B-602.3**   *602.3*

Spout Height

_____D.   Spout outlets shall be 36 inches maximum above the finish floor or ground.
11B-602.4   *602.4*

Spout Location

_____E.   The spout shall be located 15 inches minimum from the vertical support and 5 inches maximum from the front edge of the unit, including bumpers.
11B-602.5   *602.5*   **Fig. CD-25A**

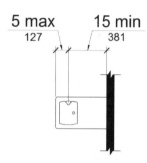

Fig. CD-25A
Drinking Fountain Spout Location
© ICC Reproduced with Permission

Water Flow

_____F.   The spout shall provide a flow of water 4 inches high minimum and shall be located 5 inches maximum from the front of the unit. The angle of the water stream shall be measured horizontally relative to the front face of the unit.
11B-602.6   *602.6*

_____G.   Where spouts are located less than 3 inches of the front of the unit, the angle of the water stream shall be 30 degrees maximum.   **11B-602.6**   *602.6*

_____H.   Where spouts are located between 3 inches and 5 inches maximum from the front of the unit, the angle of the water stream shall be 15 degrees maximum. **11B-602.6**   *602.6*

ADVISORY: Water Flow. The purpose of requiring the drinking fountain spout to produce a flow of 4 inches (100 mm) high minimum is so that a cup can be inserted under the flow of water to provide a drink of water for an individual who, because of a disability, would otherwise be incapable of using the drinking fountain.

Drinking Fountains for Standing Persons

_____I.   Spout outlets of drinking fountains for standing persons shall be 38 inches minimum and 43 inches maximum above the finish floor or ground.
11B-602.7   *602.7*   **Fig. CD-25B**

Depth

_____J.   Wall- and post-mounted cantilevered drinking fountains shall be 18 inches minimum and 19 inches maximum in depth. **11B-602.8**

Pedestrian Protection

_____K.   All drinking fountains shall either be located completely within alcoves, positioned completely between wing walls, or otherwise positioned so as not to encroach into pedestrian ways. **11B-602.9**   **Fig. CD-25B**

_____K1.   The protected area within which a drinking fountain is located shall be 32 inches wide minimum, and 18 inches deep minimum, and shall comply as detailed below: **11B-602.9**   **Fig. CD-25B**

Maneuvering Clearance

Where a clear floor or ground space is located in an alcove or otherwise confined on all or part of three sides, additional maneuvering clearance is provided as follows:

_____L.   **Forward Approach.** Alcoves are 36 inches wide minimum where the depth exceeds 24 inches. **11B-305.7.1** *305.7.1* **Fig. CD-10C, CD-25B**

_____M.   **Parallel Approach.** Alcoves are 60 inches wide minimum where The depth exceeds 15 inches. **11B-305.7.2** *305.7.2* **Fig. CD-10D**

_____N.   When used, wing walls or barriers shall project horizontally at least as far as the drinking fountain and to within 6 inches vertically from the floor or ground surface. **11B-602.9**

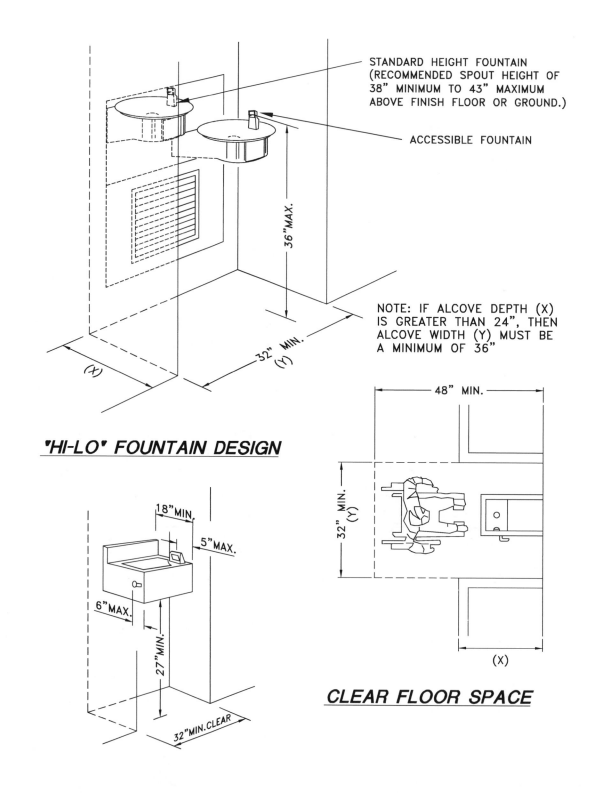

STANDARD HEIGHT FOUNTAIN
(RECOMMENDED SPOUT HEIGHT OF
38" MINIMUM TO 43" MAXIMUM
ABOVE FINISH FLOOR OR GROUND.)

ACCESSIBLE FOUNTAIN

36" MAX.

32" MIN.
(Y)

(X)

NOTE: IF ALCOVE DEPTH (X)
IS GREATER THAN 24", THEN
ALCOVE WIDTH (Y) MUST BE
A MINIMUM OF 36"

"HI-LO" FOUNTAIN DESIGN

18" MIN.

5" MAX.

6" MAX.

27" MIN.

32" MIN. CLEAR

ALCOVE INSTALLATION

48" MIN.

32" MIN.
(Y)

(X)

CLEAR FLOOR SPACE

Fig. CD-25B
Pedestrian Protection/Drinking Fountains for Standing Persons

26. <u>TOILET FACILITIES AND BATHING FACILITIES</u>

<u>General</u>
Where toilet facilities and bathing facilities are provided, they shall comply as detailed. Where toilet facilities and bathing facilities are provided in multi-story facilities permitted by an exception not to connect stories by an accessible route, toilet facilities and bathing facilities shall be provided on a story connected by an accessible route to an accessible entrance. **11B-213.1** *213.1*

<u>Toilet Facilities for Designated User Groups</u>
Where separate toilet facilities are provided for the exclusive use of separate user groups, the toilet facilities serving each user group shall comply as detailed. **11B-213.1.1**

<u>Water Closets and Toilet Compartments for Children's Use</u>
Water closets and toilet compartments for children's use are permitted to comply with the Suggested Dimensions for Children's Use detailed herein. **11B-604.1, 11B-604.9** *604.1, 604.9*

<u>Toilet Rooms and Bathing Rooms</u>
Where toilet rooms are provided, each toilet room shall comply as detailed. Where bathing rooms are provided, each bathing room shall comply as detailed. **11B-213.2** *213.2*

EXCEPTIONS:

1. In alterations, where it is technically infeasible to comply with the technical requirements for an accessible toilet room or bathing room, altering existing toilet or bathing rooms shall not be required where a single accessible unisex toilet room or bathing room is provided and located in the same area and on the same floor as existing inaccessible toilet or bathing rooms.

2. Reserved.

3. Where multiple single user portable toilet or bathing units are clustered at a single location 5%, but no fewer than one, of the toilet units and bathing units at each cluster are required to be accessible. Portable toilet units and bathing units that are accessible shall be identified by the International Symbol of Accessibility.

4. Where multiple single user toilet rooms are clustered at a single location, no more than 50% of the single user toilet rooms for each use at each cluster are required to be accessible.

5. Where toilet and bathing rooms are provided in guest rooms that are not required to provide mobility features, toilet and bathing facilities shall only be required to allow a person using a wheelchair measuring 30 inches by 48 inches to touch the wheelchair to any lavatory, urinal, water closet, tub, sauna, shower stall and any other similar sanitary installation, if provided.

> ***ADVISORY: Toilet Rooms and Bathing Rooms.** These requirements allow the use of unisex (or single-user) toilet rooms in alterations when technical infeasibility can be demonstrated. Unisex toilet rooms benefit people who use opposite sex personal care assistants. For this reason, it is advantageous to install unisex toilet rooms in addition to accessible single-sex toilet rooms in new facilities.*

> **ADVISORY:** *Toilet Rooms and Bathing Rooms Exceptions 3 and 4.* A "cluster" is a group of toilet rooms proximate to one another. Generally, toilet rooms in a cluster are within sight of, or adjacent to, one another.

Unisex (Single-Use or Family) Toilet and Unisex Bathing Rooms

Unisex toilet rooms shall contain not more than one lavatory, and two water closets without urinals or one water closet and one urinal. Unisex bathing rooms shall contain one shower or one shower and one bathtub, one lavatory, and one water closet. Doors to unisex toilet rooms and unisex bathing rooms shall have privacy latches. **11B-213.2.1** *213.2.1*

Unisex (Patient) Toilet Rooms in Medical Care and Long-Term Care Facilities

Common-use unisex toilet rooms for exclusive patient use not located within patient bedrooms shall contain a lavatory and one water closet. **11B-213.2.2**

Unisex (Patient) Bathing Rooms in Medical Care and Long-Term Care Facilities

Common-use unisex bathing rooms for exclusive patient use not located within patient bedrooms shall contain one shower or one bathtub, one lavatory, and one water closet. **11B-213.2.3**

Plumbing Fixtures and Accessories

Plumbing fixtures and accessories provided in a toilet room or bathing room required to be accessible shall comply as detailed. **11B-213.3** *213.3*

> **ADVISORY:** *Plumbing Fixtures and Accessories.* The 2013 California Plumbing Code (CPC), Table 4-1 provides minimum plumbing fixture requirements for new buildings, and changes of occupancy or use in an existing building resulting in increased occupant load. CPC Table 4-1 should be consulted in conjunction with the requirements of this section.

Construction Support Facilities

These requirements shall apply to temporary or permanent construction support facilities for uses and activities not directly associated with the actual processes of construction, including but not limited to offices, meeting rooms, plan rooms, other administrative or support functions. When provided, toilet and bathing facilities serving construction support facilities shall comply as required for accessible toilet facilities. When toilet and bathing facilities are provided by portable units, at least one of each type shall be accessible and connected to the construction support facilities it serves by an accessible route. **11B-201.4**

> **EXCEPTION:** During construction an accessible route shall not be required between site arrival points or the boundary of the area of construction and the entrance to the construction support facilities if the only means of access between them is a vehicular way not providing pedestrian access.

Construction Sites

Portable toilet units provided for use exclusively by construction personnel on a construction site are not required to be accessible or be on an accessible route. **11B-203.2** *203.2*

Toilet Compartments

Where toilet compartments are provided, at least one toilet compartment shall be a Wheelchair Accessible Compartment. In addition to the compartment required to be accessible, at least one compartment shall be an "Ambulatory Accessible Compartment", where six or more toilet compartments are provided, or where the combination of urinals and water closets totals six or more fixtures. **11B-213.3.1** *213.3.1*

> **ADVISORY: Toilet Compartments.** *A toilet compartment is a partitioned space that is located within a toilet room, and that normally contains no more than one water closet. A toilet compartment may also contain a lavatory. Full-height partitions and door assemblies can comprise toilet compartments where the minimum required spaces are provided within the compartment.*

Water Closets
Where water closets are provided, at least one shall be accessible. **11B-213.3.2**   *213.3.2*

Urinals
Where one or more urinals are provided, at least one shall be accessible. **11B-213.3.3**   *213.3.3*

Lavatories
Where lavatories are provided, at least five percent but no fewer than one shall be accessible and shall NOT be located in a toilet compartment. **11B-213.3.4**   *213.3.4*

Mirrors
Where mirrors are provided, at least one shall be accessible. **11B-213.3.5** *213.3.5*

Bathing Facilities
Where bathtubs or showers are provided, at least one accessible bathtub or one accessible shower shall be provided. Where two or more accessible showers are provided in the same functional area, at least one shower shall be opposite hand from the other or others (that is, one left-hand controls versus right-hand controls. **11B-213.3.6**   *213.3.6*

Coat Hooks and Shelves
Where coat hooks or shelves are provided in toilet rooms without toilet compartments, at least one of each type shall comply with these provisions. Where coat hooks or shelves are provided in toilet compartments, at least one of each type shall comply with these provisions. Where coat hooks or shelves are provided in bathing facilities, at least one of each type shall comply with these provisions. Coat hooks shall be located within one of the accessible reach ranges. Shelves shall be located 40 inches minimum and 48 inches maximum above the finish floor. **11B-213.3.7**, **11B-603.4**, **11B-604.8.3**, **11B-308** *213.3.7, 603.4, 604.8.3*

NOTICE TO TITLE II ENTITIES (State and Local Government Facilities Only)
California Government Code Section 7251. Toilet facilities; signs
When a building contains special toilet facilities usable by a person in a wheelchair or otherwise handicapped, a sign indicating the location of such facilities shall be posted in the building directory, in the main lobby, or at any entrance specially used by handicapped persons.

Toilet Rooms and Bathing Rooms
Doorways leading to accessible toilet rooms and bathing rooms shall be identified with the appropriate geometric symbol. Where existing toilet rooms or bathing rooms are NOT accessible, directional signs indicating the location of the nearest toilet room or bathing room that IS accessible within the facility shall be provided. These signs shall contain the International Symbol of Accessibility and compliant Visual Characters. Where a facility contains both accessible and inaccessible toilet or bathing rooms, the accessible toilet or bathing rooms shall be identified by the International Symbol of Accessibility. Where clustered single user toilet rooms or bathing units are clustered at a single location, and there are both accessible and inaccessible toilet or bathing units, the accessible toilet or bathing units shall be identified by the International Symbol of Accessibility. Existing buildings that have been remodeled to provide specific toilet rooms or bathing rooms for public use that are accessible shall have the location of and the directions to these rooms posted in or near the building lobby on a sign containing accessible Visual Characters and the International Symbol of Accessibility. **11B-216.8**   *216.8*

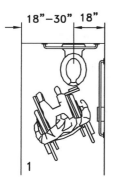
18"–30" 18"

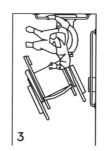

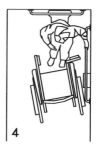

1
Takes transfer postion, swings footrest out of the way, sets brakes.

2
Removes armrest, transfers.

3
Moves wheelchair out of the way, changes position (some people fold chair or pivot it 90° to the toilet.

4
Positions on toilet, releases brakes.

DIAGONAL APPROACH

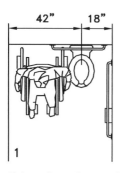
42" 18"

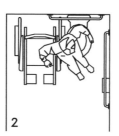

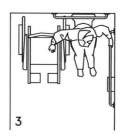

1
Takes transfer position, removes armrest, sets brakes.

2
Transfers.

3
Positions on toilet.

SIDE APPROACH

Fig. CD-26A
Wheelchair Transfers

General
Toilet and bathing rooms shall comply as detailed. **11B-603.1** *603.1*

Clearances
Clearances shall comply as detailed. **11B-603.2** *603.2*

Turning Space

_____A.   An accessible turning space is provided within the room that complies with Section 9, "TURNING SPACE". **11B-603.2.1** *603.2.1*

Overlap
Required clear floor spaces, clearances at fixtures, and turning space are permitted to overlap. **11B-603.2.2** *603.2.2*

Door Swing

_____B.   Doors do not swing into the clear floor space or clearance required for any fixture. Other than the door to the accessible water closet compartment, a door, in any position, may encroach into the turning space by 12 inches maximum. **11B-603.2.3** *603.2.3*

EXCEPTIONS:

1. **Reserved.**

2. Where the toilet room or bathing room is for individual use and a clear floor space a minimum of 30 inches wide by 48 inches long is provided within the room beyond the arc of the door swing, doors shall be permitted to swing into the clear floor space or clearance required for any fixture.

> *ADVISORY:* ***Door Swing.*** *The door to the accessible water closet compartment may swing over the turning space without limitation. Other doors may swing over the turning space up to 12 inches.*

Mirrors

_____C. If mirrors are located above lavatories or countertops, at least one is installed with the bottom edge of the reflecting surface 40 inches maximum above the finish floor or ground. **11B-603.3** *603.3*

_____D. Mirrors not located above lavatories or countertops shall be installed with the bottom edge of the reflecting surface 35 inches maximum above the finish floor or ground. **11B-213.3.5, 11B-603.3** *213.3.5, 603.3*

> *ADVISORY:* ***Mirrors.*** *A single full length mirror can accommodate a greater number of people, including children. In order for mirrors to be usable by people who are ambulatory and people who use wheelchairs, the top edge of mirrors should be 74 inches minimum from the floor or ground.*

Coat Hooks, Shelves and Medicine Cabinets

_____E. If coat hooks are provided, at least one of each type is located within an accessible reach range (see Section 12, "REACH RANGES").
11B-213.3.7, 11B-603.4 *213.3.7, 603.4*

_____F. If Shelves are provided, at least one of each type is located 40 inches minimum and 48 inches maximum above the finish floor. **11B-603.4** *603.4*

_____G. Medicine cabinets shall be located with a usable shelf no higher than 44 inches maximum above the finish floor. **11B-603.4** *603.4*

Accessories

_____H. Where towel or sanitary napkin dispensers, waste receptacles, or other accessories are provided in toilet facilities, at least one of each type is located on an accessible route. **11B-603.5**

_____I. All operable parts, including coin slots, are 40 inches maximum above the finish floor. **11B-603.5**

Guest Room Toilet and Bathing Rooms

_____J. Toilet and bathing rooms within guest rooms that are not required to provide mobility features shall provide al toilet and bathing fixtures in a location that allows a person using a wheelchair measuring 30 inches by 48 inches to touch the wheelchair to any lavatory,

---

urinal, water closet, tub, sauna, shower stall and any other similar sanitary installation, if provided.   **11B-603.6**

26A. <u>WATER CLOSETS AND WHEELCHAIR-ACCESSIBLE TOILET COMPARTMENTS</u>

General
Water closets and toilet compartments shall comply as detailed.   **11B-604.1**   *604.1*

WATER CLOSETS

Location

_____A.   The water closet is positioned with a wall or partition to the rear and to one side. **11B-604.2**   *604.2*   **Fig. CD-26AA**

_____B.   The centerline of the water closet is located at 17 inches minimum to 18 inches maximum distance from the side-wall or partition.   **11B-604.2**   *604.2*   **Fig. CD-26AA**

_____C.   The water closet is arranged for a left-hand or right-hand approach.   **11B-604.2** *604.2*

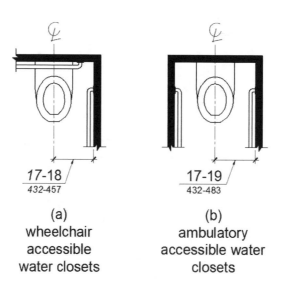

(a)
wheelchair
accessible
water closets

(b)
ambulatory
accessible water
closets

Fig. CD-26AA
Water Closet Location
© ICC Reproduced with Permission

Clearance
Clearances around water closets and in toilet compartment shall comply as detailed. **11B-604.3**   *604.3*

Size

_____D.   The clearance <u>around</u> the water closet is a minimum of 60 inches wide, measured perpendicular from the side wall and 56 inches minimum measured perpendicular from the rear wall. **11B-604.3.1** *604.3.1* **Fig. CD-26AB**

_____ E. A minimum 60 inches wide and 48 inches deep maneuvering space is provided in front of the water closet. **11B-604.3.1** *604.3.1* **Fig. CD-26AB**

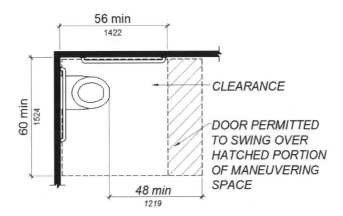

Fig. CD-26AB
Size of Clearance at Water Closets
© ICC Reproduced with Permission

Overlap

_____ F. The required clearance around the water closet is permitted to overlap the water closet, associated grab bars, dispensers, sanitary napkin disposal units, coat hooks, shelves, accessible routes, clear floor space and clearances required at other fixtures, and the turning space. No other fixtures or obstructions shall be located within the required water closet clearance. **11B-604.3.2** *604.3.2*

> **EXCEPTION:** In residential dwelling units, an accessible lavatory shall be permitted on the rear wall 18 inches minimum from the water closet centerline where the clearance at the water closet is 66 inches minimum measured perpendicular from the rear wall. **11B-604.3.2** *604.3.2*
> **Fig. CD-26AC**

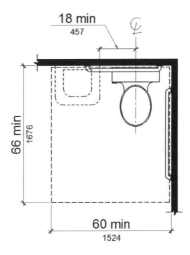

Fig. CD-26AC (Exception)
Overlap of Water Closet Clearance in Residential Dwelling Units
© ICC Reproduced with Permission

Seats

_____G.   The seat height of the water closet above the finish floor is 17 inches minimum and 19 inches maximum measured to the top of the seat.   **11B-604.4**   *604.4*

_____H.   The water closet seat is not sprung to return to a lifted position. **11B-604.4**   *604.4*

_____I.   The maximum height of the water closet seat is 2 inches. **11B-604.4** *604.4*

EXCEPTIONS:

1.   **Reserved.**

2.   In residential dwelling units, the height of water closets shall be permitted to be 15 inches minimum and 19 inches maximum above the finish floor measured to the top of the seat.

3.   A 3 inch high seat shall be permitted only in alterations where the existing fixture is less than 15 inches high.

Grab Bars

_____J.   Grab bars are provided on the side wall closest to the water closet and on the rear wall. **11B-604.5, 11B-604.8.1.5**   *604.5, 604.8.1.5*   **Fig. CD-26AD & AE**

NOTE: Where separate grab bars are required on adjacent walls at a common mounting height, an L-shaped grab bar meeting the dimensional requirements of the separate grab bars shall be permitted. **11B-609.9**.

EXCEPTIONS:

1.   **Reserved.**

2.   In residential dwelling units, grab bars shall not be required to be installed in toilet or bathrooms provided that reinforcement has been installed in walls and located so as to permit the installation of accessible toilet grab bars.

3.   In detention or correction facilities, grab bars shall not be required to be installed in housing or holding cells that are specially designed without protrusions for purposes of suicide prevention.

ADVISORY. Grab Bars Exception 2. Reinforcement must be sufficient to permit the installation of rear and side wall grab bars that fully meet all accessibility requirements including, but not limited to, required length, installation height, and structural strength.

Side Wall

_____K.   The side wall grab bar is a minimum of 42 inches in length. **11B-604.5.1** *604.5.1*   **Fig. CD-26AD**

_____L.   The side wall grab bar is located 12 inches maximum from the rear wall and

extends 54 inches minimum from the rear wall with the front end positioned 24 inches minimum in front of the water closet. **11B-604.5.1**   *604.5.1*
Fig. CD-26AD

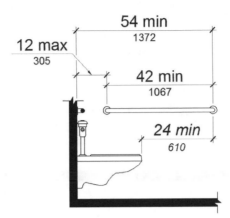

Fig. CD-26AD
Side Wall Grab Bar at Water Closets
© ICC Reproduced with Permission

Rear Wall

_____M.   The rear wall grab bar is a minimum of 36 inches in length and extends from the centerline of the water closet 12 inches minimum on one side and 24 inches minimum on the other side. **11B-604.5.2**   *604.5.2*   **Fig. CD-26AE**

EXCEPTIONS:

1. The rear grab bar shall be permitted to be 24 inches long minimum, centered on the water closet, where wall space does not permit a length of 36 inches minimum due to the location of a recessed fixture adjacent to the water closet.

2. Where an administrative authority requires flush controls for flush valves to be located in a position that conflicts with the location of the rear grab bar, then the rear grab bar shall be permitted to be split or shifted to the open side of the toilet area.

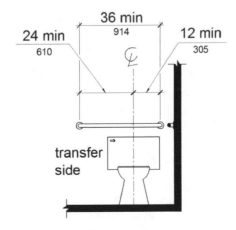

Fig. CD-26AE
Rear Wall Grab Bar at Water Closets
© ICC Reproduced with Permission

Cross Section

Grab bars shall have either a circular or non-circular cross section that meets the applicable specifications: **11B-609.2** *609.2*

_____N. **Circular Cross Section.** Grab bars with circular cross sections have an outside diameter of 1-1/4 inches minimum and 2 inches maximum. **11B-609.2.1** *609.2.1* **CD-26AF**

_____O. **Non-Circular Cross Section.** Grab bars with non-circular cross sections have a cross-section dimension of 2 inches maximum and a perimeter dimension of 4 inches minimum and 4.8 inches maximum. **11B-609.2.2** *609.2.2* **Fig. CD-26AF**

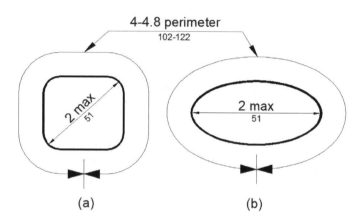

Fig. CD-26AF
Grab Bar Non-Circular Cross Section
© ICC Reproduced with Permission

Spacing

_____P. The space between the wall and the grab bar shall be 1-1/2 inches. **11B-609.3** *609.3*

_____Q. The space between the grab bar and projecting objects below and at the ends shall be 1-1/2 inches minimum. **11B-609.3** *609.3* **CD-26AG**

_____R. The space between the grab bar and projecting objects above shall be 12 inches minimum. **11B-609.3** *609.3* **CD-26AG**

EXCEPTIONS:

1. The space between the grab bars and shower controls, shower fittings, and other grab bars above shall be permitted to be 1-1/2 inches minimum.

2. For L-shaped or U-shaped grab bars the space between the walls and the grab bar shall be 1-1/2 inches minimum for a distance of 6 inches on either side of the inside corner between two adjacent wall surfaces.

ADVISORY: Spacing. *The distance between the grab bar and wall is an exact dimension. Many disabled people rely heavily upon grab bars to maintain balance and prevent serious falls. Many people brace their forearms between supports and walls to give them more leverage and stability in maintaining balance or for lifting. The grab bar clearance of 1½ inches required in this section is a safety clearance to prevent injuries resulting from arms slipping through the openings. It also provides adequate gripping room.*

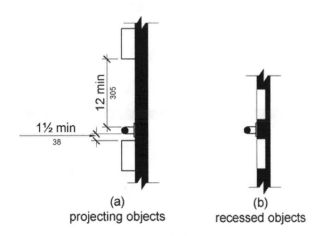

Fig. CD-26AG
Spacing of Grab Bars
© ICC Reproduced with Permission

Position of Grab Bars

_____S.   Grab bars are installed in a horizontal position, 33 inches minimum and 36 inches maximum above the finish floor, measured to the top of the gripping surface, except that at water closets for children's use, grab bars shall be installed in a horizontal position 18 inches minimum and 27 inches maximum above the finish floor measured to the top of the gripping surface. **11B-609.4** *609.4*

Surface Hazards

_____T.   The grab bar(s) and any wall or other surfaces adjacent to the grab bars is free of any sharp or abrasive elements and shall have rounded edges. **11B-609.5** *609.5*

Fittings

_____U.   Grab bars do not rotate within their fittings. **11B-609.6** *609.6*

Installation

_____V.   Grab bars shall be installed in any manner that provides a gripping surface at the specified locations and that does not obstruct the required clear floor space. **11B-609.7** *609.7*

Structural Strength

_____W.   Allowable stresses shall not be exceeded for materials used when a vertical or horizontal force of 250 pounds (1112N) is applied at any point on the grab bar, fastener, mounting device, or supporting structure. **11B-609.8** *609.8*

Alternate Configuration

_____X.   L-Shaped or U-Shaped grab bars shall be permitted. **11B-609.9**

Flush Controls

_____Y.   Flush controls are hand-operated or automatic. **11B-604.6**   604.6

_____Z.   Flush controls are located on the open side of the water closet. **11B-604.6**
                604.6

_____AA.   Hand-operated flush controls are located 44 inches maximum distance above the
                floor.   **11B-604.6**   *604.6*

_____BB.   Hand-operated controls are operable with one hand and do not require
                tight grasping, pinching or twisting of the wrist. **11B-604.6, 11B-309.4** *604.6, 309.4*

_____CC.   5 lb. maximum force is required to operate the hand-operated flush
                controls. **11B-604.6, 11B-309.4** *604.6, 309.4*

> **ADVISORY: Flush Controls.** *If plumbing valves are located directly behind the toilet seat, flush valves and related plumbing can cause injury or imbalance when a person leans back against them. To prevent causing injury or imbalance, the plumbing can be located behind walls or to the side of the toilet; or if approved by the local authority having jurisdiction, provide a toilet seat lid.*

Dispensers

_____DD.   The toilet paper dispenser is located 7 inches minimum and 9 inches maximum
                in front of the water closet measured to the centerline of the dispenser.
                1B-604.7   *604.7*   **Fig. CD-26AH**

_____EE.   The outlet of the toilet paper dispenser is below the grab bar, 19 inches minimum
                Above the finish floor and is not located behind grab bars.   **11B-604.7**   *604.7*
                Fig. CD-26AH

_____FF.   The toilet paper dispenser allows for continuous paper flow and does not control
                delivery.   **11B-604.7** *604.7*

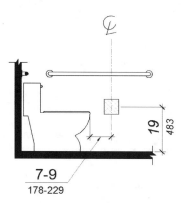

Fig. CD-26AH
Dispenser Outlet Location
© ICC Reproduced with Permission

WHEELCHAIR-ACCESSIBLE TOILET COMPARTMENTS

Size

_____GG. If a wheelchair-accessible compartment has a **wall-hung water closet**, the compartment is a minimum of 60 inches wide, measured perpendicular to the side wall and 56 inches deep minimum measured perpendicular to the rear wall.
11B-604.8.1.1   *604.8.1.1*   **Fig. CD-26AI & AJ**

_____HH. If a wheelchair-accessible compartment has a **floor-mounted water closet**, the compartment is a minimum of 60 inches wide, measured perpendicular to the side wall, and 59 inches deep minimum measured perpendicular to the rear wall. **604.8.1.1**   *604.8.1.1*   **Fig. CD-26AI & AJ**

> ***Advisory: Size.*** *The minimum space requirement in toilet compartments is provided so that a person using a wheelchair can maneuver into position at the water closet. This space cannot be obstructed by baby changing tables or other fixtures or conveniences, except as is allowed by the "overlap" provision. If toilet compartments are to be used to house fixtures other than those associated with the water closet, it will likely require the compartment to be designed to exceed the minimum space requirements. Convenience fixtures such as baby changing tables must also be accessible to people with disabilities as well as to other users. Toilet compartments that are designed to meet, and not exceed, the minimum space requirements may not provide adequate space for maneuvering into position at a baby changing table.*

Maneuvering Space with In-Swinging Door

_____II. In a wheelchair-accessible compartment with an in-swinging door, an additional minimum 60 inches wide by 36 inches deep maneuvering space is provided in front of the minimum 56 inch or 59 inch depth of the compartment, measured perpendicular to the rear wall.
11B-604.8.1.1.
Fig. CD-26AI(b), Fig. CD-26AJ(b)

Maneuvering Space with Side-Opening Door

_____JJ. In a wheelchair-accessible compartment with a side-opening door, either in-swinging or out-swinging, a minimum 60 inches wide and 60 inches deep maneuvering space is provided in front of the water closet. **11B-608.8.1.1.2**   **Fig. CD-26AI & AJ**

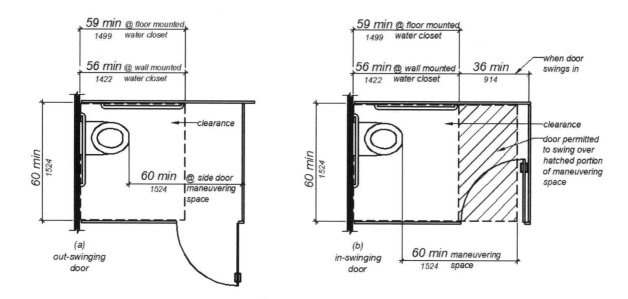

Fig. CD-26AI
Maneuvering Space with Side-Opening Door
© ICC Reproduced with Permission

Maneuvering Space with End-Opening Door

_____KK.　In a wheelchair-accessible compartment with an end-opening door (facing the water closet), either in-swinging or out-swinging, a minimum 60 inches wide and 48 inches deep maneuvering space is provided in front of the water closet.
11B-608.8.1.1.2　　Fig. CD-26AJ

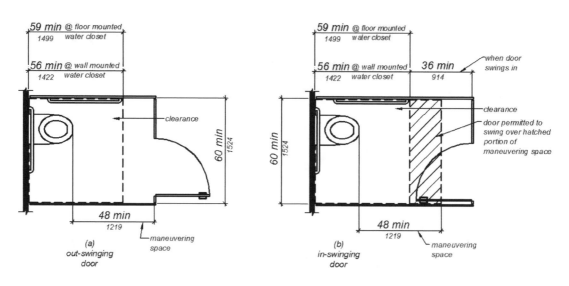

Fig. CD-26AJ
Maneuvering Space with End-Opening Door
© ICC Reproduced with Permission

Doors

_____LL.	Toilet compartment doors, including door hardware, comply with the applicable provisions of Section 14, "DOORS, DOORWAYS AND GATES", and Section 14A "MANEUVERING CLEARANCES AT DOORS, DOORWAYS AND GATES", except if the approach is to the latch side of the compartment door, clearance between the door side of the compartment and any obstruction is 48 inches minimum, measured perpendicular to the compartment door in its closed position.	**11B-604.8.1.2**	*604.8.1.2*

_____MM.	The toilet compartment door is located in the front partition or in the side wall or partition farthest from the water closet. **11B-604.8.1.2**	*604.8.1.2*	**Fig. CD-26AI & AJ**

_____NN.	Where located in the front partition, the door opening shall be 4 inches maximum from the side wall or partition farthest from the water closet. **11B-604.8.1.2** *604.8.1.2*	**Fig. CD-26AK**

_____OO.	Where located in the side wall or partition, the door opening shall be 4 inches maximum from the front partition.	**11B-604.8.1.2**	*604.8.1.2*

_____PP.	The door is self-closing.	**11B-604.8.1.2**	*604.8.1.2*

_____QQ.	A door pull is placed on both sides of the door near the latch. **11B-604.8.1.2**	*604.8.1.2*

_____RR.	The door pulls, have a shape that is operable with one hand and does not require tight grasping, pinching or twisting of the wrist.	**11B-604.8.1.2, 11B-309.4**	*604.8.1.2, 309.4*

_____SS.	The force required to activate the operable parts on doors and gates is 5 pounds maximum.	**11B-604.8.1.2, 11B-309.4**	*6048.1.2, 309.4*

_____TT.	Doors do not swing into the clear floor space or clearance required for any fixture.	Doors may swing into that portion of the maneuvering space which does not overlap the clearance required at a water closet.	**11B-604.8.1.2**	*604.8.1.2*

> **EXCEPTION:** When located at a side of the toilet compartment, the toilet compartment door opening shall provide a clear width of 34 inches minimum.

ADVISORY: Doors. This item describes requirements pertaining to the accessible water closet compartment door and the route to the door.
The door is required to have a latch which is flip-over style, sliding or which otherwise does not require the user to grasp. This is to facilitate the door by people with limited hand or finger dexterity.
The last part of this item addresses the required maneuvering space at the compartment door. This space is required to comply with the requirements for door maneuvering space contained in Section 14A, "MANEUVERING CLEARANCES AT DOORS, DOORWAYS AND GATES", except the space in front of the door shall be no less than 48 inches deep, measured perpendicular to the closed door. Where Fig. CD-14G(f) and Fig. CD-14H(j) specifically allow a 44 inch minimum maneuvering space perpendicular to door in general in the closed position, a minimum of 48 inches must be provided in order to comply with the toilet compartment door requirements.

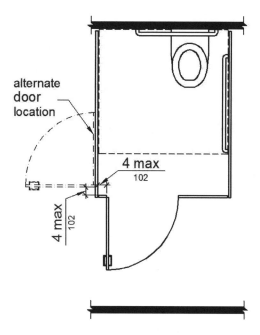

Fig. CD-26AK
Wheelchair Accessible Toilet Compartment Doors
© ICC Reproduced with Permission

Approach

_____UU.   The compartment is arranged for left-hand or right-hand approach to the water closet.   **11B-604.8.1.3**   *604.8.1.3*

Toe Clearance

_____VV.   At least one side partition provides a toe clearance of 9 inches minimum above the finish floor and 6 inches deep minimum beyond the compartment-side face of the partition, exclusive of partition support members. Compartments for children's use provide a toe clearance of 12 inches minimum. **11B-604.8.1.4**   *604.8.1.4*   **Fig. CD-26AL**

   EXCEPTION: Toe clearance at the side partition is not required in a compartment greater than 66 inches wide.

_____WW.   Partition components at toe clearances are smooth without sharp edges or abrasive surfaces. **11B-604.8.1.4**   *604.8.1.4*

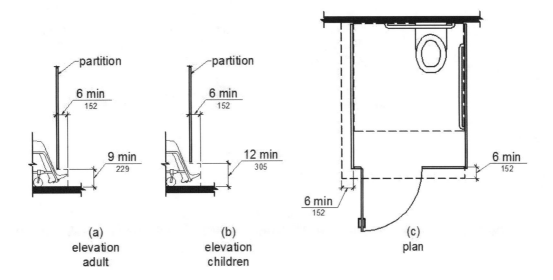

Fig. CD-26AL
Wheelchair Accessible Toilet Compartment Toe Clearance
© ICC Reproduced with Permission

26B. AMBULATORY-ACCESSIBLE TOILET COMPARTMENTS

Ambulatory-Accessible Compartment

Where toilet compartments are provided, at least one compartment shall be wheelchair-accessible. In addition to the compartment required to be accessible, where six (6) or more toilet compartments are provided, or where the combination of urinals and water closets totals six (6) or more fixtures, at least one compartment shall be an "Ambulatory Accessible Compartment". **11B-213.3.1** *213.3.1*

Size

_____A. The ambulatory accessible compartment has a depth of 60 inches minimum. **11B-604.8.2.1** *604.8.2* **Fig. CD-26BA**

_____B. The ambulatory accessible compartment has a width of 35 inches minimum and 37 inches maximum. **11B-604.8.2** *604.8.2.1* **Fig. CD-26BA**

Doors

_____C. Toilet compartment doors, including door hardware, comply with the applicable provisions of Checklist Section 13, "DOORS, DOORWAYS AND GATES", and Section 13B "MANEUVERING CLEARANCES AT DOORS, DOORWAYS AND GATES", except if the approach is to the latch side of the compartment door, clearance between the door side of the compartment and any obstruction is 44 inches minimum. **11B-604.8.1.2** *604.8.1.2*

_____D. The door is self-closing. **11B-604.8.2.2** *604.8.2.2*

_____E. A door pull is placed on both sides of the door near the latch. **11B-604.8.2.2** *604.8.2.2*

_____F.   The door pull has a shape that is operable with one hand and does not require tight grasping, pinching or twisting of the wrist.   **11B-604.8.2.2, 11B-309.4** *604.8.2.2, 309.4*

_____G.   The force required to activate the operable parts on the door is 5 pounds maximum. **11B-604.8.2.2, 11B-309.4** *604.8.2.2, 309.4*

_____H.   Toilet compartment doors do not swing into the minimum required compartment area. **11B-604.8.2.2** *604.8.2.2*

Water Closet Location

_____I.   The centerline of the water closet shall be 17 inches minimum and 19 inches maximum from the side wall or partition in the ambulatory accessible toilet compartment. **11B-604.2** *604.2*

Water Closet Seat

_____J.   The seat height of the water closet above the finish floor is 17 inches minimum and 19 inches maximum measured to the top of the seat. **11B-604.4** *604.4*

_____K.   The water closet seat is not sprung to return to a lifted position. **11B-604.4** *604.4*

_____L.   The maximum height of the water closet seat is 2 inches. **11B-604.4** *604.4*

Grab Bars

_____M.   Grab bars are provided on both sides of the compartment. **11B-604.8.2.3** *604.8.2.3* **Fig. CD-26BA**

_____N.   Grab bars shall be installed in any manner that provides a gripping surface at the specified locations and that does not obstruct the required clear floor space. **11B-609.7** *609.7*

_____O.   The side wall grab bars are a minimum of 42 inches in length. **11B-604.5.1** *604.5.1* **Fig. CD-26AD**

_____P.   The side wall grab bars are located 12 inches maximum from the rear wall and extend 54 inches minimum from the rear wall with the front ends positioned 24 inches minimum in front of the water closet. **11B-604.5.1** *604.5.1* **Fig. CD-26AD**

_____Q.   Grab bars are installed in a horizontal position, 33 inches minimum and 36 inches maximum above the finish floor, measured to the top of the gripping surface. **11B-609.4** *609.4*

_____R.   The grab bar(s) have either a circular or non-circular cross section that meets the applicable specifications: **11B-609.2** *609.2*

_____S.   **Circular Cross Section.** Grab bars with circular cross sections have an outside diameter of 1-1/4 inches minimum and 2 inches maximum. **11B-609.2.1** *609.2.1* **Fig. CD-26AF**

_____T.   **Non-Circular Cross Section.** Grab bars with non-circular cross
                sections have a cross-section dimension of 2 inches maximum and a perimeter
                dimension of 4 inches minimum and 4.8 inches maximum. **11B-609.2.2**   *609.2.2*
                Fig. CD-26AF

_____U.   The space between the wall and the grab bar(s) is 1-1/2 inches. **11B-609.3**   *609.3*

_____V.   The space between the grab bar(s) and projecting objects below and at the ends is 1-1/2 inches
                minimum. **11B-609.3**   *609.3*   **Fig. CD-26AG**

_____W.   The space between the grab bar(s) and projecting objects above is 12 inches
                minimum. **11B-609.3**   *609.3*   **Fig. CD-26AG**

_____X.   Allowable stresses shall not be exceeded for materials used when a vertical or
                horizontal force of 250 pounds (1112N) is applied at any point on the grab bar, fastener,
                mounting device, or supporting structure.   **11B-609.8** *609.8*

_____Y.   The grab bar(s) and any wall or other surfaces adjacent to the grab bars is free of
                any sharp or abrasive elements and shall have rounded edges. **11B-609.5** *609.5*

_____Z.   Grab bars do not rotate within their fittings.   **11B-609.6** *609.6*

Flush Controls

_____AA.   Flush controls are hand-operated or automatic. **11B-604.6**   604.6

_____BB.   Hand-operated flush controls are located 44 inches maximum distance above the
                floor. **11B-604.6**   *604.6*

_____CC.   Hand-operated controls are operable with one hand and do not require
                Tight grasping, pinching or twisting of the wrist. **11B-604.6, 11B-309.4**   *604.6, 309.4*

_____DD.   5 lb. maximum force is required to operate the hand-operated flush
                controls. **11B-604.6, 11B-309.4**   *604.6, 309.4*

Dispensers

_____EE.   The toilet paper dispenser is located 7 inches minimum and 9 inches maximum
                in front of the water closet measured to the centerline of the dispenser.
                11B-604.7   *604.7*   **Fig. CD-26AH**

_____FF.   The outlet of the toilet paper dispenser is below the grab bar, 19 inches minimum
                Above the finish floor and is not located behind grab bars. **11B-604.7**   *604.7*   **Fig. CD-26AH**

_____GG.   The toilet paper dispenser allows for continuous paper flow and does not control delivery.
                11B-604.7   *604.7*

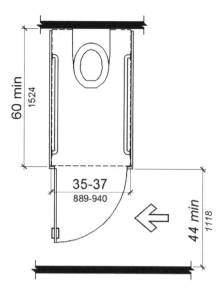

Fig. CD-26BA
Ambulatory Accessible Toilet Compartment
© ICC Reproduced with Permission

Coat Hooks and Shelves

_____HH.   If coat hooks are provided, at least one of each type is located within an accessible reach range (see Section 12, "REACH RANGES").
11B-213.3.7, **11B-604.8.3**   *213.3.7, 604.8.3*

_____II.   If Shelves are provided, at least one of each type is located 40 inches minimum and 48 inches maximum above the finish floor.   **11B-604.8.3**   *604.8.3*

26C. WATER CLOSETS AND TOILET COMPARTMENTS FOR CHILDREN'S USE

Water Closets and Toilet Compartments

_____A.   The water closets and toilet compartments intended for children's use comply With Checklist Section 26A, "WATER CLOSETS AND WHEELCHAIR-ACCESSIBLE TOILET COMPARTMENTS", substituting the alternate dimensions contained in the "TABLE OF SUGGESTED DIMENSIONS FOR CHILDREN'S USE", below, depending on the age group the facilities are intended to serve, and including the Checklist items listed below.   **11B-604.9 to 11B-604.9.7**   *604.9 to 604.9.7*

> *ADVISORY: Water Closets and Toilet Compartments for Children's Use. The specifications contained in the design and construction guidelines for disabled access are based on adult dimensions and anthropometrics. These same requirements also provide alternative suggested dimensions for water closets and toilet compartments for primary use by children. The "Table of Suggested Dimensions for Children's Use" provides additional guidance in applying the specifications to water closets for children according to the age group served and reflects the difference in the size, stature, and reach ranges for children ages 3 through 12. The specifications chosen should correspond to the*

> *age of the primary user group. The specifications of one age group should be applied consistently in the installation of a water closet and related elements.*

Water Closet Location

_____ B.   The centerline of the water closet shall be 12 inches minimum and 18 inches maximum from the side wall or partition. **11B-604.9.1**   *604.9.1*

Water Closet Height

_____ C.   The height of water closets shall be 11 inches minimum and 17 inches maximum, measured to the top of the seat. **11B-604.9.3**   *604.9.3*

Flush Controls

_____ D.   Hand-operated flush controls, if provided, shall be installed 36 inches maximum above the finish floor. **11B-604.9.5**   *604.9.5*

Dispensers

_____ E.   The outlet of the toilet paper dispenser shall be 14 inches minimum and 19 inches maximum above the finish floor. **11B-604.9.6**   *604.9.6*

TABLE OF SUGGESTED DIMENSIONS FOR CHILDREN'S USE

Suggested Dimensions for Water Closets Serving Children Ages 3 through 12			
	Ages 3 and 4	**Ages 5 through 8**	**Ages 9 through 12**
Water Closet Centerline	12 inches (305 mm)	12 to 15 inches (305 to *381* mm)	15 to 18 inches (*381* to *457* mm)
Toilet Seat Height	11 to 12 inches (279 to 305 mm)	12 to 15 inches (305 to *381* mm)	15 to 17 inches (*381* to *432* mm)
Grab Bar Height	18 to 20 inches (457 to *508* mm)	20 to 25 inches (*508* to 635 mm)	25 to 27 inches (635 to *686* mm)
Dispenser Height	14 inches (356 mm)	14 to 17 inches (356 to *432* mm)	17 to 19 inches (*432* to *483* mm)

26D.   SIGNS FOR TOILET ROOMS AND BATHING ROOMS

Toilet Rooms and Bathing Rooms
Doorways leading to accessible toilet rooms and bathing rooms shall be identified with the appropriate geometric symbol. Where existing toilet rooms or bathing rooms are NOT accessible, directional signs indicating the location of the nearest toilet room or bathing room that IS accessible within the facility shall be provided. These signs shall contain the International Symbol of Accessibility and compliant Visual Characters. Where a facility contains both accessible and inaccessible toilet or bathing rooms, the accessible toilet or bathing rooms shall be identified by the International Symbol of Accessibility. Where clustered single user toilet rooms or bathing units are clustered at a single location, and there are both accessible and inaccessible toilet or bathing units, the accessible toilet or bathing units shall be identified by the International Symbol of Accessibility. Existing buildings that have been remodeled to provide specific toilet rooms or bathing rooms for public use that are accessible shall have the location

of and the directions to these rooms posted in or near the building lobby on a sign containing accessible Visual Characters and the International Symbol of Accessibility. **11B-216.8** *216.8*

NOTICE TO TITLE II ENTITIES (State and Local Government Facilities Only)
California Government Code Section 7251. Toilet facilities; signs
When a building contains special toilet facilities usable by a person in a wheelchair or   otherwise handicapped, a sign indicating the location of such facilities shall be posted in   the building directory, in the main lobby, or at any entrance specially used by   handicapped persons.

GEOMETRIC SYMBOLS
Doorways leading to accessible toilet rooms and accessible bathing rooms shall be identified by the appropriate geometric symbol sign detailed herein. **11B-703.7.2.6**
Fig. CD-26DA

> **EXCEPTION:** Geometric symbols shall not be required at inmate toilet rooms and bathing rooms in detention and correctional facilities where only one gender is housed.   **11B-703.7.2.6**

> *Advisory: Toilet and Bathing Facilities Geometric Symbols. There is no requirement for providing gender pictograms in combination with the geometric identification symbols required at doorways leading to men's, women's and unisex toilet and bathing facilities.*
>
> *When toilet and bathing facilities have doorway openings instead of doors, such as at airports or stadiums, the geometric identification symbol should be located at the proper height adjacent to the opening or incorporated into the required tactile identification sign. For example, the geometric symbol may be used as the sign background with raised characters and Braille.*

MEN'S TOILET AND BATHING FACILITIES

_____ A.   Men's toilet and bathing facilities are identified by an equilateral triangle ¼ inch thick with edges 12 inches long and a vertex pointing upward. The triangle symbol contrasts with the door, either light on a dark background or dark on a light background. **11B-703.7.2.6.1**
Fig. CD-26DA

> **EXCEPTION:** Within secure perimeter of detention and correctional facilities, geometric symbols are not required to be ¼ inch thick.

WOMEN'S TOILET AND BATHING FACILITIES

_____ B.   Women's toilet and bathing facilities are identified by a circle, ¼ inch thick and 12 inches in diameter. The circle symbol contrasts with the door, either light on a dark background or dark on a light background.   **11B-703.7.2.6.2**
Fig. CD-26DA

> **EXCEPTION:** Within secure perimeter of detention and correctional facilities, geometric symbols are not required to be ¼ inch thick.

UNISEX TOILET AND BATHING FACILITIES

_____ C.   Unisex toilet and bathing facilities are identified by a circle, ¼ inch thick and 12 inches in diameter with a ¼ inch thick triangle with a vertex pointing upward superimposed on the circle and within the 12 inch diameter. The triangle symbol contrasts with the circle symbol, either light on a dark background or dark on a light background. The circle symbol contrasts with the door, either light on a dark background or dark on a light background.
11B-703.7.2.6.3   Fig. CD-26DA

> **EXCEPTION:** Within secure perimeter of detention and correctional facilities, geometric symbols are not required to be ¼ inch thick.

MOUNTING LOCATION

_____D.   The geometric symbol is mounted at 58 inches minimum and 60 inches maximum above the finish floor or ground surface measured from the centerline of the symbol. **11B-703.7.2.6   Fig. CD-26DA**

_____E.   Where a door is provided the symbol is mounted within 1 inch of the vertical centerline of the door.   **11B-703.7.2.6**

EDGES AND CORNERS

_____F.   The edges of the signs are rounded, chamfered or eased, and any corners have a minimum radius of 1/8 inch.   **11B-703.7.2.6.4**

DIRECTIONAL SIGNS AND INTERNATIONAL SYMBOL OF ACCESSIBILITY

IDENTIFICATION OF ACCESSIBLE FACILITIES

_____G.   Where a facility contains BOTH accessible AND inaccessible toilet or bathing rooms, the accessible toilet or bathing rooms shall be identified by the INTERNATIONAL SYMBOL OF ACCESSIBILITY, complying with Section 30F, "SYMBOLS OF ACCESSIBILITY. **11B-216.8**   *216.8*

DIRECTIONAL SIGNS TO INACCESSIBLE FACILIITIES

_____H.   Where existing toilet rooms or bathing rooms are NOT accessible, directional signs indicating the location of the nearest toilet room or bathing room that IS accessible within the facility are provided.   These signs contain the INTERNATIONAL SYMBOL OF ACCESSIBILITY, complying with Section 30F, "SYMBOLS OF ACCESSIBILITY.   **11B-216.8** *216.8*

_____I.   Existing buildings that have been remodeled to provide specific toilet rooms or bathing rooms for public use that are accessible shall have the location of, and the directions to these rooms posted in or near the building lobby or entrance on a sign containing accessible VISUAL CHARACTERS, per Section 30D, and the INTERNATIONAL SYMBOL OF ACCESSIBILITY, complying with Section 30F, "SYMBOLS OF ACCESSIBILITY. **11B-216.8**   *216.8*

INTERNATIONAL SYMBOL OF ACCESSIBILITY

_____J.   The International Symbol of Accessibility consists of a white figure on a blue background. The blue shall be equal to Color No. 15090 in Federal Standard 595C.   **11B-216.8, 11B-703.7.2.1**   *216.8, 703.7.2.1*

> **EXCEPTION:** The appropriate enforcement agency may approve other colors to compliment decor or unique design. The symbol contrast shall be light on dark or dark on light. **11B-703.7.2.1**

DESIGNATIONS
Interior and exterior signs identifying permanent rooms and spaces shall comply with the Section 30 requirements for SIGNS, RAISED CHARACTERS, BRAILLE and VISUAL CHARACTERS. Where pictograms are provided as designations of permanent rooms and spaces, the pictograms shall comply and shall have text descriptions with compliant RAISED AND VISUAL CHARACTERS. **11B-216.2** *216.2*   **Fig. CD-26DA**

EXCEPTION: Exterior signs that are not located at the door to the space they serve shall not be required to have compliant Raised Characters.

WALL—SIGNAGE (TYP.)

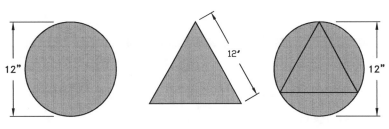

DOOR MOUNTED SIGNAGE (TYP.)

NOTE: PICTOGRAMS AND/OR LETTERING ARE NOT
REQUIRED ON DOOR—MOUNTED SIGNAGE.

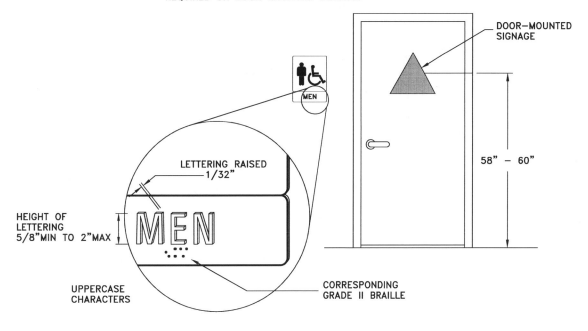

Fig. CD-26DA
Signs for Toilet Rooms and Bathing Rooms

26E. URINALS

General

Where one or more urinals are provided, at least one is accessible and shall comply as follows. **11B-213.3.3**   *213.3.3*

> **ADVISORY:** **General.** *Stall-type urinals provide greater accessibility for a broader range of persons, including people of short stature.*

Height and Depth

_____A.   The accessible urinal is the stall-type or wall hung type. **11B-605.2**   *605.2*

_____B.   The accessible urinal is a minimum of 13-1/2 inches deep measured from the Outer face of the urinal rim to the back of the fixture. **11B-605.2**   *605.2*   **Fig. CD-26EA**

_____C.   The urinal rim is located at a maximum height of 17 inches above the finished floor or ground.   **11B-605.2**   *605.2*   **Fig. CD-26EA**

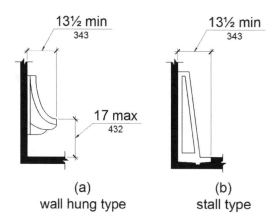

(a)
wall hung type

(b)
stall type

Fig. CD-26EA
Height and Depth of Urinals
© ICC Reproduced with Permission

Clear Floor Space

_____D.   A clear floor or ground space complying with Section 10, "CLEAR FLOOR OR GROUND SPACE", positioned for forward approach, shall be provided. **11B-605.3**   *605.3*

Flush Controls

_____E.   Flush controls are hand-operated or automatic.   **11B-605.4**   *605.4*

_____F.   Hand-operated flush controls are located 44 inches maximum distance above the floor.   **11B-604.6**   *604.6*

_____G.   Hand-operated controls are operable with one hand and do not require

---

tight grasping, pinching or twisting of the wrist. **11B-605.4, 11B-309.4**   *605.4, 309.4*

_____H.   5 lb. maximum force is required to operate the hand-operated flush controls. **11B-605.4, 11B-309.4**   *605.4, 309.4*

26F.   <u>LAVATORIES AND SINKS</u>

<u>GENERAL</u>
Lavatories and sinks shall comply as detailed.   **11B-606.1**   *606.1*

> **ADVISORY: General.**   *If soap and towel dispensers are provided, they must be located within the reach ranges specified in Section 12, "REACH RANGES" except for those in toilet and bathing rooms which must comply with the requirement for the highest operable parts to be located a maximum of 40 inches above the finish floor. Locate soap and towel dispensers so that they are conveniently usable by a person at the accessible lavatory.*

<u>GENERAL</u>
Where lavatories are provided, at least five percent (5%), but no fewer than one lavatory, shall be accessible and not be located in a toilet compartment.   **11B-213.3.4**   *213.3.4*

<u>GENERAL</u>
Where sinks are provided, at least five percent (5%), but no fewer than one, of each type provided in each accessible room or space shall be accessible.   **11B-212.3.**   *212.3.*

EXCEPTIONS:

1.   Mop, service or scullery sinks shall not be required to comply.

2.   Scrub sinks, as defined in California Plumbing Code Section 221.0, shall not be required to comply.

<u>Clear Floor Space</u>

_____A.   A clear floor space complying with Section 10, "CLEAR FLOOR OR GROUND SPACE", positioned for a forward approach, and knee and toe clearance complying with Section 11, "KNEE AND TOE CLEARANCE", shall be provided .   **11B-606.2, 11B-305.2, 11B-305.3**   *606.2, 305.2, 305.3*

EXCEPTIONS:

1.   A parallel approach that complies with Section 10, "CLEAR FLOOR OR GROUND SPACE" shall be permitted to a kitchen sink in a space where a cook top or conventional range is not provided and to wet bars.

2.   *Reserved.*

3.   In residential dwelling units, cabinetry shall be permitted under lavatories and kitchen sinks provided that all of the following conditions are met:

(a)   the cabinetry can be removed without removal or replacement of the fixture;

(b)   the finish floor extends under the cabinetry; and

(c)   the walls behind and surrounding the cabinetry are finished.

4.   A knee clearance of 24 inches (610 mm) minimum above the finish floor or ground shall be permitted at lavatories and sinks used primarily by children 6 through 12 years where the rim or counter surface is 31 inches (787 mm) maximum above the finish floor or ground.

5.   A compliant parallel approach shall be permitted to lavatories and sinks used primarily by children 5 years and younger.

6.   The dip of the overflow shall not be considered in determining knee and toe clearances.

7.   No more than one bowl of a multi-bowl sink shall be required to provide accessible knee and toe clearance.

Height

_____Q.   Lavatories and sinks shall be installed with the front of the higher of the rim or counter surface 34 inches maximum above the finish floor or ground. **11B-606.3**   *606.3*

EXCEPTIONS:

1.   **Reserved.**

2.   In residential dwelling unit kitchens, sinks that are adjustable to variable heights, 29 inches minimum and 36 inches maximum, shall be permitted where rough-in plumbing permits connections of supply and drain pipes for sinks mounted at the height of 29 inches.

Faucets

_____R.   Hand-operated metering faucets shall remain open for 10 seconds minimum. **11B-606.4**   *606.4*

_____S.   Controls for faucets are operable with one hand and do not require tight grasping, pinching or twisting of the wrist. The force required to activate operable parts is no greater than 5 pounds maximum. Hand-operated controls for faucets shall be located within accessible reach ranges. **11B-309.4, 11B-606.4**   *309.4, 606.4*

Exposed Pipes and Surfaces

_____T.   Water supply and drain pipes under lavatories and sinks shall be installed or otherwise configured to protect against contact. There shall be no sharp or abrasive surfaces under lavatories and sinks. **11B-606.5**   *606.5* **Fig. CD-26FA**

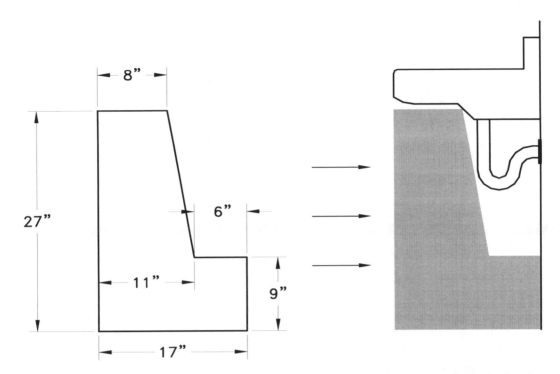

MINIMUM LAVATORY KNEE AND TOE CLEARANCES

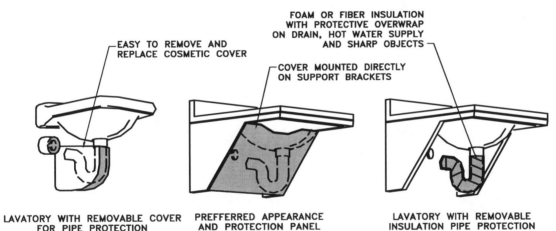

EASY TO REMOVE AND
REPLACE COSMETIC COVER

FOAM OR FIBER INSULATION
WITH PROTECTIVE OVERWRAP
ON DRAIN, HOT WATER SUPPLY
AND SHARP OBJECTS

COVER MOUNTED DIRECTLY
ON SUPPORT BRACKETS

LAVATORY WITH REMOVABLE COVER
FOR PIPE PROTECTION

PREFFERRED APPEARANCE
AND PROTECTION PANEL

LAVATORY WITH REMOVABLE
INSULATION PIPE PROTECTION

HOT WATER AND DRAIN PIPES UNDER LAVATORIES SHALL BE INSULATED OR OTHERWISE
CONFIGURED TO PROTECT AGAINST CONTACT. THERE SHALL BE NO SHARP OR ABRASIVE
SURFACES UNDER LAVATORIES.

LAVATORY INSULATION / CONTACT PROTECTION

Fig. CD-26FA
Exposed Pipes and Surfaces/Knee and Toe Clearance

Adjacent Side Wall or Partition

_____ U.   Lavatories, when located adjacent to a side wall or partition, shall be a minimum
of 18 inches to the centerline of the fixture.   **11B-606.6**   **Fig. CD-26FB**

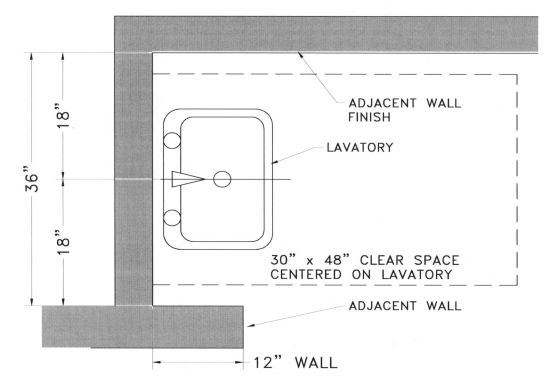

Fig. CD-26FB
Adjacent Side Wall or Partition

Sink Depth

_____V.   Sinks shall be 6-1/2 inches deep maximum.   **11B-606.7**

26G.   BATHTUBS

General
Bathtubs shall comply as detailed.   **11B-607.1**   *607.1*

Clearance

_____A.   Clearance in front of bathtubs shall extend the length of the bathtub and shall be 48 inches wide minimum for forward approach and 30 inches wide minimum for parallel approach. An accessible lavatory shall be permitted at the control end of the clearance. Where a permanent seat is provided at the head end of the bathtub, the clearance shall extend 12 inches minimum beyond the wall at the head end of the bathtub.   **11B-607.2**   *607.2*   **Fig. CD-26GA**

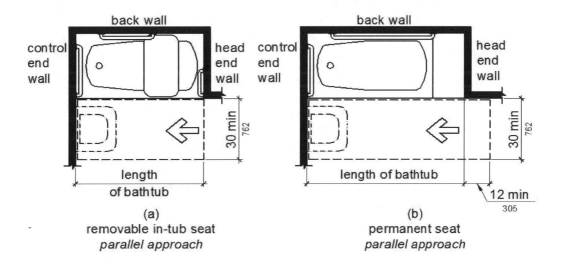

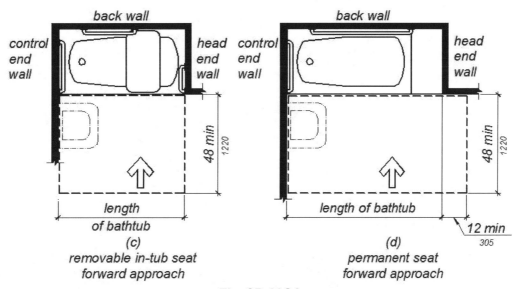

Fig. CD-26GA
Clearance for Bathtubs
© ICC Reproduced with Permission

Seat

_____B.   A permanent seat at the head end of the bathtub or a removable in-tub seat shall be provided.   **11B-607.3**   *607.3*   **Fig. CD-26GB**

Bathtub Seats

_____C.   The top of bathtub seats shall be 17 inches minimum and 19 inches maximum above the bathroom finish floor.   **11B-610.2** *610.2* **Fig. CD-26GB**

_____D.   The depth of a removable in-tub seat shall be 15 inches minimum and 16 Inches maximum. The seat shall be capable of secure placement. **11B-610.2**   *610.2*   **Fig. CD-26GB**

_____E.   Permanent seats at the head end of the bathtub shall be 15 inches deep minimum and shall extend from the back wall to or beyond the outer edge of the bathtub. **11B-610.2**   *610.2*   **Fig. CD-26GB**

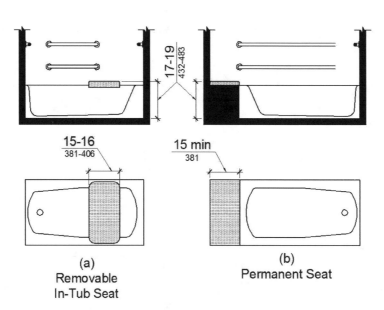

Fig. CD-26GB
Bathtub Seats
© ICC Reproduced with Permission

Bathtubs WITH Permanent Seats

Back Wall Grab Bars

_____F.   Two grab bars shall be installed on the back wall in a horizontal position, One located 33 inches minimum and 36 inches maximum above the finish floor measured to the top of the gripping surface, and the other located 8 inches minimum and 10 inches maximum above the rim of the bathtub. **11B-607.4.1.1, 11B-609.4** *607.4.1.1, 609.4* **Fig. CD-26GC**

_____G.   Each grab bar shall be installed 15 inches maximum from the head end
wall and 12 inches maximum from the control end wall. **11B-607.4.1.1, 11B-609.4**
607.4.1.1, 609.4 **Fig. CD-26GC**

Control End Wall Grab Bar

_____H.   A grab bar 24 inches long minimum shall be installed on the control end
wall at the front edge of the bathtub. **11B-607.4.1.2**   *607.4.1.2*   **Fig. CD-28GC**

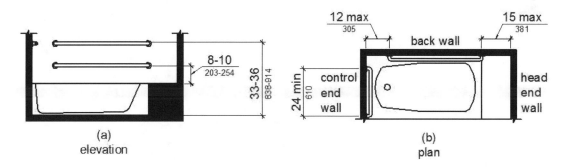

Fig. CD-26GC
Grab Bars for Bathtubs with Permanent Seats
© ICC Reproduced with Permission

Bathtubs WITHOUT Permanent Seats

Back Wall Grab Bars

_____I.   Two grab bars shall be installed on the back wall in a horizontal position,
one located 33 inches minimum and 36 inches maximum above the finish floor
measured to the top of the gripping surface, and the other located 8 inches minimum
and 10 inches maximum above the rim of the bathtub. **11B-607.4.2.1, 11B-609.4**
607.4.2.1, 609.4 **Fig. CD-26GD**

_____J.   Each grab bar shall be 24 inches long minimum and shall be installed 24
inches maximum from the head end wall and 12 inches maximum from the control end
wall. **11B-607.4.2.1, 11B-609.4** *607.4.2.1, 609.4*
Fig. CD-26GD

Control End Wall Grab Bar

_____K.   A grab bar 24 inches long minimum shall be installed on the control end
wall at the front edge of the bathtub. **11B-607.4.2.2**   *607.4.2.2*   **Fig. CD-26GD**

Head End Wall Grab Bar

_____L.   A grab bar 12 inches long minimum shall be installed on the head end
wall at the front edge of the bathtub. **11B-607.4.1.2**   *607.4.1.2*   **Fig. CD-26GD**

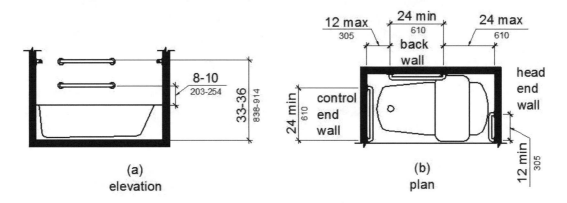

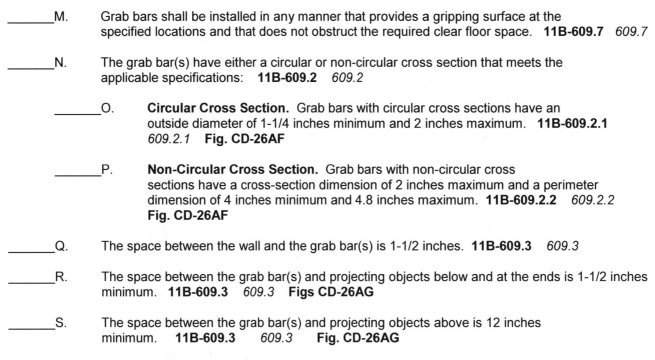

Fig. CD-26GD
Grab Bars for Bathtubs with Removable In-Tub Seats
© ICC Reproduced with Permission

ALL Bathtub Grab Bars

_____M.   Grab bars shall be installed in any manner that provides a gripping surface at the specified locations and that does not obstruct the required clear floor space. **11B-609.7** *609.7*

_____N.   The grab bar(s) have either a circular or non-circular cross section that meets the applicable specifications: **11B-609.2** *609.2*

_____O.   **Circular Cross Section.** Grab bars with circular cross sections have an outside diameter of 1-1/4 inches minimum and 2 inches maximum. **11B-609.2.1** *609.2.1* **Fig. CD-26AF**

_____P.   **Non-Circular Cross Section.** Grab bars with non-circular cross sections have a cross-section dimension of 2 inches maximum and a perimeter dimension of 4 inches minimum and 4.8 inches maximum. **11B-609.2.2** *609.2.2* **Fig. CD-26AF**

_____Q.   The space between the wall and the grab bar(s) is 1-1/2 inches. **11B-609.3** *609.3*

_____R.   The space between the grab bar(s) and projecting objects below and at the ends is 1-1/2 inches minimum. **11B-609.3** *609.3* **Figs CD-26AG**

_____S.   The space between the grab bar(s) and projecting objects above is 12 inches minimum. **11B-609.3** *609.3* **Fig. CD-26AG**

> **NOTE: Alternate Configuration.** L-shaped or U-shaped grab bars shall be permitted. **EXCEPTION:** For L-shaped or U-shaped grab bars, the space between the walls and the grab bar shall be 1-1/2 inches minimum for a distance of 6 inches on either side of the inside corner between two adjacent wall surfaces. **11B-609.9, 11B-609.3** *609.9, 609.3*
>
> > **EXCEPTIONS: 1.** Reserved.
> > **2.** In residential dwelling units, grab bars shall not be required to be installed in bathtubs located in bathing facilities provided that reinforcement has been installed in walls and located so as to permit the installation of compliant grab bars. **11B-604.5** *604.5*

_____T.   Allowable stresses shall not be exceeded for materials used when a vertical or horizontal force of 250 pounds (1112N) is applied at any point on the grab bar, fastener, mounting device, or supporting structure.   **11B-609.8** *609.8*

_____U.   The grab bar(s) and any wall or other surfaces adjacent to the grab bars is free of any sharp or abrasive elements and shall have rounded edges. **11B-609.5** *609.5*

_____V.   Grab bars do not rotate within their fittings.   **11B-609.6** *609.6*

Controls

_____W.   Controls, other than drain stoppers, shall be located on an end wall.
11B-607.5   *607.5*   **Fig CD-26GE**

_____X.   Controls shall be between the bathtub rim and grab bar, and between the open side of the bathtub and the centerline of the width of the bathtub. **11B-607.5**   *607.5*   **Fig CD-26GE**

_____Y.   Operable parts of controls are operable with one hand and do not require tight grasping, pinching or twisting of the wrist. The force required to activate operable parts is no greater than 5 pounds maximum.
11B-607.5, 11B-309.4   *607.5, 309.4*

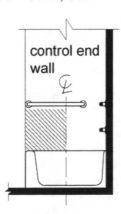

Fig. CD-26GE
Bathtub Control Location
© ICC Reproduced with Permission

Shower Spray Unit and Water

_____Z.   A shower spray unit with a hose 59 inches long minimum that can be used both as a fixed position shower head and as a handheld shower shall be provided.   **11B-607.6**   *607.6*

_____AA.   The shower spray unit shall have an on/off control with a non-positive shut-off.
11B-607.6   *607.6*

_____BB.   If an adjustable-height shower head on a vertical bar is used, the bar shall be installed so as not to obstruct the use of grab bars. **11B-607.6**   *607.6*

_____CC.   Bathtub shower spray units shall deliver water that is 120 degrees Fahrenheit (49 degrees Celsius) maximum.   **11B-607.6**   *607.6*

> **ADVISORY: Shower Spray Unit and Water.** *Ensure that hand-held shower spray units are capable of delivering water pressure substantially equivalent to fixed shower heads.*

Bathtub Enclosures

_____DD.   Enclosures for bathtubs shall not obstruct controls, faucets, shower and spray units or obstruct transfer from wheelchairs onto bathtub seats or into bathtubs.   **11B-607.7** *607.7*

_____EE.   Enclosures on bathtubs shall not have tracks installed on the rim of the open face of the bathtub.   **11B-607.7**   *607.7*

26H. <u>SHOWER COMPARTMENTS</u>

General
Shower compartments shall comply as detailed.   **11B-608.1**   *608.1*

Size and Clearance for Shower Compartments
Shower compartments shall have sizes and clearances as detailed. **11B-608.2**   *608.2*

> **ADVISORY: General.** *Shower stalls that are 60 inches (1525 mm) wide and have no curb may increase the usability of a bathroom because the shower area provides additional maneuvering space.*

<u>STANDARD ROLL-IN TYPE SHOWER COMPARTMENTS</u>

_____A.   Standard roll-in type shower compartments shall be 30 inches wide minimum by 60 inches deep minimum clear inside dimensions measured at center points of opposing sides with a full opening width on the long side.   **11B-608.2.2** *608.2.2*   **Fig. CD-26HA**

Clearance

_____B.   A 36 inch wide minimum by 60 inch long minimum clearance shall be provided adjacent to the open face of the shower compartment.   **11B-608.2.2.1** *608.2.2.1* **Fig. CD-26HA**

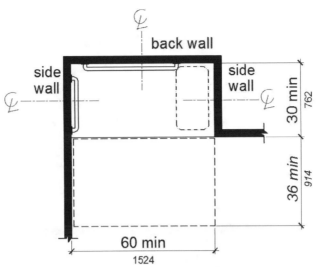

Fig. CD-26HA
Standard Roll-In Type Shower Compartment Size and Clearance
© ICC Reproduced with Permission

Grab Bars

Grab bars shall comply as detailed. Where multiple grab bars are used,
required horizontal grab bars shall be installed at the same height above the finish floor. Where separate grab bars are required on adjacent walls at a common mounting height, an L-shaped or U-shaped grab bar meeting the dimensional requirements detailed shall be permitted. **11B-608.3** *608.3*

EXCEPTIONS:

1. **Reserved.**

2. In residential dwelling units, grab bars shall not be required to be installed in showers, located in bathing facilities provided that reinforcement has been installed in walls and located so as to permit the installation of compliant grab bars.

_____C. Grab bars shall be provided on the back wall and the side wall opposite the seat. Grab bars shall not be provided above the seat. **11B-608.3.2** *608.3.2* **Fig. CD-26HB**

_____D. Grab bars shall be installed 6 inches maximum from adjacent walls. **11B-608.3.2** *608.3.2* **Fig. CD-26HB**

_____E. Grab bars shall be installed in any manner that provides a gripping surface at the specified locations and that does not obstruct the required clear floor space. **11B-609.7** *609.7*

_____F. The grab bar(s) have either a circular or non-circular cross section that meets the applicable specifications. **11B-609.2** *609.2*

_____G. **Circular Cross Section.** Grab bars with circular cross sections have an outside diameter of 1-1/4 inches minimum and 2 inches maximum. **11B-609.2.1** *609.2.1* **Fig. CD-26AF**

_____H. **Non-Circular Cross Section.** Grab bars with non-circular cross sections have a cross-section dimension of 2 inches maximum and a perimeter dimension of 4 inches minimum and 4.8 inches maximum. **11B-609.2.2** *609.2.2* **Fig. CD-26AF**

_____I. The space between the wall and the grab bar(s) is 1-1/2 inches. **11B-609.3** *609.3*

_____J. The space between the grab bar(s) and projecting objects below and at the ends is 1-1/2 inches minimum. **11B-609.3** *609.3* **Fig. CD-26AG**

_____K. The space between the grab bar(s) and projecting objects above is 12 Inches minimum. **11B-609.3** *609.3* **Fig. 26-AG**

_____L. Grab bars shall be installed in a horizontal position, 33 inches minimum and 36 inches maximum above the finish floor measured to the top of the gripping surface. **11B-609.4** *609.4*

_____M. Allowable stresses shall not be exceeded for materials used when a vertical or horizontal force of 250 pounds (1112N) is applied at any point on the grab bar, fastener, mounting device, or supporting structure. **11B-609.8** *609.8*

_____N. The grab bar(s) and any wall or other surfaces adjacent to the grab bars is free of any sharp or abrasive elements and shall have rounded edges. **11B-609.5** *609.5*

_____O. Grab bars do not rotate within their fittings. **11B-609.6** *609.6*

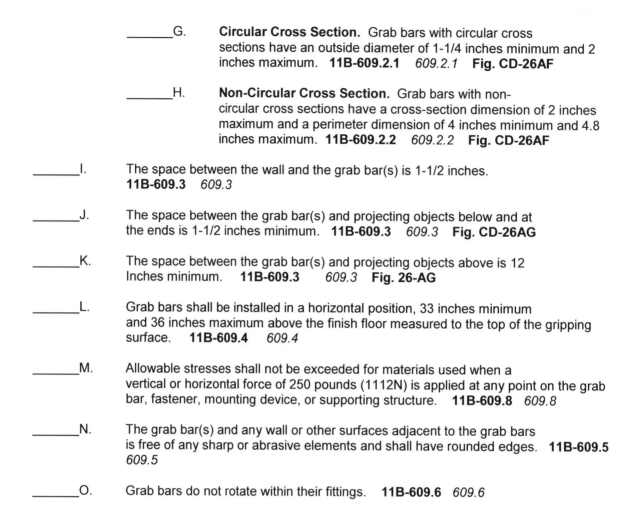

With Seat

Fig. CD-26HB
Grab Bars for Standard Roll-In Type Showers
© ICC Reproduced with Permission

Shower Compartment Seat
A folding seat shall be provided in roll-in type showers and shall comply as detailed. **11B-608.4, 11B-610.1** *608.4, 610.1*

EXCEPTION: In residential dwelling units, seats shall not be required in shower compartments provided that reinforcement has been installed in walls so as to permit the installation of compliant seats.

_____P.   A seat in a standard roll-in shower compartment shall be a folding type, shall be installed on the side wall adjacent to the controls, and shall extend from the back wall to a point within 3 inches of the compartment entry. **11B-610.3**   *610.3*   **Fig. CD-26HC**

_____Q.   The top of the seat shall be 17 inches minimum and 19 inches maximum above the bathroom finish floor.   **11B-610.3**   *610.3*

_____R.   When folded, the seat shall extend 6 inches maximum from the mounting wall.   **11B-610.3**   *610.3*

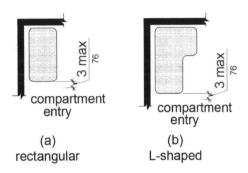

(a)
rectangular

(b)
L-shaped

Fig. CD-26HC
Extent of Seat
© ICC Reproduced with Permission

Rectangular Seats

_____S.   The rear edge of a rectangular seat shall be 2-1/2 inches maximum and the front edge 15 inches minimum and 16 inches maximum from the seat wall.   **11B-610.3.1**   *610.3.1*   **Fig. CD-26HD**

_____T.   The side edge of the seat shall be 1-1/2 inches maximum from the adjacent wall. **11B-610.3.1**   *610.3.1*   **Fig. CD-26HD**

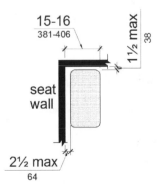

Fig. CD-26HD
Rectangular Shower Seat
© ICC Reproduced with Permission

L-Shaped Seats

_____U.　The rear edge of an L-shaped seat shall be 2-1/2 inches maximum and the front edge 15 inches minimum and 16 inches maximum from the seat wall. **11B-610.3.2**　*610.3.2*
Fig. CD-26HE

_____V.　The rear edge of the "L" portion of the seat shall be 1-1/2 inches maximum from the wall and the front edge shall be 14 inches minimum and 15 inches maximum from the wall. **11B-610.3.2**　*610.3.2*　**Fig. CD-26HE**

_____W.　The end of the "L" shall be 22 inches minimum and 23 inches maximum from the main seat wall. **11B-610.3.2**　*610.3.2*　**Fig. CD-26HE**

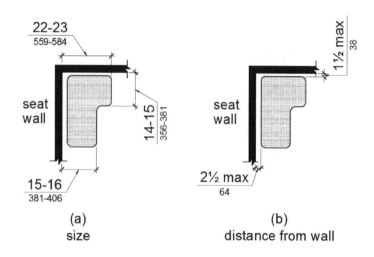

Fig. CD-26HE
L-Shaped Shower Seat
© ICC Reproduced with Permission

Structural Strength

_____X.　Allowable stresses shall not be exceeded for materials used when a vertical or horizontal force of 250 pounds is applied at any point on the seat, fastener, mounting device, or supporting structure. **11B-610.4**　*610.4*

Controls

_____Y.　In standard roll-in type shower compartments, operable parts of controls and faucets shall be installed on the back wall of the compartment adjacent to the seat wall 19 inches minimum and 27 inches maximum from the seat wall; and shall be located above the grab bar, but no higher than 48 inches above the shower floor, with their centerline at 39 inches to 41 inches above the shower floor. **11B-608.5.2**　*608.5.2*
Fig. CD-26HF

_____Z.　Operable parts of the shower spray unit including the handle, shall be installed on the back wall adjacent to the seat wall 19 inches minimum and 27 inches maximum from the seat wall; and shall be located above the grab bar, but no higher than 48 inches above the shower floor.   **11B-608.5.2**   *608.5.2*   **Fig. CD-26HF**

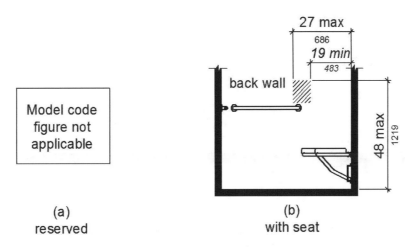

Fig. CD-26HF
Standard Roll-In Type Shower Compartment Control Location
© ICC Reproduced with Permission

Shower Spray Unit and Water

_____AA.   A shower spray unit with a hose 59 inches long minimum that can be used both as a fixed position shower head and as a handheld shower shall be provided. **11B-608.6**   *608.6*

_____BB.   The shower spray unit shall have an on/off control with a non-positive shut-off.   **11B-608.6**   *608.6*

_____CC.   If an adjustable-height shower head on a vertical bar is used, the bar shall be installed so as not to obstruct the use of grab bars.   **11B-608.6**   *608.6*

_____DD.   Shower spray units shall deliver water that is 120 degrees Fahrenheit (49 degrees Celsius) maximum.   **11B-608.6**   *608.6*

> **EXCEPTION:** Where subject to excessive vandalism two fixed shower heads shall be installed instead of a handheld spray unit in facilities that are not transient lodging guest rooms. Each shower head shall be installed so it can be operated independently of the other and shall have swivel angle adjustments, both vertically and horizontally.

ADVISORY: *Shower Spray Unit and Water.* *Ensure that hand-held shower spray units are capable of delivering water pressure substantially equivalent to fixed shower heads.*

Shower Thresholds

_____EE.   Changes in level between 1/4" high minimum and 1/2" high maximum are beveled with a slope not steeper than 1:2.   **11B-608.7, 11B-303.2** *608.7, 303.2* **Fig. CD-8B**

_____FF. Changes in level of 1/4" high maximum are permitted to be vertical and without edge treatment. **11B-608.7, 11B-303.2** *608.7, 303.2* **Fig. CD-8A**

Shower Enclosures

_____GG. Enclosures for shower compartments shall not obstruct controls, faucets, and shower spray units or obstruct transfer from wheelchairs onto shower seats. **11B-608.8** *608.8*

Shower Floor or Ground Surface

_____HH. Floor or ground surfaces of showers shall be stable, firm, and slip resistant and shall be sloped 1:48 maximum in any direction. Where drains are provided, grate openings shall be ¼ inch maximum and flush with the floor surface. **11B-609.9, 11B-302.1**

Soap Dish

_____II. Where a soap dish is provided, it shall be located on the control wall at 40 inches maximum above the shower floor, and within the reach limits from the seat. **11B-608.10**

ALTERNATE ROLL-IN TYPE SHOWER COMPARTMENTS

Clearance

_____A. Alternate roll-in type shower compartments shall be 36 inches wide and 60 inches deep minimum clear inside dimensions measured at center points of opposing sides with a full opening width on the long side. **11B-608.2.3** *608.2.3* **Fig. CD-26HG**

_____B. A 36 inch wide minimum entry shall be provided at one end on the long side of the compartment. **11B-608.2.3** *608.2.3* **Fig. CD-26HG**

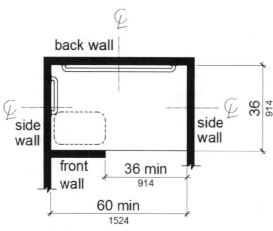

Note: inside finished dimensions measured at the center points of opposing sides

Fig. CD-26HG
Alternate Roll-In Type Shower Compartment Size and Clearance
© ICC Reproduced with Permission

Grab Bars

Grab bars shall comply as detailed. Where multiple grab bars are used, required horizontal grab bars shall be installed at the same height above the finish floor. Where separate grab bars are required on adjacent walls at a common mounting height, an L-shaped or U-shaped grab bar meeting the dimensional requirements detailed shall be permitted. **11B-608.3, 11B-609.9**   *608.3, 609.9*

EXCEPTIONS:

1. **Reserved.**

2. In residential dwelling units, grab bars shall not be required to be installed in showers, located in bathing facilities provided that reinforcement has been installed in walls and located so as to permit the installation of compliant grab bars.

_____C. Grab bars shall be provided on the back wall and the side wall farthest from the compartment entry. Grab bars shall not be provided above the seat. **11B-608.3.3**   *608.3.3* **Fig. CD-26HH**

_____D. Grab bars shall be installed 6 inches maximum from adjacent walls. **11B-608.3.3**   *608.3.3*   **Fig. CD-26HH**

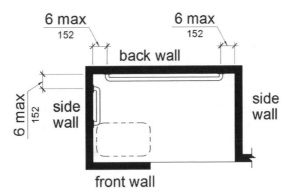

Fig. CD-26HH
Grab Bars for Alternate Roll-In Type Showers
© ICC Reproduced with Permission

_____E. Grab bars shall be installed in any manner that provides a gripping surface at the specified locations and that does not obstruct the required clear floor space. **11B-609.7**   *609.7*

_____F. The grab bar(s) have either a circular or non-circular cross section that meets the applicable specifications. **11B-609.2**   *609.2*

_____G.   **Circular Cross Section.** Grab bars with circular cross sections have an outside diameter of 1-1/4 inches minimum and 2 inches maximum. **11B-609.2.1** *609.2.1* **Fig. CD-26AF**

_____H.   **Non-Circular Cross Section.** Grab bars with non-circular cross sections have a cross-section dimension of 2 inches maximum and a perimeter dimension of 4 inches minimum and 4.8 inches maximum. **11B-609.2.2** *609.2.2* **Fig. CD-26AF**

_____I.   The space between the wall and the grab bar(s) is 1-1/2 inches. **11B-609.3** *609.3*

_____J.   The space between the grab bar(s) and projecting objects below and at the ends is 1-1/2 inches minimum. **11B-609.3** *609.3* **Fig. CD-26AG**

_____K.   The space between the grab bar(s) and projecting objects above is 12 Inches minimum. **11B-609.3** *609.3* **Fig. 26AG**

_____L.   Grab bars shall be installed in a horizontal position, 33 inches minimum and 36 inches maximum above the finish floor measured to the top of the gripping surface. **11B-609.4** *609.4*

_____M.   Allowable stresses shall not be exceeded for materials used when a vertical or horizontal force of 250 pounds (1112N) is applied at any point on the grab bar, fastener, mounting device, or supporting structure. **11B-609.8** *609.8*

_____N.   The grab bar(s) and any wall or other surfaces adjacent to the grab bars is free of any sharp or abrasive elements and shall have rounded edges. **11B-609.5** *609.5*

_____O.   Grab bars do not rotate within their fittings. **11B-609.6** *609.6*

Shower Compartment Seat
A folding seat shall be provided in roll-in type showers and shall comply as detailed. **11B-608.4, 11B-610.1** *608.4, 610.1*

> **EXCEPTION:** In residential dwelling units, seats shall not be required in shower compartments provided that reinforcement has been installed in walls so as to permit the installation of compliant seats.

_____P.   A seat in an alternate type roll-in shower compartment shall be a folding type, shall be installed on the front wall opposite the back wall, and shall extend from the adjacent side wall to a point within 3 inches of the compartment entry. **11B-610.3** *610.3* **Fig. CD-26HC**

_____Q.   The top of the seat shall be 17 inches minimum and 19 inches maximum above the bathroom finish floor. **11B-610.3** *610.3*

_____R.   When folded, the seat shall extend 6 inches maximum from the mounting wall. **11B-610.3** *610.3*

Rectangular Seats

_____S.   The rear edge of a rectangular seat shall be 2-1/2 inches maximum and the front edge 15 inches minimum and 16 inches maximum from the seat wall. **11B-610.3.1** *610.3.1* **Fig. CD-26HD**

_____T.   The side edge of the seat shall be 1-1/2 inches maximum from the adjacent wall. **11B-610.3.1**   *610.3.1*   **Fig. CD-26HD**

L-Shaped Seats

_____U.   The rear edge of an L-shaped seat shall be 2-1/2 inches maximum and the front edge 15 inches minimum and 16 inches maximum from the seat wall.   **11B-610.3.2**   *610.3.2*
Fig. CD-26HE

_____V.   The rear edge of the "L" portion of the seat shall be 1-1/2 inches maximum from the wall and the front edge shall be 14 inches minimum and 15 inches maximum from the wall. **11B-610.3.2**   *610.3.2*   **Fig. CD-26HE**

_____W.   The end of the "L" shall be 22 inches minimum and 23 inches maximum from the main seat wall. **11B-610.3.2**   *610.3.2*   **Fig. CD-26HE**

Structural Strength

_____X.   Allowable stresses shall not be exceeded for materials used when a vertical or horizontal force of 250 pounds is applied at any point on the seat, fastener, mounting device, or supporting structure.   **11B-610.4**   *610.4*

Controls

_____Y.   In alternate type roll-in type shower compartments, operable parts of controls and faucets shall be installed on the side wall of the compartment adjacent to the seat wall 19 inches minimum and 27 inches maximum from the seat wall; and shall be located above the grab bar, but no higher than 48 inches above the shower floor, with their centerline at 39 inches to 41 inches above the shower floor.   **11B-608.5.3**   *608.5.3*   **Fig. CD-26HI**

_____Z.   Operable parts of the shower spray unit, including the handle, shall be installed on the side wall adjacent to the seat wall 17 inches minimum and 19 inches maximum from the seat wall or on the back wall opposite the seat 15 inches maximum, left or right, of the centerline of the seat; and shall be located above the grab bar, but no higher than 48 inches above the shower floor.   **11B-608.5.3**   *608.5.3*   **Fig. CD-26HI**

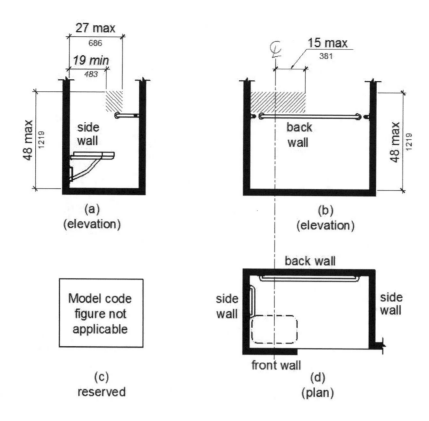

Fig. CD-HI
Alternate Roll-In Type Shower Compartment Control Location
© ICC Reproduced with Permission

Shower Spray Unit and Water

_____AA. A shower spray unit with a hose 59 inches long minimum that can be used both as a fixed position shower head and as a handheld shower shall be provided. **11B-608.6** *608.6*

_____BB. The shower spray unit shall have an on/off control with a non-positive shut-off. **11B-607.6** *607.6*

_____CC. If an adjustable-height shower head on a vertical bar is used, the bar shall be installed so as not to obstruct the use of grab bars. **11B-608.6** *608.6*

_____DD. Bathtub shower spray units shall deliver water that is 120 degrees Fahrenheit (49 degrees Celsius) maximum. **11B-608.6** *608.6*

> **EXCEPTION:** Where subject to excessive vandalism two fixed shower heads shall be installed instead of a handheld spray unit in facilities that are not transient lodging guest rooms. Each shower head shall be installed so it can be operated independently of the other and shall have swivel angle adjustments, both vertically and horizontally.

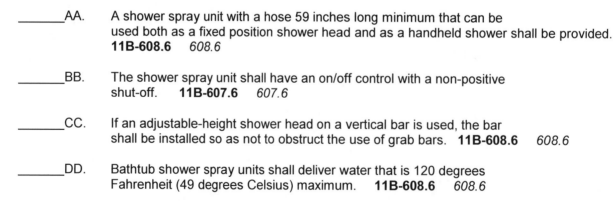

ADVISORY: Shower Spray Unit and Water. *Ensure that hand-held shower spray units are capable of delivering water pressure substantially equivalent to fixed shower heads.*

Shower Thresholds

_____EE.   Changes in level between 1/4" high minimum and 1/2" high maximum
are beveled with a slope not steeper than 1:2.   **11B-608.7, 11B-303.2**
608.7, 303.2 **Fig. CD-8B**

_____FF.   Changes in level of 1/4" high maximum are permitted to be vertical and
without edge treatment.   **11B-608.7, 11B-303.2**   *608.7, 303.2*
Fig. CD-8A

Shower Enclosures

_____GG.   Enclosures for shower compartments shall not obstruct controls, faucets,
and shower spray units or obstruct transfer from wheelchairs onto shower seats.
11B-608.8   *608.8*

Shower Floor or Ground Surface

_____HH.   Floor or ground surfaces of showers shall be stable, firm, and slip
resistant and shall be sloped 1:48 maximum in any direction.   Where drains are
provided, grate openings shall be ¼ inch maximum and flush with the floor surface.
11B-609.9, 11B-302.1

Soap Dish

_____II.   Where a soap dish is provided, it shall be located on the control wall at
40 inches maximum above the shower floor, and within the reach limits from the seat.
11B-608.10

27. <u>WASHING MACHINES AND CLOTHES DRYERS</u>

General
Where provided, washing machines and clothes dryers shall comply as detailed.
11B-214.1, 611.1   *214.1. 611.1*

Washing Machines
Where three or fewer washing machines are provided, at least one shall comply as detailed.   Where
more than three washing machines are provided, at least two shall comply as detailed.   **11B-214.2,
11B-611**   *214.2, 611*

Clothes Dryers
Where three or fewer clothes dryers are provided, at least one shall comply as detailed.   Where more
than three clothes dryers are provided, at least two shall comply as detailed.   **11B-214.3, 11B-611**
214.3, 611

Clear Floor Space

_____A.   A clear floor or ground space complying with Section 10, "CLEAR FLOOR OR
GROUND SPACE", positioned for parallel approach shall be provided. The clear floor
or ground space shall be centered on the appliance. **11B-611.2**   *611.2*

Operating Parts

_____B.   Operable parts, including doors, lint screens, and detergent and bleach compartments shall comply with Section 13, "OPERABLE PARTS". **11B-611.3** *611.3*

Height

_____C.   Top loading machines shall have the door to the laundry compartment located 36 Inches maximum above the finish floor. Front loading machines shall have the bottom of the opening to the laundry compartment located 15 inches minimum and 36 inches maximum above the finish floor. **11B-611.4** *611.4* **Fig CD-27A**

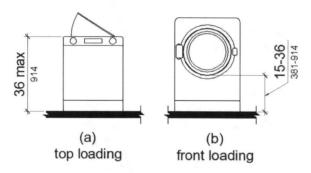

Fig. CD-27A
Height of Laundry Compartment Opening
© ICC Reproduced with Permission

28. <u>SAUNAS AND STEAM ROOMS</u>

<u>Saunas and Steam Rooms</u>   **11B-241**   *241*

General
Where provided, saunas and steam rooms shall comply as detailed. **11B-241** *241*

> **EXCEPTION:** Where saunas or steam rooms are clustered at a single location, no more than 5 percent of the saunas and steam rooms, but no fewer than one, of each type in each cluster shall be required to comply as detailed.

General
Saunas and steam rooms shall comply as detailed. **11B-612.1**   *612.1*

Bench

_____A.   Where seating is provided in saunas and steam rooms, at least one bench shall comply with Section 47, "BENCHES". Doors shall not swing into the clear floor space at the end of the bench seat.   **11B-612.2**   *612.2*

> **EXCEPTION:** A readily removable bench shall be permitted to obstruct the required turning space and the required clear floor or ground space.

Turning Space

_____B.   A turning space that complies with Section 9, "TURNING SPACE" shall be provided within saunas and steam rooms. **11B-612.3**   *612.3*

29. <u>FIRE ALARM SYSTEMS</u>

General
Where fire alarm systems provide audible alarm coverage, alarms shall comply as detailed. **11B-215.1** *215.1*

> **EXCEPTION:** In existing facilities, visible alarms shall not be required except where an existing fire alarm system is upgraded or replaced, or a new fire alarm system is installed.

> *Advisory: General. Unlike audible alarms, visible alarms must be located within the space they serve so that the signal is visible. Facility alarm systems (other than fire alarm systems) such as those used for tornado warnings and other emergencies are not required to comply with the technical criteria for alarms in this section. Every effort should be made to ensure that such alarms can be differentiated in their signal from fire alarms systems and that people who need to be notified of emergencies are adequately safeguarded. Consult local fire departments and prepare evacuation plans taking into consideration the needs of every building occupant, including people with disabilities.*

Public and Common Use Areas
Alarms in public use areas and common use areas shall comply as detailed. **11B-215.2**, **CBC §907.5.3.2.1**

Visible Alarms
Visible alarm notification appliances shall be provided in public use areas and common use areas, including, but not limited to:

1. Sanitary facilities including restrooms, bathrooms and shower rooms
2. Corridors
3. Music practice rooms
4. Band rooms
5. Gymnasiums
6. Multipurpose rooms
7. Occupational shops
8. Occupied rooms where ambient noise impairs hearing of the fire alarm
9. Lobbies
10. Meeting rooms
11. Classrooms. **CBC §907.5.2.3.1**

> **EXCEPTIONS:**
>
> 1. In other than Group I-2 and I-2.1, visible alarm notification appliances are not required in alterations, except where an existing fire alarm system is upgraded or replaced, or a new fire alarm system is installed. **CBC §907.5.2.3**
>
> 2. Visible alarm notification appliances shall not be required in enclosed exit stairways, exterior exit stairs, and exterior exit ramps. **CBC §907.5.2.3**
>
> 3. Visible alarm notification appliances shall not be required in elevator cars. **CBC §907.5.2.3**

Employee Work Areas
Where employee work areas have audible alarm coverage, the wiring system shall be designated so that compliant visible alarms can be integrated into the alarm system as detailed. **11B-215.3** *215.3*

Employee Work Areas

Where employee work areas have audible alarm coverage, the notification appliance circuits serving the employee work areas shall be initially designed with a minimum of 20 percent spare capacity to account for the potential of adding visible notification appliances in the future to accommodate hearing impaired employees.
CBC §907.5.2.3.2

Transient Lodging

Guest rooms required to have communication features shall provide alarms that comply as detailed.
11B-215.4 *215.4*

Groups R-1 and R-2.1. Group R-1 and R-2.1 dwelling units or sleeping units in accordance with Table 29A (below) shall be provided with a visible alarm notification appliance, activated by both the in-room smoke alarm and the building fire alarm system. **CBC §907.5.2.3.3**

TABLE 29A
VISIBLE AND AUDIBLE ALARMS

NUMBER OF SLEEPING UNITS	SLEEPING UNITS WITH VISIBLE AND AUDIBLE ALARMS
6 to 25	2
26 to 50	4
51 to 75	7
76 to 100	9
101 to 150	12
151 to 200	14
201 to 300	17
301 to 400	20
401 to 500	22
501 to 1,000	5% of total
1,001 and over	50 plus 3 for each 100 over 1,000

Residential Facilities

Where provided in residential dwelling units required to comply, alarms shall comply as detailed.
11B-215.5 *215.5*

Group R-2. In group R-2 occupancies required by CBC Section 907 to have a fire alarm system, all dwelling units and sleeping units shall be provided with the capability to support visible alarm notification appliances in accordance with NFPA 72. Such capability shall be permitted to include the potential for future interconnection of the building fire alarm system with the unit smoke alarms, replacement of audible appliances, or future extension of the existing wiring from the unit smoke alarm locations to required locations for visible appliances. **CBC §907.5.2.3.4**

General

Fire alarm systems shall have permanently installed audible and visible alarms complying with NFPA 72 and the sections below from CBC Chapter 9. **11B-702.1** *702.1*

EXCEPTION: Reserved.

Audible Alarms

Audible alarm notification appliances shall be provided and shall emit a distinctive sound that is not to be used for any purpose other than that of a fire alarm. In Group I-2 occupancies, audible appliances placed in patient areas shall be only chimes or similar sounding devices for alerting staff. See Section 907.6.5. **CBC §907.5.2.1**

EXCEPTIONS:

1.   Visible alarm notification appliances shall be allowed in lieu of audible alarm notification appliances in patient areas of Group I-2 occupancies. **CBC §907.5.2.1**

2.   Where provided, audible notification appliances located in each occupant evacuation elevator lobby in accordance with Section 3008.5.1 shall be connected to a separate notification zone for manual paging only. **CBC §907.5.2.1**

Average Sound Pressure

The audible alarm notification appliances shall be provide a sound pressure level of 15 decibels (dBA) above the average ambient sound level or 5 dBA above the maximum sound level having a duration of at least 60 seconds, whichever is greater, in every occupiable space within the building.   **CBC §907.5.2.1.1**

Maximum Sound Pressure

The maximum sound pressure level for audible alarm notification appliances shall be 110 dBA at the minimum hearing distance from the audible appliance. Where the average ambient noise is greater than 95 dBA, visible alarm notification appliances shall be provided in accordance with NFPA 72 and audible alarm notification appliances shall not be required. **CBC §907.5.2.1.2**

Audible Alarm Signal

The audible signal shall be the standard fire alarm evacuation signal, ANSI S3.41 Audible Emergency Evacuation Signal, "three pulse temporal pattern", as described in NFPA 72. **CBC §907.5.2.1.3**

EXCEPTION: The use of the existing evacuation signaling scheme shall be permitted where approved by the enforcing agency.

Manual Fire Alarm Boxes   CBC §907.4.2

Height

The height of the manual fire alarm boxes shall be a minimum of 42 inches (1067 mm) and a maximum of 48 inches (*1219* mm) measured vertically, from the floor level to the highest point of the activating handle or lever of the box. Manual fire alarm boxes shall be operable with one hand and shall not require tight grasping, pinching, or twisting of the wrist. The force required to activate operable parts shall not exceed 5 pounds (22.2 N) maximum.   **CBC §907.4.2.2**

EXCEPTION: [DSA-AC] In existing buildings there is no requirement to retroactively relocate existing manual fire alarm boxes to a minimum of 42 inches (1067 mm) and a maximum of 48 inches (1219 mm) from the floor level to the activating handle or lever of the box.

> ***ADVISORY: Height.*** *The requirements contained in Section 13 of this guidebook, titled "OPERABLE PARTS", requires controls and operating mechanisms to be operable with one hand without tight grasping, pinching or twisting of the wrist (must be operable by persons with limited manual dexterity). The maximum effort to activate controls shall be no greater than 5 pounds-force.*

30. <u>SIGNS</u>

<u>General</u>
New or altered signs shall be provided as detailed and shall comply. The addition of or replacement of signs shall not trigger any additional path of travel requirements. **11B-216.1** *216.1*

> **EXCEPTIONS:**
>
> 1. Building directories, menus, seat and row designations in assembly areas, occupant names, building addresses, and company names and logos shall not be required to comply.
>
> 2. In parking facilities, signs shall not be required to comply with the requirements for Designations, Directional and Informational Signs or for other improvements.
>
> 3. Temporary, 7 days or less, signs shall not be required to comply.
>
> 4. In detention and correctional facilities, signs not located in public use areas shall not be required to comply.

<u>Designations</u>
Interior and exterior signs identifying permanent rooms and spaces shall comply with the Section 30 requirements for SIGNS, RAISED CHARACTERS, BRAILLE and VISUAL CHARACTERS. Where pictograms are provided as designations of permanent rooms and spaces, the pictograms shall comply and shall have text descriptions with compliant RAISED AND VISUAL CHARACTERS. **11B-216.2** *216.2*

> **EXCEPTION:** Exterior signs that are not located at the door to the space they serve shall not be required to have compliant Raised Characters.

> *Advisory: Designations. This section applies to signs that provide designations, labels, or names for interior rooms and spaces where the sign is not likely to change over time. Examples include interior signs labeling restrooms, room and floor numbers or letters, and room names. Tactile text descriptors are required for pictograms that are provided to label or identify a permanent room or space. Pictograms that provided information about a room or space, such as "no smoking", occupant logos, and the International Symbol of Accessibility, are not required to have text descriptors.*
>
> *People with visual impairments benefit from tactile signs containing raised characters and/or accompanying Braille. They also benefit from an orderly scheme of consecutive room numbers for way-finding, though way-finding is not required by code. Tactile exit signs complying with Chapter 10 contribute to a safe environment for people with visual impairments.*

<u>Directional and Informational Signs</u>
Signs that provide direction to or information about interior spaces and facilities of the site shall have compliant Visual Characters. **11B-216.3** *216.3*

> *Advisory: Directional and Informational Signs. Information about interior and exterior spaces and facilities include rules of conduct, occupant load, and similar signs. Signs providing direction to rooms or spaces include those that identify egress routes.*

Means of Egress
Signs for means of egress shall comply as detailed. **11B-216.4**   *216.4*

Exit Doors
Signs required by CBC Chapter 10, Section 1011.4 at doors to passageways, exit discharge, and exit stairways shall comply with the General requirements for Signs, Raised Characters, Braille and Visual Characters. **11B-216.4.1**   *216.4.1*

Areas of Refuge and Exterior Areas for Assisted Rescue
Signs required by CBC Chapter 10, Section 1007.11 to provide instructions in areas of refuge shall comply with the requirements for Visual Signs. Signs required by CBC Chapter 10, Section 1007.9 at doors to areas of refuge and exterior areas for assisted rescue shall comply with the requirements for Visual Signs and include an International Symbol of Accessibility. **11B-216.4.2**   *216.4.2*

Directional Signs
Signs required by CBC Chapter 10, Section 1007.10 to provide directions to accessible means of egress shall comply as detailed. **11B-216.4.3**   *216.4.3*

Delayed Egress Locks
Signs required by CBC Chapter 10, Section 1008.1.9.7, Item 5.1 at doors with delayed egress locks shall comply with the General requirements for Signs, Raised Characters, Braille and Visual Characters. **11B-216.4.4**   *216.4.4*

Parking
Parking spaces that comply with the requirements of Section 21, "PARKING SPACES", shall be identified by compliant signage. **11B-216.5**   *216.5*

EXCEPTIONS:

1.   **Reserved**.

2.   In residential facilities, where parking spaces are assigned to specific residential dwelling units, identification of accessible parking spaces shall not be required.

Entrances
In existing buildings and facilities where not all entrances comply with Section 14, "DOORS, DOORWAYS AND GATES", entrances that comply with Section 14, "DOORS, DOORWAYS AND GATES" shall be identified by the International Symbol of Accessibility. Directional signs complying with the requirements for accessible Visual Characters shall be provided at entrances that do not comply. Directional signs with accessible Visual Characters, Including the International Symbol of Accessibility, indicating the accessible route to the nearest accessible entrance shall be provided at junctions when the accessible route diverges from the regular circulation path.
11B-216.6   *216.6*

EXCEPTIONS:

1.   An International Symbol of Accessibility is not required at entrances to individual rooms, suites, offices, sales or rental establishments, or other such spaces when all entrances to the building or facility are accessible and persons entering the building or facility have passed through one or more entrances with signage complying with this section.

2.   An International Symbol of Accessibility is not required at entrances to machinery spaces frequented only by service personnel for maintenance, repair, or occasional monitoring of equipment; for

example, elevator pits or elevator penthouses; mechanical, electrical or communications equipment rooms; piping or equipment catwalks; electric substations and transformer vaults; and highway and tunnel utility facilities.

Advisory: Entrances. Where a directional sign is required, it should be located to minimize backtracking. In some cases, this could mean locating a sign at the beginning of a route, not just at the inaccessible entrances to a building.

Directional signs are needed where the accessible route diverges from the route for the general public and should be located at decision points (for example where the path to the stairs diverges from the path to an elevator or ramp). Directional signs are not needed where paths are equal and/or readily apparent. The signage program should be designed to consider differing uses of a facility which occur at different times of the day. For example, portions of a facility may be closed in the evening; appropriate signage should be provided to give adequate direction during the hours of use in addition to the typical operational hours.

The sign program should be designed to provide the appropriate level of signage at points necessary for convenient navigation around the site. Too many signs can be confusing to everyone utilizing the site.

Elevators
Where existing elevators do not comply with Section 17, "ELEVATORS", elevators that do comply shall be clearly identified with the International Symbol Accessibility. Existing buildings that have been remodeled to provide specific elevators for public use that comply with these buildings standards shall have the location of and the directions to these elevators posted in the building lobby on a sign complying with accessible Visual Characters, including the International Symbol of Accessibility. **11B-216.7** *216.7*

Toilet Rooms and Bathing Rooms
Doorways leading to accessible toilet rooms and bathing rooms shall be identified with the appropriate geometric symbol. Where existing toilet rooms or bathing rooms are NOT accessible, directional signs indicating the location of the nearest toilet room or bathing room that IS accessible within the facility shall be provided. These signs shall contain the International Symbol of Accessibility and compliant Visual Characters. Where a facility contains both accessible and inaccessible toilet or bathing rooms, the accessible toilet or bathing rooms shall be identified by the International Symbol of Accessibility. Where clustered single user toilet rooms or bathing units are clustered at a single location, and there are both accessible and inaccessible toilet or bathing units, the accessible toilet or bathing units shall be identified by the International Symbol of Accessibility. Existing buildings that have been remodeled to provide specific toilet rooms or bathing rooms for public use that are accessible shall have the location of and the directions to these rooms posted in or near the building lobby on a sign containing accessible Visual Characters and the International Symbol of Accessibility. **11B-216.8** *216.8*

TTYs
Identification and directional signs for public TTYs shall be provided as detailed. **11B-216.9** *216.9*

Identification Signs
Public TTYs shall be identified by the International Symbol of TTY that complies with Item LL of this Section, below. **11B-216.9.1** *216.9.1*

Directional Signs
Directional signs indicating the location of the nearest public TTY shall be provided at all banks of public pay telephones not containing a public TTY. In addition, where signs provide direction to public pay telephones, they shall also provide direction to public TTYs. If a facility has no banks of telephones, the directional signage shall be provided at the entrance or in a building

---

directory. Directional signs shall comply with the requirements for accessible Visual Characters, and shall include the International Symbol of Access for Hearing Loss that complies with Item LL of this Section, below.   **11B-216.9.2**   *216.9.2*

Assistive-Listening Systems

Each assembly area required to provide assistive listening systems shall provide signs informing patrons of the availability of the assistive listening system. The sign shall include wording that states "Assistive-Listening System Available" and shall be posted in a prominent place at or near the assembly area entrance. Assistive listening signs shall comply with the requirements for accessible Visual Characters and shall include the International "Symbol of Access for Hearing Loss that complies with Item MM of this Section, below.   **11B-216.10**   *216.10*

> *Advisory: Assistive Listening Systems. The term "prominent place" means a place that arriving persons would easily notice. It is helpful, though not required, to identify the location or person to contact for obtaining the system on the sign. Note that a tactile sign is not required by this section.*

> **EXCEPTION:** Where ticket offices or windows are provided, signs shall not be required at each assembly area provided that signs are displayed at each ticket office or window informing patrons of the availability of assistive listening systems.

Check-Out Aisles

Where more than one check-out aisle is provided, check-out aisles that comply with the requirements for accessible check-out aisles in Section 48, "SALES AND SERVICE", shall be identified with a sign clearly visible to a person in a wheelchair displaying the International Symbol of Accessibility. The sign shall be a minimum of 4 inches by 4 inches. Where check-out aisles are identified by numbers, letters, or functions, signs identifying accessible check-out aisles shall display the International Symbol of Accessibility in the same location as the check-out aisle identification.
11B-216.11, 11B-904.3.4   *216.11*

> **EXCEPTION:** Where all check-out aisles are accessible, signs identifying the accessible checkout aisles shall not be required.

Amusement Rides

Signs identifying the type of access provided on amusement rides shall be provided at entries to queues and waiting lines. In addition, where accessible unload areas also serve as accessible load areas, signs indicating the location of the accessible load and unload areas shall be provided at entries to queues and waiting lines. Signs shall comply with the requirements for accessible Visual Characters, and shall include the International Symbol of Accessibility.   **11B-216.12**   *216.12*

> *Advisory: Amusement Rides. Amusement rides designed primarily for children, amusement rides that are controlled or operated by the rider, and amusement rides without seats, are not required to provide wheelchair spaces, transfer seats, or transfer systems, and need not meet the sign requirements detailed. The load and unload areas of these rides must, however, be on an accessible route and provide turning space.*

Cleaner Air Symbol

Use of the cleaner air symbol is **VOLUNTARY**. Where publicly funded facilities or any facilities leased or rented by the State of California, not concessionaires, comply with the Conditions of Use identified below, a compliant Cleaner Air Symbol is permitted to be posted as detailed to indicate rooms, facilities, and paths of travel that are accessible to and usable by people who are adversely impacted by airborne chemicals or particulate(s) and/or the use of electrical fixtures and/or devices.   **11B-216.13.1**

Removal of Cleaner Air Symbol

If the path of travel, room and/or facility identified by the Cleaner Air Symbol should temporarily or permanently cease to meet the minimum conditions as set forth above, the cleaner air symbol shall be removed and shall not be replaced until the minimum conditions are again met. **11B-216.13.2**

Conditions of Use

1. Floor or wall coverings, floor or wall covering adhesives, carpets, formaldehyde-emitting particleboard cabinetry, cupboards or doors have not been installed or replaced in the previous 12 months.

2. Incandescent lighting provided in lieu of fluorescent or halogen lighting, and electrical systems and equipment shall be operable by or on behalf of the occupant or user of the room, facility or path of travel.

3. Heating, ventilation, air conditioning and their controls shall be operable by or on behalf of the occupant or user.

4. To maintain "cleaner-air" designation only nonirritating, nontoxic products will be used in cleaning, maintenance, disinfection, pest management or for any minimal touch-ups that are essential for occupancy of the area. Deodorizers or Fragrance Emission Devises and Systems (FEDS) shall not be used in the designated area. Pest control practices for cleaner-air areas shall include the use of bait stations using boric acid, sticky traps and silicon caulk for sealing cracks and crevices. Areas shall be routinely monitored for pest problems. Additional nontoxic treatment methods, such as temperature extremes for termites, may be employed in the event of more urgent problems. These pest control practices shall not be used 48 hours prior to placement of the sign, and the facility shall be ventilated with outside air for a minimum of 24 hours following use or application.

5. Signage shall be posted requesting occupants or users not to smoke or wear perfumes, colognes or scented personal care products. Fragranced products shall not be used in the designated cleaner-air room, facility or path of travel.

6. A log shall be maintained on site, accessible to the public either in person or by telephone, e-mail, fax or other accessible means as requested. One or more individuals shall be designated to maintain the log. The log shall record any product or practice used in the cleaner-air designated room, facility or path of travel, as well as scheduled activities, that may impact the Cleaner-Air designation. The log shall also include the product label as well as the material Safety Data Sheets (MSDS). **11B-216.13.3**

General Sign Requirements
Signs shall comply as detailed. Where both visual and tactile characters are required, either one sign with both visual and tactile characters, or two separate signs, one with visual, and one with tactile characters, shall be provided. **11B-703.1**   *703.1*

Plan Review and Inspection
Signs as specified in this Section, or in other sections of this code, when included in the construction of new buildings or facilities, or when included, altered or replaced due to additions, alterations or renovations to existing buildings or facilities, and when a permit is required, shall comply as follows: **11B-703.1.1**

Plan Review
Plans, specifications or other information indicating compliance with these regulations shall be submitted to the enforcing agency for review and approval. **11B-703.1.1.1**

Inspection
Signs and identification devices shall be field inspected after installation and approved by the enforcing agency prior to the issuance of a final certificate of occupancy per Chapter 1, Division II, Section 111, or final approval where no certificate of occupancy is issued. The inspection shall include, but not be limited to, verification that Braille dots and cells are properly spaced and the size, proportion and type of raised characters are in compliance with these regulations. **11B-703.1.1.2**

30A. RAISED CHARACTERS

Raised characters shall comply as detailed and shall be duplicated in Braille that complies as detailed. Raised characters shall be installed as detailed. **11B-703.2** *703.2*

> ***Advisory: Raised Characters.*** *Signs that are designed to be read by touch should not have sharp or abrasive edges.*

Depth

_____A. Raised characters shall be 1/32 inch minimum above their background. **11B-703.2.1** *703.2.1*

Case

_____B. Characters shall be uppercase. **11B-703.2.2** *703.2.2*

Style

_____C. Characters shall be sans serif. Characters shall not be italic, oblique, script, highly decorative, or other unusual forms. **11B-703.2.3** *703.2.3*

Character Proportions

_____D. Characters shall be selected from fonts where the width of the uppercase letter "O" is 60 percent minimum and 110 percent maximum of the height of the uppercase letter "I". **11B-703.2.4** *703.2.4*

Character Height

_____E. Character height measured vertically from the baseline of the character shall be 5/8 inch minimum and 2 inches maximum based on the height of the uppercase letter "I". **11B-703.2.5** *703.2.5* **Fig. CD-30A**

EXCEPTION: Reserved.

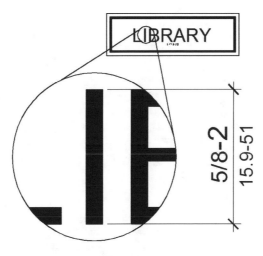

Fig. CD-30A
Height of Raised Characters
© ICC Reproduced with Permission

Stroke Thickness

_____ F. Stroke thickness of the uppercase letter "I" shall be 15 percent maximum of the height of the character. **11B-703.2.6** *703.2.6*

Character Spacing

Character spacing shall be measured between the two closest points of adjacent raised characters within a message, excluding word spaces.

_____ G. Where characters have rectangle cross sections, spacing between individual raised characters shall be 1/8 inch minimum and 4 times the raised characters stroke width maximum. **11B-703.2.7** *703.2.7*

_____ H. Where characters have other cross sections, spacing between individual raised characters shall 1/16 inch minimum and 4 times the raised character stroke width maximum at the base of the cross sections, and 1/8 inch minimum and 4 times the raised character stroke width maximum at the top of the cross sections. **11B-703.2.7** *703.2.7*

_____ I. Characters shall be separated from raised borders and decorative elements 3/8 inch minimum **11B-703.2.7** *703.2.7*

Line Spacing

_____ J. Spacing between the baselines of separate lines of raised characters within a message shall be 135 percent minimum and 170 percent maximum of the raised character height. **11B-703.2.8** *703.2.8*

Format

_____ K. Text shall be in a horizontal format. **11B-703.2.9** *703.2.9*

30B. BRAILLE

BRAILLE

Braille shall be contracted (Grade 2) and shall comply as detailed. **11B-703.3** *703.3*

> *Advisory: Braille. Contracted Braille uses special characters called contractions to make words shorter. Standard English uses contractions like "don't" as a short way of writing two words, such as "do" and "not". In Braille there are many additional contractions. Some contractions stand for a whole word and other contractions stand for a group of letters within a word. In addition to contractions, the Braille code includes short-form words which are abbreviated spellings of common longer words. For example, "tomorrow" is spelled "tm", "friend" is spelled "fr", and "little" is spelled "ll" in Braille.*

Dimensions and Capitalization

_____ A. Braille dots shall have a domed or rounded shape and shall comply with Table 30B. The indication of an uppercase letter or letters shall only be used before the first word of sentences, proper nouns and names, individual letters of the alphabet, initials, acronyms. **11B.703.3.1** *703.3.1* **Fig. CD-30BA**

TABLE 30B
BRAILLE DIMENSIONS

Measurement Range	Minimum in Inches Maximum in Inches
Dot base diameter	0.059 (1.5 mm) to 0.063 (1.6 mm)
Distance between two dots in the same cell[1]	0.100 (2.5 mm)
Distance between corresponding dots in adjacent cells[1]	0.300 (7.6 mm)
Dot height	0.025 (0.6 mm) to 0.037 (0.9 mm)
Distance between corresponding dots from one cell directly below[1]	0.395 (10 mm) to 0.400 (10.2 mm)

1. Measured center to center.

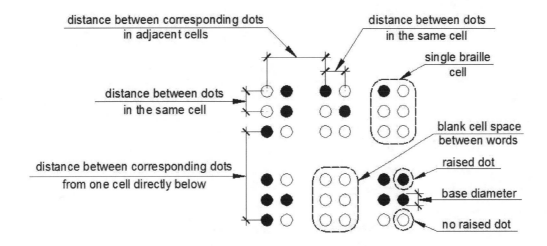

Fig. CD-30BA
Braille Measurement
© ICC Reproduced with Permission

Position

_____ B. Braille shall be positioned below the corresponding text in a horizontal format, flush left or centered. **11B-703.3.2** *703.3.2* **Fig. CD-30BC**

_____ C. If text is multi-lined, Braille shall be placed below the entire text. Braille shall be separated 3/8 inch below the entire text **11B-703.3.2** *703.3.2* **Fig. CD-30BC**

_____ D. Braille shall be separated 3/8 inch minimum and ½ inch maximum from any other tactile characters and 3/8 inch minimum from raised borders and decorative elements. **11B-703.3.2** *703.3.2* **Fig. CD-30BC**

EXCEPTION: Braille provided on elevator car controls shall be separated 3/16 inch and shall be located either directly below the corresponding raised characters or symbols.

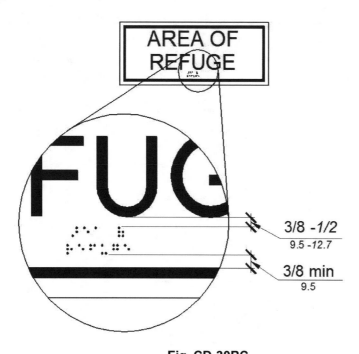

Fig. CD-30BC
Position of Braille
© ICC Reproduced with Permission

30C. INSTALLATION HEIGHT AND LOCATION

Installation Height and Location
Signs with tactile characters shall comply as detailed. **11B-703.4**   *703.4*

Height Above Finish Floor or Ground

_____A.   Tactile characters on signs shall be located 48 inches above the finish floor or ground surface, measured from the baseline of the lowest Braille cells and 60 inches maximum above the finish floor or ground surface, measured from the baseline of the highest line of raised characters. **11B-703.4.1**   *703.4.1* **Fig. CD-30CA**

   EXCEPTION: Tactile characters for elevator car controls shall not be required to comply with this provision.

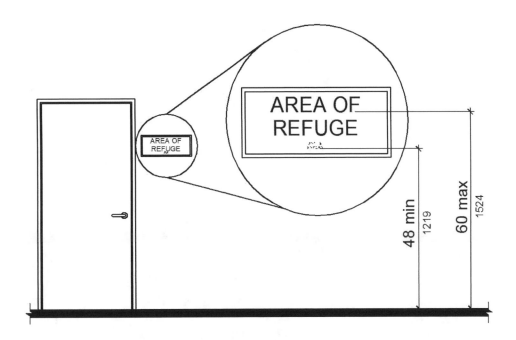

Fig. 30CA
Height of Tactile Characters Above Finish Floor or Ground
© ICC Reproduced with Permission

Location

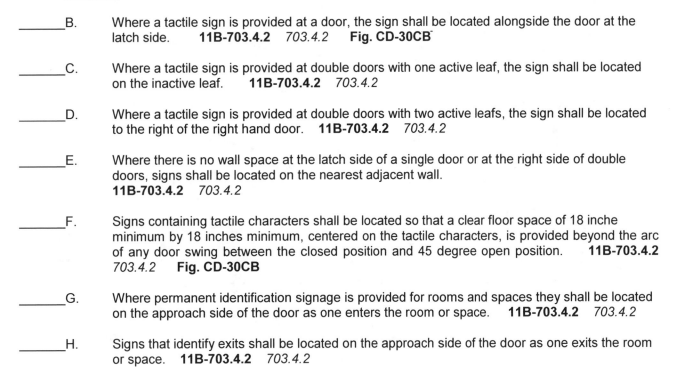

_____ B. Where a tactile sign is provided at a door, the sign shall be located alongside the door at the latch side. **11B-703.4.2** *703.4.2* **Fig. CD-30CB**

_____ C. Where a tactile sign is provided at double doors with one active leaf, the sign shall be located on the inactive leaf. **11B-703.4.2** *703.4.2*

_____ D. Where a tactile sign is provided at double doors with two active leafs, the sign shall be located to the right of the right hand door. **11B-703.4.2** *703.4.2*

_____ E. Where there is no wall space at the latch side of a single door or at the right side of double doors, signs shall be located on the nearest adjacent wall.
11B-703.4.2 *703.4.2*

_____ F. Signs containing tactile characters shall be located so that a clear floor space of 18 inch minimum by 18 inches minimum, centered on the tactile characters, is provided beyond the arc of any door swing between the closed position and 45 degree open position. **11B-703.4.2** *703.4.2* **Fig. CD-30CB**

_____ G. Where permanent identification signage is provided for rooms and spaces they shall be located on the approach side of the door as one enters the room or space. **11B-703.4.2** *703.4.2*

_____ H. Signs that identify exits shall be located on the approach side of the door as one exits the room or space. **11B-703.4.2** *703.4.2*

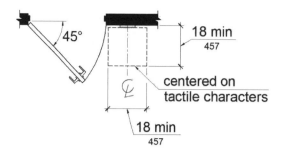

Fig. 30CB
Location of Tactile Signs at Doors
© ICC Reproduced with Permission

> **Advisory: Location.** *Persons with visual impairments are trained to look in a consistent location for tactile signs. When a tactile sign is provided at a door, the sign must be located alongside the door, preferably at the latch side. When tactile signs are mounted inconsistently on surrounding wall surfaces, or mounted on the door itself, they are difficult to find and may create a safety hazard. In this section, the term "wall space" refers to any type of partition assembly or adjacent surface of sufficient dimension where a sign can lay flat, including glass partitions or sidelites.*

30D. <u>VISUAL CHARACTERS</u>

<u>VISUAL CHARACTERS</u>
Visual characters shall comply as detailed.   **11B-703.5**   *703.5*

> **EXCEPTION:** Where visual characters comply with the requirements for Raised Lettering and are accompanied by compliant Braille, they shall not be required to comply with Items B through F, H and I of this section.

<u>Finish and Contrast</u>

_____A.   Characters and their background shall have a non-glare finish.

_____A1.   Characters shall contrast with their background with either light characters on a dark background or dark characters on a light background.   **11B-703.5.1**   *703.5.1*

> **Advisory: Finish and Contrast.** *Signs are more legible for persons with low vision when characters contrast as much as possible with their background. Additional factors affecting the ease with which the text can be distinguished from its background include shadows cast be lighting source, surface glare, and the uniformity of the text and its background colors and textures.*

<u>Case</u>

_____B.   Characters shall be uppercase and lowercase or a combination of both.   **11B-703.5.2**   *703.5.2*

<u>Style</u>

_____C.   Characters shall be conventional in form. Characters shall not be italic, oblique, script, highly decorative, or of other unusual forms. **11B-703.5.3**   *703.5.3*

Character Proportions

_____D.   Characters shall be selected from fonts where the width of the uppercase letter "O" is 60 percent minimum and 110 percent maximum of the height of the uppercase letter "I". **11B-703.5.4**   *703.5.4*

Character Height

_____E.   Minimum character height shall comply with Table 30B. Viewing distance shall be measured as the horizontal distance between the character and an obstruction preventing further approach towards the sign. Character height shall be based on the uppercase letter "I".   **11B-703.5.5** *703.5.5*

> **EXCEPTION:**   Where provided, floor plans providing emergency procedures information in accordance with CCR Title 19 shall not be required to comply with minimum character heights.

TABLE 30B
VISUAL CHARACTER HEIGHT

Height to Finish Floor or Ground From Baseline of Character	Horizontal Viewing Distance	Minimum Character Height
40 inches (*1016* mm) to less than or equal to 70 inches (*1778* mm)	less than 72 inches (*1829* mm)	5/8 inch (*15.9* mm)
	72 inches (*1829* mm) and greater	5/8 inch (*15.9* mm), plus 1/8 inch (*3.2* mm) per foot (*305* mm) of viewing distance above 72 inches (*1829* mm)
Greater than 70 inches (*1778* mm) to less than or equal to 120 inches (*3048* mm)	less than 180 inches (*4572* mm)	2 inches (*51* mm)
	180 inches (*4572* mm) and greater	2 inches (*51* mm), plus 1/8 inch (*3.2* mm) per foot (*305* mm) of viewing distance above 180 inches (*4572* mm)
Greater than 120 inches (*3048* mm)	less than 21 feet (*6401* mm)	3 inches (*76* mm)
	21 feet (*6401* mm) and greater	3 inches (*76* mm), plus 1/8 inch (*3.2* mm) per foot (*305* mm) of viewing distance above 21 feet (*6401* mm)

Height from Finish Floor or Ground

_____F.   Visual characters shall be 40 inches minimum above the finish floor or ground. **11B-703.5.6**   *703.5.6*

> **EXEPTIONS:**
>
> 1.   Visual characters indicating elevator car controls shall not be required to comply with this provision.
>
> 2.   Floor-level exit signs complying with CBC Chapter 10, Section 1011.7 shall not be required to comply with this provision.

3.   Where provided, floor plans providing emergency procedures information in accordance with Title 19 shall not be required to comply with this provision.

Stroke Thickness

_____G.   Stroke thickness of the uppercase letter "I" shall be 10 percent minimum and 20 percent maximum of the height of the character.   **11B-703.5.7**   *703.5.7*

Character Spacing

_____H.   Character spacing shall be measured between the two closest points of adjacent characters, excluding word spaces. Spacing between individual characters shall be 10 percent minimum and 35 percent maximum of character height.
11B-703.5.8   *703.5.8*

Line Spacing

_____I.   Spacing between the baselines of separate lines of characters within a message shall be 135 percent minimum and 170 percent maximum of the character height.   **11B-703.5.9**   *703.5.9*

Format

_____J.   Text shall be in a horizontal format.   **11B-703.5.10**   *703.5.10*

30E. PICTOGRAMS

Pictograms
Pictograms shall comply as detailed. **11B-703.6**   *703.6*

> *Advisory: **Pictograms.** Pictograms and other symbols, such as the International Symbol of Accessibility (ISA), which are included on signs with raised characters and Braille are not required to be raised. The ISA. When included on a tactile sign, does not require any accompanying text, either visual or tactile.*

Pictogram Field

_____A.   Pictograms shall have a field height of 6 inches minimum.   **11B-703.6.2**   *703.6.2*

_____A1.   Characters and Braille shall not be located in the pictogram field.
11B-703.6.1   *703.6.1* **Fig. CD-30EA**

> *Advisory: **Pictogram Field.** Pictograms include both a symbol and the field (or background) on which it is displayed. The 6-inch vertical dimension applies to the field, not the symbol. The symbol may be smaller than the field. The required equivalent verbal description must be placed below the pictogram, and may not intrude into the 6-inch field.*

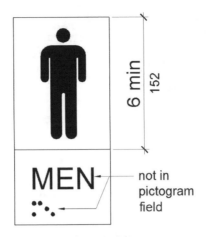

Fig. CD-30EA
Pictogram Field
© ICC Reproduced with Permission

Finish and contrast

_____B.   Pictograms and their field shall have a non-glare finish. **11B-703.6.2**   *703.6.2*

_____B1.   Pictograms shall contrast with their field with either a light pictogram on a dark field or a dark pictogram on a light field.   **11B-703.6.2**   *703.6.2*

> *Advisory: Finish and Contrast. Signs are more legible for persons with low vision when characters contrast as much as possible with their background. Additional factors affecting the ease with which the text can be distinguished from its background include shadows cast be lighting source, surface glare, and the uniformity of the text and its background colors and textures.*

Text Descriptors

_____C.   Pictograms shall have text descriptors located directly below the pictogram field. Text descriptors shall comply with the requirements for Raised Characters, Braille, and Installation Height and Location.   **11B-703.6.3**   *703.6.3*

30F. <u>SYMBOLS OF ACCESSIBILITY</u>

Symbols of Accessibility
Symbols of accessibility shall comply as detailed. **11B-703.7**   *703.7*

> *Advisory: Symbols of Accessibility. Symbols of accessibility, such as the International Symbol of Accessibility (ISA), which are included on signs with raised characters and Braille are not required to be raised. The ISA. When included on a tactile sign, does not require any accompanying text, either visual or tactile.*

Finish and Contrast

_____A.   Symbols of accessibility and their background shall have a non-glare finish. Symbols of accessibility shall contract with their background with either a light symbol on a dark background or a dark symbol on a light background.   **11B-703.7.1**   *703.7.1*

> **Advisory: Finish and Contrast.** *Signs are more legible for persons with low vision when characters contrast as much as possible with their background. Additional factors affecting the ease with which the text can be distinguished from its background include shadows cast be lighting source, surface glare, and the uniformity of the text and its background colors and textures.*

Symbols
11B-703.7.2 *703.7.2*

International Symbol of Accessibility

_____KK.   The International Symbol of Accessibility shall comply with Fig. CD-30FA. The symbol shall consist of a white figure on a blue background. The blue shall be Color No. 15090 in Federal Standard 595C. **11B-703.7.2.1** *703.7.2.1* **Fig. CD-30FA**

> **EXCEPTION:** The appropriate enforce agency may approve other colors to complement décor or unique design. The symbol contrast shall be light on dark or dark on light.

Fig. CD-30FA
International Symbol of Accessibility
© ICC Reproduced with Permission

> **Advisory: International Symbol of Accessibility Exception.** *This exception includes latitude in the use of other colors, but artistic license in the graphic representation of the symbol itself is not permitted.*

International Symbol of TTY

_____LL.   The International Symbol of TTY. The International Symbol of TTY shall comply with Fig. CD-30H. **11B-703.7.2.2** *703.7.2.2* **Fig. CD-30FB**

Fig, CD-30FB
International Symbol of TTY
© ICC Reproduced with Permission

<u>**Volume Control Telephones**</u>

_____MM.   Telephones with a volume control shall be identified by a pictogram of a telephone handset with radiating sound waves on a square field such as shown in Fig. CD-30FC.   **11B-703.7.2.3**   *703.7.2.3*   **Fig. CD-30FC**

Fig. CD-30FC
Volume Control Telephone
© ICC Reproduced with Permission

<u>**Assistive Listening Systems**</u>

_____NN.   Assistive listening systems shall be identified by the International Symbol for Hearing Loss complying with Fig. CD-30FD. **11B-703.7.2.4**   *703.7.2.4*
Fig. CD-30FD

Fig. CD-30FD
International Symbol of Access for Hearing Loss
© ICC Reproduced with Permission

<u>**Cleaner Air Symbol**</u>
Rooms, facilities and paths of travel that are accessible to and usable by people who are adversely impacted by airborne chemicals or particulate(s) and/or the use of electrical fixtures and/or devices shall be identified by the Cleaner Air Symbol complying with Fig. CD-30FE. This symbol is to be used strictly for publicly funded facilities or any facilities leased or rented by State of California, not concessionaires. **11B-703.7.2.5**

The symbol, which shall include the text "Cleaner Air" as shown, shall be displayed either as a negative or positive image within a square that is a minimum of 6 inches on each side. The symbol may be shown in black and white or in color. When color is used, it shall be Federal Blue (Color No. 15090 Federal Standard 595C) on white, or white on Federal Blue. There shall be at least 70-percent color contrast between the background of the sign from the surface that it is mounted on.   **11B-703.7.2.5**
Fig. CD-30FE

Fig, CD-30FE
Cleaner Air Symbol
© ICC Reproduced with Permission

30G. TOILET AND BATHING FACILITY SIGNS

Toilet and Bathing Facilities Geometric Symbols
Doorways leading to accessible toilet rooms and accessible bathing rooms shall be identified by the appropriate geometric symbol sign detailed herein. **11B-216.8**
Fig. CD-30GA

> **EXCEPTION:** Geometric symbols shall not be required at inmate toilet rooms and bathing rooms in detention and correctional facilities where only one gender is housed. **11B-703.7.2.6**

> *Advisory: Toilet and Bathing Facilities Geometric Symbols. There is no requirement for providing gender pictograms in combination with the geometric identification symbols required at doorways leading to men's, women's and unisex toilet and bathing facilities.*
>
> *When toilet and bathing facilities have doorway openings instead of doors, such as at airports or stadiums, the geometric identification symbol should be located at the proper height adjacent to the opening or incorporated into the required tactile identification sign. For example, the geometric symbol may be used as the sign background with raised characters and Braille.*

MEN'S TOILET AND BATHING FACILITIES

_____A.   Men's toilet and bathing facilities are identified by an equilateral triangle ¼ inch thick with edges 12 inches long and a vertex pointing upward. The triangle symbol contrasts with the door, either light on a dark background or dark on a light background. **11B-703.7.2.6.1**
Fig. CD-30GA

> **EXCEPTION:** Within secure perimeter of detention and correctional facilities, geometric symbols are not required to be ¼ inch thick.

WOMEN'S TOILET AND BATHING FACILITIES

_____B.   Women's toilet and bathing facilities are identified by a circle, ¼ inch thick and 12 inches in diameter. The circle symbol contrasts with the door, either light on a dark background or dark on a light background. **11B-703.7.2.6.2   Fig. CD-30GA**

> **EXCEPTION:** Within secure perimeter of detention and correctional facilities, geometric symbols are not required to be ¼ inch thick.

UNISEX TOILET AND BATHING FACILITIES

_____C.   Unisex toilet and bathing facilities are identified by a circle, ¼ inch thick and 12 inches in diameter with a ¼ inch thick triangle with a vertex pointing upward superimposed on the circle and within the 12 inch diameter. The triangle symbol contrasts with the circle symbol, either light on a dark background or dark on a light background. The circle symbol contrasts with the door, either light on a dark background or dark on a light background. **11B-703.7.2.6.3 Fig. CD-30GA**

> **EXCEPTION:** Within secure perimeter of detention and correctional facilities, geometric symbols are not required to be ¼ inch thick.

MOUNTING LOCATION

_____D.   The geometric symbol is mounted at 58 inches minimum and 60 inches maximum above the finish floor or ground surface measured from the centerline of the symbol. **11B-703.7.2.6 Fig. CD-30GA**

_____E.   Where a door is provided the symbol is mounted within 1 inch of the vertical centerline of the door. **11B-703.7.2.6**

EDGES AND CORNERS

_____F.   The edges of the signs are rounded, chamfered or eased, and any corners have a minimum radius of 1/8 inch. **11B-703.7.2.6.4**

DIRECTIONAL SIGNS AND INTERNATIONAL SYMBOL OF ACCESSIBILITY

IDENTIFICATION OF ACCESSIBLE FACILITIES

_____G.   Where a facility contains BOTH accessible AND inaccessible toilet or bathing rooms, the accessible toilet or bathing rooms shall be identified by the INTERNATIONAL SYMBOL OF ACCESSIBILITY, complying with Section 30F, "SYMBOLS OF ACCESSIBILITY". **11B-216.8** *216.8*

DIRECTIONAL SIGNS TO INACCESSIBLE FACILIITIES

_____H.   Where existing toilet rooms or bathing rooms are NOT accessible, directional signs indicating the location of the nearest toilet room or bathing room that IS accessible within the facility are provided. These signs contain the INTERNATIONAL SYMBOL OF ACCESSIBILITY, complying with Section 30F, "SYMBOLS OF ACCESSIBILITY. **11B-216.8** *216.8*

_____I.   Existing buildings that have been remodeled to provide specific toilet rooms or bathing rooms for public use that are accessible shall have the location of, and the directions to these rooms posted in or near the building lobby or entrance on a sign containing accessible VISUAL CHARACTERS, per Section 30D, and the INTERNATIONAL SYMBOL OF ACCESSIBILITY, complying with Section 30F, "SYMBOLS OF ACCESSIBILITY. **11B-216.8** *216.8*

INTERNATIONAL SYMBOL OF ACCESSIBILITY

_____J.   The International Symbol of Accessibility consists of a white figure on a blue background. The blue shall be equal to Color No. 15090 in Federal Standard 595C. **11B-216.8, 11B-703.7.2.1** *216.8, 703.7.2.1*

> **EXCEPTION:** The appropriate enforcement agency may approve other colors to compliment decor or unique design. The symbol contrast shall be light on dark or dark on light. **11B-703.7.2.1**

DESIGNATIONS

Interior and exterior signs identifying permanent rooms and spaces shall comply with the Section 30 requirements for SIGNS, RAISED CHARACTERS, BRAILLE and VISUAL CHARACTERS. Where pictograms are provided as designations of permanent rooms and spaces, the pictograms shall comply and shall have text descriptions with compliant RAISED AND VISUAL CHARACTERS. **11B-216.2** *216.2* **Fig. CD-30GA**

> **EXCEPTION:** Exterior signs that are not located at the door to the space they serve shall not be required to have compliant Raised Characters.

WALL–SIGNAGE (TYP.)

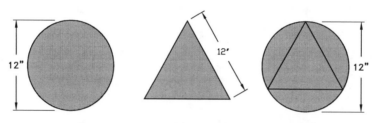

DOOR MOUNTED SIGNAGE (TYP.)

NOTE: PICTOGRAMS AND/OR LETTERING ARE NOT REQUIRED ON DOOR–MOUNTED SIGNAGE.

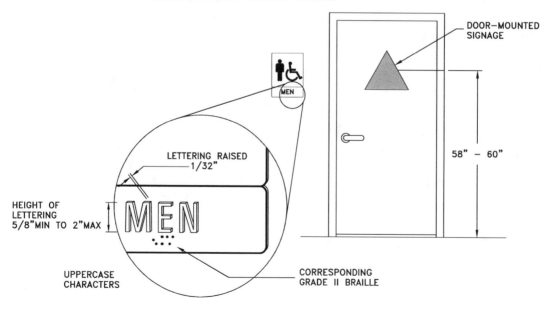

Fig, CD-30GA
Toilet and Bathing Facility Signs

30H. PEDESTRIAN TRAFFIC-CONTROL BUTTONS

Pedestrian Traffic-Control Buttons

_____A. Pole-supported pedestrian traffic-control buttons shall be identified with color coding consisting of a textured horizontal yellow band 2 inches (51 mm) in width encircling the pole, and a 1-inch-wide (25 mm) dark border band above and below this yellow band. Color coding shall be placed immediately above the control button. **11B-703.7.2.7   Fig. CD-30HA**

_____A1. Control buttons shall be located no higher than 48 inches (1219 mm) above the ground surface adjacent to the pole.   **11B-703.7.2.7 Fig. CD-30HA**

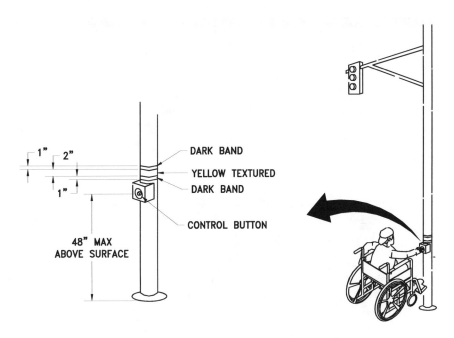

Fig, CD-30HA
Pedestrian Traffic-Control Buttons

30I. TACTILE EXIT SIGNS

Raised Character and Braille Exit Signs
Raised Character and Braille exit signs shall comply with Sections 30A, 30B, 30C, and 30E, above. Tactile exit signs shall be required at the following locations: **CBC §1011.4**

_____A. Each grade-level exterior exit door that is required to comply with CBC Section 1011.1, "EXIT SIGNS, (WHERE REQUIRED) shall be identified by a tactile exit sign with the word, "EXIT". **CBC §1011.4**

_____B. Each exit door that is required to comply with CBC Section 1011.1, "EXIT SIGNS, (WHERE REQUIRED), and leads directly to a grade-level exterior exit by means of a stairway or ramp shall be identified by a tactile exit sign with the following words as appropriate: **CBC §1011.4**

_____B1. "EXIT STAIR DOWN"

_____B2.   "EXIT RAMP DOWN"

_____B3.   "EXIT STAIR UP"

_____B4.   "EXIT RAMP UP"

_____C.   Each exit door that is required to comply with CBC Section 1011.1, "EXIT SIGNS, (WHERE REQUIRED), leads directly to a grade-level exterior exit by means of an exit enclosure or an exit passageway shall be identified by tactile exit sign with the words, "EXIT ROUTE". **CBC §1011.4**

_____D.   Each exit access door from an interior room or area to a corridor or hallway that is required to comply with CBC Section 1011.1, "EXIT SIGNS, (WHERE REQUIRED), shall be identified by a tactile exit sign with the words "EXIT ROUTE."

_____E.   Each exit door through a horizontal exit that is required to comply with CBC Section 1011.1, "EXIT SIGNS, (WHERE REQUIRED), shall be identified by a sign with the words, "TO EXIT."

General

Signs shall comply as detailed. Where both visual and tactile characters are required, either one sign with both visual and tactile characters, or two separate signs, one with visual, and one with tactile characters, shall be provided. **11B-703.1**   *703.1*

Plan Review and Inspection

Signs as specified in this Section, or in other sections of this code, when included in the construction of new buildings or facilities, or when included, altered or replaced due to additions, alterations or renovations to existing buildings or facilities, and when a permit is required, shall comply as follows: **11B-703.1.1**

Plan Review

Plans, specifications or other information indicating compliance with these regulations shall be submitted to the enforcing agency for review and approval. **11B-703.1.1.1**

Inspection

Signs and identification devices shall be field inspected after installation and approved by the enforcing agency prior to the issuance of a final certificate of occupancy per Chapter 1, Division II, Section 111, or final approval where no certificate of occupancy is issued. The inspection shall include, but not be limited to, verification that Braille dots and cells are properly spaced and the size, proportion and type of raised characters are in compliance with these regulations. **11B-703.1.1.2**

30J. <u>DELAYED EGRESS LOCKS</u>

Door Operation   CBC §1008.1.9

Delayed Egress Locks

A sign shall be provided on the door located above and within 12 inches (305 mm) of the release device reading: "KEEP PUSHING. THIS DOOR WILL OPEN IN 15 [30] SECONDS. ALARM WILL SOUND." Sign lettering shall be at least 1 inch (25 mm) in height and shall have a stroke of not less than 1/8 inch (3.2 mm). **CBC §1008.1.9.7(5)**

A tactile sign shall also be provided in Braille and raised characters, which complies with Sections 30A, 30B, 30C, and 30E, above. **CBC §1008.1.9.7(5.1)**

30K. <u>STAIRWAY IDENTIFICATION SIGNS</u>

<u>Stairway Identification Signs</u>

A sign shall be provided at each floor landing in an interior exit stairway and ramp connecting more than three stories designating the floor level, the terminus of the top and bottom of the interior exit stairway and ramp and the identification of the stair or ramp. The signage shall also state the story of, and the direction to, the exit discharge and the availability of roof access from the interior exit stairway and ramp for the fire department. The sign shall be located 5 feet (1524 mm) above the floor landing in a position that is readily visible when the doors are in the open and closed positions.   **CBC §1022.9**

In addition to the stairway identification sign, raised characters and braille floor identification signs that comply with Sections 30A, 30B, 30C, and 30E, above, shall be located at the landing of each floor level, placed adjacent to the door on the latch side, in all enclosed stairways in buildings two or more stories in height to identify the floor level. At the exit discharge level, the sign shall include a raised five pointed star located to the left of the identifying floor level. The outside diameter of the star shall be the same as the height of the raised characters.

31. <u>TELEPHONES</u>

<u>Telephones</u>   **11B-217, 11B-704**   *217, 704*

<u>General</u>
Public telephones shall comply as detailed.   **11B-704.1**   *704.1*

<u>General</u>
Where coin-operated public pay telephones, coinless public pay telephones, public closed-circuit telephones, public courtesy phones, or other types of public telephones are provided, public telephones shall be provided in accordance with the requirements listed for each type of public telephone provided. For purposes of this section, a bank of telephones shall be considered to be two or more adjacent telephones.   **11B-217.1**   *217.1*

> **ADVISORY: General.** *These requirements apply to all types of public telephones including courtesy phones at airports and rail stations that provide a free direct connection to hotels, transportation services, and tourist attractions.*

<u>VOLUME CONTROLS</u>

<u>Volume Controls</u>

_____A.   All public telephones shall have volume controls complying as detailed.
11B-217.3   *217.3*

<u>Volume Control Telephones</u>
Public telephones required to have volume controls shall be equipped with a receive volume control that provides a gain adjustable up to 20 dB minimum. For incremental volume control, provide at least one intermediate step of 12 dB of gain minimum. An automatic reset shall be provided. Volume control telephones shall be equipped with a receiver that generates a magnetic field in the area of the receiver cap. Public telephones with volume control shall be hearing aid compatible.   **11B-704.3**   *704.3*

> **ADVISORY:** *Volume Control Telephones.* *Amplifiers on pay phones are located in the base or the handset or are built into the telephone. Most are operated by pressing a button or key. If the microphone in the handset is not being used, a mute button that temporarily turns off the microphone can also reduce the amount of background noise which the person hears in the earpiece. If a volume adjustment is provided that allows the user to set the level anywhere from the base volume to the upper requirement of 20 dB, there is no need to specify a lower limit. If a stepped volume control is provided, one of the intermediate levels must provide 12 dB of gain. Consider compatibility issues when matching an amplified handset with a phone or phone system. Amplified handsets that can be switched with pay telephone handsets are available. Portable and in-line amplifiers can be used with some phones but are not practical at most public phones covered by these requirements.*

WHEELCHAIR ACCESSIBLE TELEPHONES

Wheelchair Accessible Telephones
Where public telephones are provided, wheelchair accessible telephones complying as detailed shall be provided in accordance with Table 31A.   **11B-217.2**   *217.2*

> **EXCEPTION:** Drive-up only public telephones shall not be required to comply .

Table 31A
Wheelchair Accessible Telephones

Number of Telephones Provided on a Floor, Level, or Exterior Site	Minimum Number of Required Wheelchair Accessible Telephones
1 or more single units	At least 50 percent of telephone units, but not less than 1 per floor, level, and exterior site
1 bank	At least 50 percent of telephone units per bank, but not less than 1 per floor, level, and exterior site
2 or more banks	At least 50 percent of telephone units per bank, but not less than 1 per bank At least 1 telephone per floor shall meet the requirements for a forward reach telephone.

Wheelchair Acessible Telephones
Wheelchair accessible telephones shall comply as detailed.   **11B-704.2**   *11B-704.2*

Clear Floor or Ground Space

_____B.   A clear floor or ground space complying with Section 10, "CLEAR FLOOR OR GROUND SPACE" shall be provided. The clear floor or ground space shall not be obstructed by bases, enclosures, or seats.   **11B-704.2.1**   *704.2.1*

> **ADVISORY: Clear Floor or Ground Space.** *Because clear floor and ground space is required to be unobstructed, telephones, enclosures and related telephone book storage cannot encroach on the required clear floor or ground space and must comply with the provisions for protruding objects.*

Parallel Approach

_____C.   Where a parallel approach is provided, the distance from the edge of the telephone enclosure to the face of the telephone unit shall be 10 inches (*254* mm) maximum   **11B-704.2.1.1**   *11B-704.2.1.1*   **Fig. CD-31A**

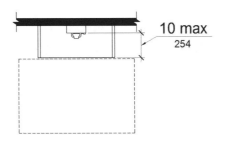

Fig. CD-31A
Parallel Approach to Telephone
© ICC Reproduced with Permission

Forward Approach

_____D.   Where a forward approach is provided at a telephone within an enclosure, the counter may extend beyond the face of the telephone 20 inches (508 mm) into the required clear floor or ground space and the enclosure may extend beyond the face of the telephone 24 inches (610 mm). If an additional 6 inches (152 mm) in width of clear floor space is provided, creating a clear floor space of 36 inches by 48 inches (914 mm by 1219 mm), the enclosure may extend more than 24 inches (610 mm) beyond the face of the telephone.   **11B-704.2.1.2** *704.2.1.2*   **Fig. CD-31B**

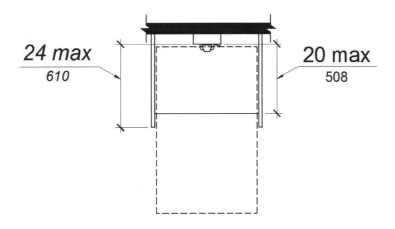

Fig. CD-31B
Forward Approach to Telephone
© ICC Reproduced with Permission

Operable Parts

_____E.   Operable parts shall comply with Section 13, "OPERABLE PARTS".
11B-704.2.2   *704.2.2*

_____E1.   Telephones shall have push-button controls where such service is available.
11B-704.2.2   *704.2.2*

Telephone Directories

_____F.   Telephone directories, where provided, shall be located in accordance with Section 13, "OPERABLE PARTS".   **11B-704.2.3**   *704.2.3*

Cord Length

_____G.   The cord from the telephone to the handset shall be 29 inches (*737 mm*) long minimum. **11B-704.2.4** *704.2.4*

TTYs

TTYs

TTYs complying with these provisions shall be provided as detailed. **11B-217.4** *217.4*

> **ADVISORY: TTYs.** *Separate requirements are provided based on the number of public pay telephones provided at a bank of telephones, within a floor, a building, or on a site. In some instances one TTY can be used to satisfy more than one of these requirements. For example, a TTY required for a bank can satisfy the requirements for a building. However, the requirement for at least one TTY on an exterior site cannot be met by installing a TTY in a bank inside a building. Consideration should be given to phone systems that can accommodate both digital and analog transmissions for compatibility with digital and analog TTYs.*

Bank Requirement
Where four or more public pay telephones are provided at a bank of telephones, at least one public TTY that complies as detailed shall be provided at that bank.
11B-217.4.1 *217.4.1*

      Exception: *Reserved.*

Floor Requirement
TTYs in public buildings shall be provided as detailed. TTYs in private buildings shall be provided as detailed.   **11B-217.4.2** *217.4.2*

Public Buildings
Where at least one public pay telephone is provided on a floor of a public building, at least one public TTY shall be provided on that floor.   **11B-217.4.2.1** *217.4.2.1*

Private Buildings
Where four or more public pay telephones are provided on a floor of a private building, at least one public TTY shall be provided on that floor.   **11B-217.4.2.2** *217.4.2.2*

Building Requirement
TTYs in public buildings shall be provided as detailed. TTYs in private buildings shall be provided as detailed.   **11B-217.4.3** *217.4.3*

Public Buildings
Where at least one public pay telephone is provided in a public building, at least one public TTY shall be provided in the building. Where at least one public pay telephone is provided in a public use area of a public building, at least one public TTY shall be provided in the public building in a public use area.   **11B-217.4.3.1** *217.4.3.1*

Private Buildings
Where four or more public pay telephones are provided in a private building, at least one public TTY shall be provided in the building.
11B-217.4.3.2 *217.4.3.2*

> **EXCEPTION:** In a stadium or arena, in a convention center, in a hotel with a convention center or in a covered mall, if an interior public pay telephone is provided at least one interior public TTY shall be provided in the facility.

Exterior Site Requirement
Where four or more public pay telephones are provided on an exterior site, at least one public TTY shall be provided on the site. **11B-217.4.4**   *217.4.4*

Transportation Facilities
In transportation facilities, in addition to the requirements for TTYs listed above, where at least one public pay telephone serves a particular entrance to a bus or rail facility, at least one public TTY shall be provided to serve that entrance. In airports, in addition to the requirements for TTYs listed above, where four or more public pay telephones are located in a terminal outside the security areas, a concourse within the security areas, or a baggage claim area in a terminal, at least one public TTY shall be provided in each location. **11B-217.4.7**   *217.4.7*

Rest Stops, Emergency Roadside Stops, and Service Plazas
Where at least one public pay telephone is provided at a public rest stop, emergency roadside stop, or service plaza, at least one public TTY shall be provided. **11B-217.4.5**   *217.4.5*

Hospitals
Where at least one public pay telephone is provided serving a hospital emergency room, hospital recovery room, or hospital waiting room, at least one public TTY shall be provided at each location. **11B-217.4.6**   *217.4.6*

Detention and Correctional Facilities
In detention and correctional facilities, where at least one pay telephone is provided in a secured area used only by detainees or inmates and security personnel, at least one TTY shall be provided in at least one secured area. **11B-217.4.8**   *217.4.8*

TTYs

_____H. TTYs provided at a public pay telephone shall be permanently affixed within, or adjacent to, the telephone enclosure. **11B-704.4**   *704.4*

_____I. Where an acoustic coupler is used, the telephone cord shall be sufficiently long to allow connection of the TTY and the telephone receiver. **11B-704.4**   *704.4*

> **ADVISORY: TTYs.** *Ensure that sufficient electrical service is available where TTYs are to be installed.*

Height

_____J. When in use, the touch surface of TTY keypads shall be 34 inches (*864* mm) minimum above the finish floor. **11B-704.4.1**   *704.4.1*

> **EXCEPTION:** Where seats are provided, TTYs shall not be required to comply with this provision.

> **ADVISORY: Height.** *A telephone with a TTY installed underneath cannot also be a wheelchair accessible telephone because the required 34 inches (865 mm) minimum keypad height can causes the highest operable part of the telephone, usually the coin slot, to exceed the maximum permitted side and forward reach ranges.*
>
> **ADVISORY: Height Exception.** *While seats are not required at TTYs, reading and typing at a TTY is more suited to sitting than standing. Facilities that often provide seats at TTY's include, but are not limited to, airports and other passenger terminals or stations, courts, art galleries, and convention centers.*

Shelves for Portable TTYs

Where a bank of telephones in the interior of a building consists of three or more public pay telephones, at least one public pay telephone at the bank shall be provided with a shelf and an electrical outlet as detailed. **11B-217.5** *217.5*

EXCEPTIONS:

1. Secured areas of detention and correctional facilities where shelves and outlets are prohibited for purposes of security or safety shall not be required to comply with this requirement.

2. The shelf and electrical outlet shall not be required at a bank of telephones with a TTY.

TTY Shelf

_____K. Public pay telephones required to accommodate portable TTYs shall be equipped with a shelf and an electrical outlet within or adjacent to the telephone enclosure. **11B-704.5** *704.5*

_____L. The telephone handset shall be capable of being placed flush on the surface of the shelf. **11B-704.5** *704.5*

_____M. The shelf shall be capable of accommodating a TTY and shall have 6 inches (*152* mm) minimum vertical clearance above the area where the TTY is to be placed. **11B-704.5** *704.5*

32. DETECTABLE WARNINGS AND DETECTABLE DIRECTIONAL TEXTURE

Detectable Warnings and Detectable Directional Texture **11B-247** *247*

Product Approval

Only approved DSA-AC detectable warning products and directional surfaces shall be installed as provided in the California Code of Regulations (CCR), Title 24, Part 1, Chapter 5, Articles 2, 3 and 4. Refer to CCR Title 24, Part 12, Chapter 11B, Section 12, 11B.205 for building and facility access specifications for product approval for detectable warning products and directional surfaces. **11B-705.3**

Detectable Warnings   **11B-247.1** *247.1*

General

Detectable warnings shall be provided as detailed and shall comply as detailed.
11B-247.1.1 *247.1.1*

> **ADVISORY: General.** *Detectable warnings are provided for the benefit of persons with visual impairments to indicate transitions to potentially hazardous areas.*

Where Required/Locations
Detectable warnings shall be provided as required below: **11B-247.1.2, 11B-705.1.2**

Platform Edges

_____A.   Detectable warning surfaces at platform boarding edges shall be 24 inches (610 mm) wide and shall extend the full length of the public use areas of the platform. **11B247.1.1, 11B--705.1.2.1**   *705.2*

Curb Ramps

_____B.   Detectable warnings at curb ramps shall extend 36 inches (914 mm) in the direction of travel. Detectable warnings shall extend the full width of the ramp run excluding any flared sides. Detectable warnings shall be located so the edge nearest the curb is 6 inches (152 mm) minimum and 8 inches (203 mm) maximum from the line at the face of the curb marking the transition between the curb and the gutter, street or highway. **11B-247.1.2, 11B-705.1.2.2**

> **EXCEPTION:** On parallel curb ramps, detectable warnings shall be placed on the turning space at the flush transition between the street and sidewalk.

Islands or Cut-Through Medians

_____C.   Detectable warnings at pedestrian islands or cut-through medians shall be 36 inches (914 mm) minimum in depth extending the full width of the pedestrian path or cut-through, placed at the edges of the pedestrian island or cut-through median, and shall be separated by 24 inches (610 mm) minimum of walking surface without detectable warnings. **11B-247.1.2.3, 11B-705.1.2.3**

> **EXCEPTION:** Detectable warnings shall be 24 inches (610 mm) minimum in depth at pedestrian islands or cut-through medians that are less than 96 inches (2438 mm) in length in the direction of pedestrian travel.

Bus Stops

_____D.   Bus stop pads shall provide a square curb surface or detectable warnings. When detectable warnings are provided at bus stop pads, it shall be 36 inches (914 mm) in width. **11B-247.1.2.4, 11B-705.1.2.4**

Hazardous Vehicular Areas

_____E.   If a walk crosses or adjoins a vehicular way, and the walking surfaces are not separated by curbs, railings or other elements between the pedestrian areas and vehicular areas, the boundary between the areas shall be defined by a continuous detectable warning that shall be 36 inches (914 mm) in width. **11B-247.1.2.5, 11B-705.1.2.5**

Reflecting Pools

_____F.   The edges of reflecting pools shall be protected by railings, walls, warning curbs or detectable warnings. When detectable warnings are provided at reflecting pools, it shall be 24 inches (610 mm) minimum and 36 inches (914 mm) maximum in width. **11B-247.1.2.6, 11B-705.1.2.6**

Track Crossings

_____G.   Where it is necessary to cross tracks to reach transit boarding platforms, detectable warnings shall be provided and shall be 36 inches (914 mm) in the direction of pedestrian travel and extend the full width of the circulation path. **11B-247.1.2.7, 11B-705.1.2.7**

Detectable Warnings   11B-705.1

General
Detectable warnings shall consist of a surface of truncated domes and shall comply with Section 11B-705. **11B-705.1.1**   *705.1.1*

Dome Size

_____H.   Truncated domes in a detectable warning surface shall have a base diameter of 0.9 inch (22.9 mm) minimum and 0.92 inch (23.4 mm) maximum, a top diameter of 0.45 inch (11.4 mm) minimum and 0.47 inch (11.9 mm) maximum, and a height of 0.18 inch (4.6 mm) minimum and 0.22 inch (5.6 mm) maximum.   **11B-705.1.1.1**   *705.1.1*   **Fig. CD-32A**

Dome Spacing

_____I.   Truncated domes in a detectable warning surface shall have a center-to-center spacing of 2.3 inches (58 mm) minimum and 2.4 inches (61 mm) maximum, and a base-to-base spacing of 0.65 inch (16.5 mm) minimum, measured between the most adjacent domes on a square grid. **11B-705.1.1.2**   *705.1.2*   **Fig. CD-32A**

> **EXCEPTION:** Where installed in a radial pattern, truncated domes shall have a center-to-center spacing of 1.6 inches (41 mm) minimum to 2.4 inches (61 mm) maximum.

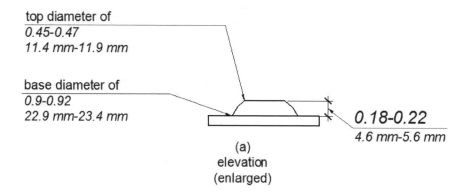

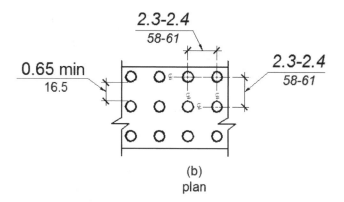

Fig. CD-32A
Size and Spacing of Truncated Domes

Contrast

____ J. Detectable warning surfaces shall contrast visually with adjacent walking surfaces either light on-dark, or dark-on-light. The material used to provide contrast shall be an integral part of the surface. Contrast shall be determined by: **11B-705.1.1.3**   *705.1.3*

> Contrast = [(B1-B2)/B1] x 100 percent where
> B1 = light reflectance value (LRV) of the lighter area and
> B2 = light reflectance value (LRV) of the darker area.

EXCEPTION: Where the detectable warning surface does not adequately contrast with adjacent surfaces, a 1 inch (25 mm) wide black strip shall separate yellow detectable warning from adjacent surfaces.

Resiliency

____ K. Detectable warning surfaces shall differ from adjoining surfaces in resiliency or sound-on-cane contact.   **11B-705.1.1.4**

> **EXCEPTION:** Detectable warning surfaces at curb ramps, islands or cut-through medians shall not be required to comply.

Color

____ L. Detectable warning surfaces shall be yellow conforming to FS 33538 of Federal Standard 595C. **11B-705.1.1.5**

> **EXCEPTION:** Detectable warning surfaces at curb ramps, islands or cut-through medians shall not be required to comply.

Detectable Directional Texture   **11B-705.2**

____ M. At transit boarding platforms, the pedestrian access shall be identified with a detectable directional texture complying as detailed.   **11B-247.2**   **Fig. CD-32B**

Detectable Directional Texture
Detectable directional texture at transit boarding platforms shall comply as detailed and shall be 0.1 inch (2.5 mm) in height that tapers off to 0.04 inch (1.0 mm), with bars raised 0.2 inch (5.1 mm) from the surface. The raised bars shall be 1.3 inches (33 mm) wide and 3 inches (76 mm) from center-to-center of each bar. This surface shall differ from adjoining walking surfaces in resiliency or sound-on-cane contact. The color shall be yellow conforming to Federal Color No. 33538. This surface will be placed directly behind the yellow detectable warning texture specified, aligning with all doors of the transit vehicles where passengers will embark. The width of the directional texture shall be equal to the width of the transit vehicle's door opening. The depth of the texture shall not be less than 36 inches (914 mm). **11B-705.2**

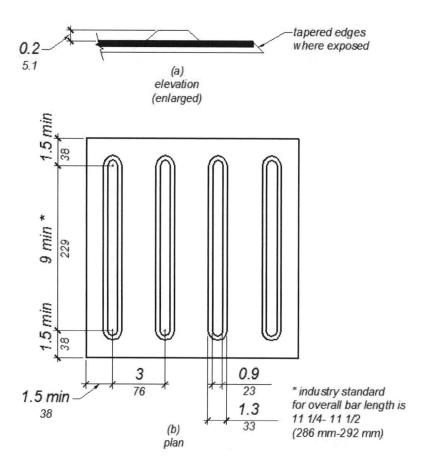

0.2
5.1

tapered edges
where exposed

(a)
elevation
(enlarged)

1.5 min
38

9 min *
229

1.5 min
38

3
76

0.9
23

1.3
33

1.5 min
38

* industry standard
for overall bar length is
11 1/4- 11 1/2
(286 mm-292 mm)

(b)
plan

Fig. CD-32B
Detectable Directional Texture
© ICC Reproduced with Permission

33. <u>ASSISTIVE LISTENING SYSTEMS</u>

<u>Assistive Listening Systems</u>   **11B-219**   *219*

General
Assistive listening systems shall be provided and comply as detailed. **11B-219.1**   *219.1*

Required Systems
An assistive listening system shall be provided in assembly areas, including conference and meeting rooms. **11B-219.2**   *219.2*

> **EXCEPTION:** This section does not apply to systems used exclusively for paging, background music, or a combination of these two uses.

Receivers
The minimum number of receivers to be provided shall be equal to 4 percent of the total number of seats, but in no case less than two. Twenty-five percent minimum of receivers provided, but no fewer than two, shall be hearing-aid compatible as detailed. **11B-219.3**   *219.3*

EXCEPTIONS:

1. Where a building contains more than one assembly area and the assembly areas required to provide assistive listening systems are under one management, the total number of required receivers shall be permitted to be calculated according to the total number of seats in the assembly areas in the building provided that all receivers are usable with all systems.

2. Where all seats in an assembly area are served by an induction loop assistive listening system, the minimum number of receivers required by to be hearing-aid compatible shall not be required to be provided.

Location

If the assistive-listening system provided is limited to specific areas or seats, then such areas or seats shall be within a 50-foot (15240 mm) viewing distance of the stage or playing area and shall have a complete view of the stage or playing area. **11B-219.4**

> *Advisory: Location. Sitting in close proximity to the performing area benefits persons with hearing impairments by allowing them to lip-read and better see the facial expressions of performers.*

Permanent and Portable Systems

Permanently installed assistive-listening systems are required in areas if (1) they accommodate at least 50 persons or if they have audio-amplification systems, and (2) they have fixed seating. If portable assistive-listening systems are used for conference or meeting rooms, the system may serve more than one room. An adequate number of electrical outlets or other supplementary wiring necessary to support a portable assistive-listening system shall be provided (See flow chart "ASSISTIVE LISTENING SYSTEM REQUIREMENTS IN ASSEMBLY AREAS"). **11B-219.5**

> *Advisory: Permanent and Portable Systems. The California Building Code (CBC) requires permanently installed assistive listening systems in those assembly areas where audible communication is integral to the use of a space (movie theaters, concert and lecture halls, playhouses, meeting rooms, etc.); where fixed seating is provided and where there may be an audio-amplification system. For other assembly areas, such as those without fixed seating, the CBC requires either a permanently installed system or a portable system. If a portable system is provided an adequate number of electrical outlets or other supplementary wiring to support the system is required. While this provision does not necessarily require the addition of electrical outlets, consideration should be given to locating outlets to support dispersion of seating available for individuals using the assistive listening systems.*

General

Assistive listening systems required in assembly areas, conference and meeting rooms shall comply as detailed. **11B-706.1** *706.1*

> *Advisory: General. Assistive listening systems are generally categorized by their mode of transmission. There are hard-wired systems and three types of wireless systems: induction loop, infrared, and FM radio transmission. Each has different advantages and disadvantages that can help determine which system is best for a given application. For example, an FM system may be better than an infrared system in some open-air assemblies since infrared signals are less effective in sunlight. On the other hand, an infrared system is typically a better choice than an FM system where confidential transmission is important because it will be contained within a given space.*
>
> *The technical standards for assistive listening systems describe minimum performance levels for volume, interference, and distortion. Sound pressure levels (SPL), expressed in decibels, measure output sound volume. Signal-to-noise ratio (SNR or S/N), also expressed in decibels,*

> *represents the relationship between the loudness of a desired sound (the signal) and the background noise in a space or piece of equipment. The higher the SNR, the more intelligible the signal. The peak clipping level limits the distortion in signal output produced when high-volume sound waves are manipulated to serve assistive listening devices.*
>
> *Selecting or specifying an effective assistive listening system for a large or complex venue requires assistance from a professional sound engineer. The federal Access Board has published technical assistance on assistive listening devices and systems.*

_____A.   Each assembly area in the building or facility, including conference and meeting rooms, provides an assistive-listening system.   **11B-219.2**   *219.2*

_____B.   The required number or personal receivers is provided.   **11B-219.3**   *219.3*

_____C.   If the assistive-listening system provided serves individual fixed seats, then such seats are located within a fifty-foot viewing distance of the stage or playing area and have a complete view of the stage or playing area.   **11B-219.4**

_____D.   Permanent and portable assistive listening systems are provided as required (See flow chart "ASSISTIVE LISTENING SYSTEM REQUIREMENTS IN ASSEMBLY AREAS".   **11B-219.5**

Receiver Jacks

_____E.   Receivers required for use with an assistive listening system shall include a 1/8 inch (3.2 mm) standard mono jack.   **11B-706.2**   *706.2*

Receiver hearing-Aid Compatibility

_____F.   Receivers required to be hearing-aid compatible shall interface with telecoils in hearing aid through the provision of neckloops.   **11B-706.3**   *706.3*

> *Advisory: Receiver Hearing-Aid Compatibility.   Neckloops and headsets that can be worn as neckloops are compatible with hearing aids. Receivers that are not compatible include earbuds, which may require removal of hearing aids, earphones, and headsets that must be worn over the ear, which can create disruptive interference in the transmission and can be uncomfortable for people wearing hearing aids.*

Sound Pressure Level

_____G.   Assistive listening systems shall be capable of providing a sound pressure level of 110 dB minimum and 118 dB maximum with a dynamic range on the volume control of 50 dB.   **11B-706.4**   *706.4*

Signal-to-Noise Ratio

_____H.   The signal-to-noise ratio for internally generated noise in assistive listening systems shall be 18 dB minimum.   **11B-706.5**   *706.5*

Peak Clipping Level

_____I.   Peak clipping shall not exceed 18 dB of clipping relative to the peaks of speech.   **11B-706.6**   *706.6*

Required Signs

_____J.   Each assembly area required to provide assistive listening systems shall provide signs informing patrons of the availability of the assistive listening system. The sign shall include wording that states "Assistive-Listening System Available" and shall be posted in a prominent place at or near the assembly area entrance. Assistive listening signs shall comply with the requirements for accessible Visual Characters and shall include the International "Symbol of Access for Hearing Loss". **11B-216.10** *216.10*

> *Advisory: Assistive Listening Systems. The term "prominent place" means a place that arriving persons would easily notice. It is helpful, though not required, to identify the location or person to contact for obtaining the system on the sign. Note that a tactile sign is not required by this section.*

EXCEPTION: Where ticket offices or windows are provided, signs shall not be required at each assembly area provided that signs are displayed at each ticket office or window informing patrons of the availability of assistive listening systems.

ASSISTIVE LISTENING SYSTEM
REQUIREMENTS IN ASSEMBLY AREAS

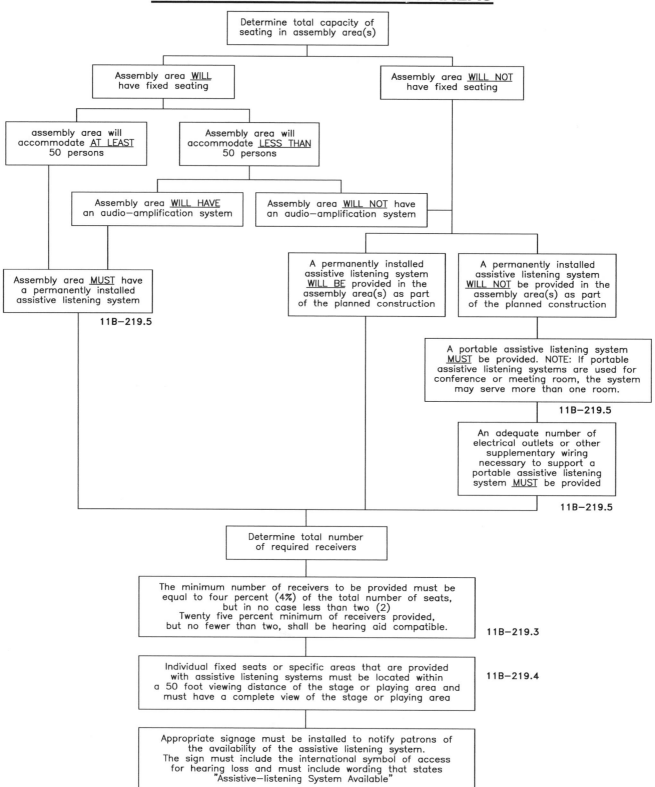

FM
ASSISTIVE LISTENING
SYSTEM AVAILABLE
— PLEASE ASK —

INFRARED
ASSISTIVE LISTENING
SYSTEM AVAILABLE
— PLEASE ASK —

AUDIO LOOP IN USE
TURN T—SWITCH FOR
BETTER HEARING
— OR ASK FOR HELP —

Summary of Assistive Listening Devices and Systems
COMPARISON OF LARGE AREA ASSISTIVE LISTENING SYSTEMS

System Description	Advantages	Disadvantages	Typical Applications
FM RADIO TRANSMISSION (40 frequencies available on narrow band transmission systems. Ten frequencies available on wideband transmission systems.) ***Transmitters:*** FM base station or personal transmitter broadcasts signal to listening area. ***Receiver:*** Pocket size with: a) Earphone(s), or b) Headset, or c) Induction neck-loop or silhouette coil coupling to personal hearing aid equipped with telecoil, or d) Direct audio input (DAI) to personal hearing aid.	• Highly portable when used with body-worn, personal transmitter • Easy to install. • May be used separately or integrated with existing PA-systems. • Multiple frequencies allow for use by different groups within same area (e.g., multi-language translation).	• Signal spill-over to adjacent rooms or listening areas (can prevent interference by using different trans-mission frequencies for each room/listening area). Choose infrared if privacy is essential. • Receivers required for everyone. Requires administration and maintenance of receivers. • Susceptible to electrical interference when used with induction neck-loop/silhouette (Provision of DAI audio shoes and cords is impractical for public applications). • Some systems more susceptible to radio wave interference and signal drift than others.	Service counters Outdoor guided tours Tour busses Meeting rooms Conference rooms Auditoriums Classrooms Courtrooms Churches and Temples Theaters Museums Theme parks Arenas Sport stadiums Retirement/nursing homes Hospitals
INFRARED ***Transmitter:*** Amplifier drives emitter panel(s) covering listening area. ***Receivers:*** Under-chin or Pendant type receiver with: a) Headset, or b) Earphone(s), or c) Induction neck-loop or silhouette coil coupling to personal hearing aid equipped with telecoil, or d) direct audio input (DAI) to personal hearing aid.	• Unlike induction or FM transmission. IR transmission does not travel through walls or other solid surfaces. • Insures confidentiality. • Infrared receivers compatible with most infrared emitters. • May be used separately or integrated with existing PA-systems. • Can be used for multi-language translation (must use special multi-frequency receivers).	• Receivers required for everyone. Requires administration and maintenance of receivers. • Ineffective in direct sunlight. • Careful installation required to insure entire listening area will receive IR signal. • Susceptible to electrical interference when used with induction neckloop/silhouette (Provision of DAI audio shoes and cords is impractical for public applications). • Lifetime of emitters varies with company. • Historical buildings may pose installation problems.	Indoor service counters Meetings requiring Confidentiality Meeting rooms Conference rooms Auditoriums Classrooms Courtrooms Churches and Temples Theaters Museums Arenas (indoors only) Sports stadiums (indoors only) Retirement/nursing Homes Hospitals

Source: Architectural and Transportation Barriers Compliance Board, 36 CFR Part 1191, Tuesday, January 13, 1998

CONTINUED ON NEXT PAGE

Summary of Assistive Listening Devices and Systems			
COMPARISON OF LARGE AREA ASSISTIVE LISTENING SYSTEMS (continued)			
System Description	**Advantages**	**Disadvantages**	**Typical Applications**
INDUCTION LOOP ***Transmitter:*** Amplifier drives an induction loop that surrounds listening area. ***Receivers:*** a) Personal hearing aid with telecoil. b) Pocket size induction receiver with earphone or headset. c) Self-contained wand. d) Telecoil inside plastic chassis which looks like a BTE, ITE, or canal hearing aid.	• Requires little, or no administration of receivers. If most people have telecoil-equipped hearing aids. Induction receivers must be used where hearing aids in use are not equipped with telecoils. • Induction receivers are compatible with all loop systems. • Unobtrusive with telecoil hearing aid. • May be used separately or integrated with existing PA-systems. • Portable systems are available for use with small groups of listeners. These portable systems can be stored in a carrying case and set up temporarily, as needed.	• Signal spill-over to adjacent rooms. • Susceptible to electrical interference. • Limited portability unless areas are pre-looped or small, portable system is used (see advantages). • Requires installation of loop wire. Installation may be difficult in pre-existing buildings. Skilled installation essential in historical buildings (and may not be permitted at all). • If listener does not have telecoil-equipped hearing aid then requires administration and maintenance of receivers.	Service counters Ports of transportation Public transportation Vehicles Tour buses Meeting rooms Conference rooms Auditoriums Classrooms Courtrooms Churches and Temples Theaters Museums Theme parks Arenas Sports stadiums Retirement/nursing homes Hospitals
Source: Architectural and Transportation Barriers Compliance Board, 36 CFR Part 1191, Tuesday, January 13, 1998			

34. <u>AUTOMATIC TELLER MACHINES, FARE MACHINES AND POINT-OF-SALE DEVICES</u>

Automatic Teller Machines and Fare Machines

Where automatic teller machines or self-service fare vending, collection, or adjustment machines are provided they shall comply as detailed. Where bins are provided for envelopes, waste paper, or other purposes, at least one of each type shall comply with Section 45, "STORAGE". **11B-220.1** *220.1*

> ***Advisory: General.*** *If a bank provides both interior and exterior ATMs, each such installation is considered a separate location. Accessible ATMs, including those with speech and those that are within reach of people who use wheelchairs, must provide all the functions provided to customers at that location at all times. For example, it is unacceptable for the accessible ATM only to provide cash withdrawals while inaccessible ATMs also sell theater tickets.*

One Automatic Teller Machine or Fare Machine

Where one automatic teller machine or fare machine is provided at a location, it shall comply with items A through M, below.   **11B-220.1.1**

Two Automatic Teller Machines or Fare Machines

Where two automatic teller machines or fare machines are provided at a location, one shall comply with items A through M, below, and one shall comply with Section 13, "OPERABLE PARTS, and items A, C through J, and L through L2, below. **11B-220.1.2**

Three or More Automatic Teller Machines or Fare Machines

Where three or more automatic teller machines or fare machines are provided at a location, at least 50 percent shall comply with items A through M, below, and the rest shall comply with Section 13, "OPERABLE PARTS, and items A, C through J, and L through L2, below.   **11B-220.1.3**

ADVISORY: *When only one ATM is provided at a location, it can provide for either forward or parallel approach, or both, and the associated reach range and height limitations that go with that approach.*

When two ATMs are provided, one of them must comply with either forward or parallel approach, or both, with the exception that the maximum allowed height to the operable parts of this ATM cannot exceed 48". The second ATM must still provide for clear floor space on an accessible route, controls which do not require more than 5 lbf or tight grasping, pinching or twisting of the wrist to operate, and accessible information and instructions for use of the ATM by persons with vision impairments, however there are no other requirements associated with this second ATM.

When three or more ATMs are provided, two of them must comply with the same requirements for two ATMs as explained above. For the additional ATMs beyond these first two, at least 50% of these must provide for either forward or parallel approach, or both, and the associated reach range and height limitations that go with that approach. The remaining ATMs are only required to provide for clear floor space on an accessible route, controls which do not require more than 5 lbf. or tight grasping, pinching or twisting of the wrist to operate, and accessible information and instructions for use of the ATM by persons with vision impairments.

Point-of-Sale Devices

Where point-of-sale devices are provided, all devices at each location shall comply with Section 13, "OPERABLE PARTS", and items B and L through L2, below. In addition, point-of-sale systems that include a video touch screen or any other non-tactile keypad shall comply with either item N or N1, below. Where point-of-sale devices are provided at check stands and sales and service counters, they shall comply with items N through N1, below, and shall also comply with items A through C, below.   **11B-220.2**

> **EXCEPTION:** Where a single point-of-sale device is installed for use with any type of motor fuel, it shall comply with the requirements for "Point-of-Sale Devices", above, and Section 13, "OPERABLE PARTS". Where more than one point-of-sale device is installed for use with a specific type of motor fuel, a minimum of two for that type shall comply with the requirements for "Point-of-Sale Devices", above, and Section 13, "OPERABLE PARTS" . Types of motor fuel include, but are not limited to, gasoline, diesel, compressed natural gas, methanol, ethanol or electricity.

Automatic Teller Machines, Fare Machines and Point-of-Sale Devices   **11B-707**   *707*

> *Advisory: Automatic Teller Machines and Fare Machines. Interactive transaction machines (ITMs), other than ATMs, are not covered by this Section. However, for entities covered by the ADA, the Department of Justice regulations that implement the ADA provide additional guidance regarding the relationship between these requirements and elements that are not directly addressed by these requirements. Federal procurement law requires that ITMs purchased by the Federal government comply with standards issued by the Access Board under Section 508 of the Rehabilitation Act of 1973, as amended. This law covers a variety of products, including computer hardware and software, websites, phone systems, fax machines, copiers, and similar technologies. For more information on Section 508 consult the Access Board's website at www.access-board.gov.*

General
Automatic teller machines, fare machines and point-of-sale devices shall comply as detailed. **11B-707.1**   *707.1*

> *Advisory: General.   If farecards have one tactually distinctive corner they can be inserted with greater accuracy. Token collection devices that are designed to accommodate tokens which are perforated can allow a person to distinguish more readily between tokens and common coins. Place accessible gates and fare vending machines in close proximity to other accessible elements when feasible so the facility is easier to use.*

Clear Floor or Ground Space

_____A.   A clear floor or ground space complying with Section 10, "CLEAR FLOOR OR GROUND SPACE" shall be provided.   **11B-707.2**   *707.2*

> EXCEPTION: Clear floor or ground space shall not be required at drive-up only automatic teller machines and fare machines.

Operable Parts

_____B.   Operable parts shall comply with Section 13, "OPERABLE PARTS". Unless a clear or correct key is provided, each operable part shall be able to be differentiated by sound or touch, without activation.   **11B-707.3**   *707.3*

> EXCEPTION: Drive-up only automatic teller machines and fare machines shall not be required to comply with the requirements for Clear Floor or Ground Space or Reach Ranges.

Privacy

_____C.   Automatic teller machines shall provide the opportunity for the same degree of privacy of input and output available to all individuals.   **11B-707.4**   *707.4*

> *Advisory: Privacy. In addition to people who are blind or visually impaired, people with limited reach who use wheelchairs or have short stature, who cannot effectively block the ATM screen with their bodies, may prefer to use speech output. Speech output users can benefit from an option to render the visible screen blank, thereby affording them greater personal security and privacy.*

Speech Output

_____D.   Machines shall be speech enabled. Operating instructions and orientation, visible transaction prompts, user input verification, error messages, and all displayed information for full use shall be accessible to and independently usable by individuals with vision impairments. **11B-707.5** *707.5*

_____D1.   Speech shall be delivered through a mechanism that is readily available to all users, including but not limited to, an industry standard connector or a telephone handset. **11B-707.5** *707.5*

_____D2.   Speech shall be recorded or digitized human, or synthesized. **11B-707.5** *707.5*

EXCEPTIONS:

1.   Audible tones shall be permitted instead of speech for visible output that is not displayed for security purposes, including but not limited to, asterisks representing personal identification numbers.

2.   Advertisements and other similar information shall not be required to be audible unless they convey information that can be used in the transaction being conducted.

3.   Where speech synthesis cannot be supported, dynamic alphabetic output shall not be required to be audible.

> *Advisory: Speech Output. If an ATM provides additional functions such as dispensing coupons, selling theater tickets, or providing copies of monthly statements, all such functions must be available to customers using speech output. To avoid confusion at the ATM, the method of initiating the speech mode should be easily discoverable and should not require specialized training. For example, if a telephone handset is provided, lifting the handset can initiate the speech mode.*

User Control

_____E.   Speech shall be capable of being repeated or interrupted. **11B-707.5.1** *707.5.1*

_____E1.   Volume control shall be provided for the speech function. **11B-707.5.1** *707.5.1*

EXCEPTION: Speech output for any single function shall be permitted to be automatically interrupted when a transaction is selected.

Receipts

_____F.   Where receipts are provided, speech output devices shall provide audible balance inquiry information, error messages, and all other information on the printed receipt necessary to complete or verify the transaction. **11B-707.5.2** *707.5.2*
EXCEPTIONS:

1. Machine location, date and time of transaction, customer account number, and the machine identifier shall not be required to be audible.

2. Information on printed receipts that duplicates information available on-screen shall not be required to be presented in the form of an audible receipt.

3. Printed copies of bank statements and checks shall not be required to be audible.

Input

Input devices shall comply as detailed.   **11B-707.6**   *707.6*

Input Controls

_____G.   At least one tactilely discernible input control shall be provided for each function.   **11B-707.6.1**   *707.6.1*

_____G1.   Where provided, key surfaces not on active areas of display screens, shall be raised above surrounding surfaces. **11B-707.6.1**   *707.6.1*

_____G2.   Where membrane keys are the only method of input, each shall be tactilely discernible from surrounding surfaces and adjacent keys.   **11B-707.6.1**   *707.6.1*

Numeric Keys

_____H.   Numeric keys shall be arranged in a 12-key ascending or descending telephone keypad layout.   **11B-707.6.2**   *707.6.2*   **Fig. CD-34A**

_____H1.   The number five key shall be tactilely distinct from the other keys.   **11B-707.6.2**   *707.6.2*   **Fig. CD-34A**

Advisory: Numeric Keys. Telephone keypads and computer keyboards differ in one significant feature, ascending versus descending numerical order. Both types of keypads are acceptable, provided the computer-style keypad is organized similarly to the number pad located at the right on most computer keyboards, and does not resemble the line of numbers located above the computer keys.

(a)
12-key
ascending

(b)
12-key
descending

Fig. CD-34A
Numeric Key Layout
© ICC Reproduced with Permission

Function Keys

Function keys shall comply as detailed.   **11B-707.6.3**   *707.6.3*

Contrast

_____ I.   Function keys shall contrast visually from background surfaces. **11B-707.6.3.1**   *707.6.3.1*

_____ I1.   Characters and symbols on key surfaces shall contrast visually from key surfaces. **11B-707.6.3.1**   *707.6.3.1*

_____ I2.   Visual contrast shall be either light-on-dark or dark-on-light. **11B-707.6.3.1**   *707.6.3.1*

EXCEPTION: Tactile symbols required by item G through H, above, shall not be required to comply with the requirement for visual contrast.

Tactile Symbols

_____ J.   Function key surfaces shall have tactile symbols as follows: Enter or Proceed key: raised circle; Clear or Correct key: raised left arrow; Cancel key: raised letter ex; Add Value key: raised plus sign; Decrease Value key: raised minus sign. **11B-707.6.3.2**   *707.6.3.2*

Display Screen
The display screen shall comply as detailed.   **11B-707.7**   *707.7*

EXCEPTION: Drive-up only automatic teller machines and fare machines shall not be required to comply with items K through K3, below.

Visibility

_____ K.   The display screen shall be visible from a point located 40 inches (*1016* mm) above the center of the clear floor space in front of the machine. **11B-707.7.1**   *707.7.1*

Vertically Mounted Display Screen

_____ K1.   Where display screens are mounted vertically or no more than 30 degrees tipped away from the viewer, the center line of the display screen and other display devices shall be no more than 52 inches (1321 mm) above the floor or ground surface. **11B-707.7.1.1**   **Fig. CD-34B**

Angle-Mounted Display Screen

_____ K2.   Where display screens are mounted between 30 degrees and 60 degrees tipped away from the viewer, the center line of the display screen and other display devices shall be no more than 44 inches (1118 mm) above the floor or ground surface. **11B-707.7.1.2**   **Fig. CD-34B**

Horizontally Mounted Display Screen

_____K3.   Where display screens are mounted no less than 60 degrees and no more than 90 degrees (horizontal) tipped away from the viewer, the center line of the display screen and other display devices shall be no more than 34 inches (864 mm) above the floor or ground surface. **11B-707.7.1.3   Fig. CD-34B**

Characters

_____L.   Characters displayed on the screen shall be in a sans serif font. **11B-707.7.2**   *707.7.2*

_____L1.   Characters shall be 3/16 inch (4.8 mm) high minimum based on the uppercase letter "I".   **11B-707.7.2**   *707.7.2*

_____L2.   Characters shall contrast with their background with either light characters on a dark background or dark characters on a light background.   **11B-707.7.2**   *707.7.2*

Braille Instructions

_____M.   Braille instructions for initiating the speech mode shall be provided. Braille shall comply with Section 30B, "BRAILLE".   **11B-707.8**   *707.8*

Point-of-Sale Devices
Point-of-sale devices shall comply as detailed.   **11B-707.9**

General

_____N.   Where point-of-sale devices are provided, all devices at each location shall comply with item B and items L through L2, above. In addition, point-of-sale systems that include a video touch screen or any other non-tactile keypad shall be equipped with either of the following: **11B-707.9.1**

Tactilely Discernible Numerical Keypad

_____N1.   A tactilely discernible numerical keypad similar to a telephone keypad containing a raised dot with a dot base diameter between 1.5 mm and 1.6 mm and a height between 0.6 mm and 0.9 mm on the number 5 key that enables a visually impaired person to enter his or her own personal identification number or any other personal information necessary to process the transaction in a manner that provides the opportunity for the same degree of privacy input and output available to all individuals. **11B-707.9.1.1**

Other Technology

_____N2.   Other technology, such as a radio frequency identification device, fingerprint biometrics, or some other mechanism that enables a visually impaired person to access the video touch screen device with his or her personal identifier and to process his or her transaction in a manner that provides the opportunity for the same degree of privacy input and output available to all individuals. Where a video screen overlay is provided it shall be equipped with a tactilely discernible numerical keypad that complies with item N1, above. **11B-707.9.1.2**

Point-of-Sale Devices at Check Stands and Sales or Service Counters

_____O.　　Where point-of-sale devices are provided at check stands and sales or service counters, they shall comply with items N through N2 and L through L2, above. **11B-707.9.2**

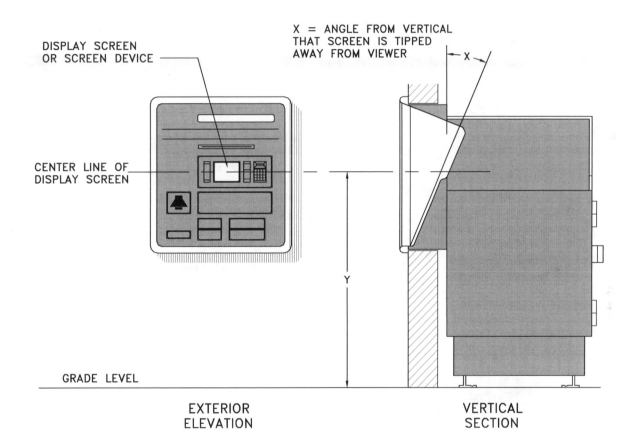

REQUIRED DIMENSIONS WHEN SCREENS ARE TIPPED AWAY FROM VIEWER:

WHEN X = 30 DEGREES OR LESS, THEN Y = 52" MAXIMUM ABOVE GRADE.
WHEN X IS BETWEEN 30 DEGREES AND 60 DEGREES, THEN Y = 44" MAXIMUM ABOVE GRADE.
WHEN X IS BETWEEN 60 DEGREES AND 90 DEGREES, THEN Y = 34" MAXIMUM ABOVE GRADE.

Fig. CD-34B
Display Screen

35. <u>TWO-WAY COMMUNICATIONS</u>

<u>Two-Way Communication Systems</u>   **11B-230, 11B-708**   *230, 708*

General

Where a two-way communication system is provided to gain admittance to a building or facility or to restricted areas within a building or facility, the system shall comply as detailed. **11B-230.1**   *230.1*

> *ADVISORY: **General.** This requirement applies to facilities such as office buildings, courthouses, and other facilities where admittance to the building or restricted spaces is dependent on two-way communication systems.*

General

Two-way communication systems shall comply as detailed. **11B-708.1**   *708.1*

> *ADVISORY: **General.** Devices that do not require handsets are easier to use by people who have a limited reach.*

Audible and Visual Indicators

_____A.   The system shall provide both audible and visual signals. **11B-708.2**   *708.2*

> *ADVISORY: **Audible and Visual Indicators.** A light can be used to indicate visually that assistance is on the way. Signs indicating the meaning of visual signals should be provided.*

Handsets

_____B.   Handset cords, if provided, shall be 29 inches (*737 mm*) long minimum. **11B-708.3**   *708.3*

Residential Dwelling Unit Communication Systems

Communications systems between a residential dwelling unit and a site, building, or floor entrance shall comply as detailed below. **11B-708.4**   *708.4*

Common Use or Public Use System Interface

_____C.   The common use or public use system interface shall include the capability of supporting voice and TTY communication with the residential dwelling unit interface. **11B-708.4.1**   *708.4.1*

Residential Dwelling Unit Interface

_____D.   The residential dwelling unit system interface shall include a telephone jack capable of supporting voice and TTY communication with the common use or public use system interface. **11B-708.4.2**   *708.4.2*

36. ASSEMBLY AREAS

Definition of "Assembly Area"
Assembly Area. A building or facility, or portion thereof, used for the purpose of entertainment, educational or civic gatherings, or similar purposes. For the purposes of these requirements, assembly areas include, but are not limited to, classrooms, lecture halls, courtrooms, public meeting rooms, public hearing rooms, legislative chambers, motion picture houses, auditoria, theaters, playhouses, dinner theaters, concert halls, centers for the performing arts, amphitheaters, arenas, stadiums, grandstands, or convention centers. **11B-106.5, 202** *106.5*

Press Boxes
Press boxes in assembly areas shall be on an accessible route. **11B-206.2.7** *206.2.6.7*

> **EXCEPTIONS:**
>
> 1. An accessible route shall not be required to press boxes in bleachers that have points of entry at only one level provided that the aggregate area of all press boxes if 500 square feet maximum.
>
> 2. An accessible route shall not be required to free-standing press boxes that are elevated above grade 12 feet minimum provided that the aggregate area of all press boxes is 500 square feet maximum.

Stairways
In assembly areas, aisle stairs shall not be required to comply with these requirements except that compliant tread striping shall be provided. **11B-210** *210*

General
Assembly areas shall provide wheelchair spaces, companion seats, designated aisle seats and semi-ambulant seats as specified herein. **11B-221.1** *221.1*

ADVISORY: General. Several different types of accessible seating are required in an assembly seating area.

Wheelchair seating areas, integrated into the general seating plan, are required so that people using wheelchairs are not isolated from other spectators or their friends and family. These seating areas must comply as detailed.

Companion seats are required next to each wheelchair seating location. The companion seat is a conventional seat that accommodates a friend or companion. These seats must comply as detailed.

Aisle seating is required to be provided in addition to the wheelchair seating areas. At least five percent of aisle seats (but not less than one) are required to either have no armrest on the aisle side or to have a removable or folding armrest on the aisle side. These seats accommodate people who have a mobility disability but who wish to use a seat that is not in a wheelchair seating location. These seats must comply as detailed.

Semi-ambulant seating is required in addition to the spaces provided for wheelchair users. At least one percent of all seats (but no fewer than two) are required to provide 24 inches clear leg room from the front edge of the seat to the nearest obstruction or to the back of the seat immediately in front. These seats accommodate people who have a mobility disability but who wish to use a seat that is not in a wheelchair seating location. These seats must comply as detailed.

Spaces and Elements
At least one accessible route shall connect accessible building or facility entrances with all accessible spaces and elements within the building or facility which are otherwise connected by a circulation path. **11B-206.2.4** *206.2.4*

> **EXCEPTION:** In assembly areas with fixed seating required to provide accessible wheelchair spaces, companion seats, designated aisle seats and semi-ambulant seats, an accessible route shall not be required to serve fixed seating where wheelchair spaces required to be on an accessible route are not provided. **Exception 2**

Accessible Route to Performance Areas
Where a circulation path directly connects a performance area to an assembly seating area, an accessible route shall directly connect the assembly seating area with the performance area. An accessible route shall be provided from performance areas to ancillary areas or facilities used by performers, unless exempted by exceptions 1 through 7 of the "Multi-Story Buildings and Facilities" provisions. **11B-206.2.6** *206.2.6*

Platform Lifts for Performance Areas and Speakers' Platforms
Platform lifts shall be permitted to provide accessible routes to performance areas and speakers' platforms. **11B-206.7.1** *206.7.1*

Lawn Seating
Lawn seating areas and exterior overflow seating areas, where fixed seats are not provided, shall connect to an accessible route. **11B-221.5** *221.5*

Wheelchair Spaces
Wheelchair spaces as specified herein shall be provided in assembly areas with fixed seating. **11B-221.2** *221.2*

ADVISORY: Wheelchair Spaces. Additional information regarding wheelchair accessible seating in venues that sell tickets for assigned seats is available on the US Department of Justice website at http://www.ada.gov/ticketing_2010.htm.

Platform Lifts for Wheelchair Spaces
Platform lifts shall be permitted to provide an accessible route to comply with the wheelchair space dispersion and line-of-sight requirements for Assembly Areas, Wheelchair Spaces, Companion Seats, Designated Aisle Seats and Semi-Ambulant Seats. **11B-206.7.2** *206.7.2*

Number and Location
Wheelchair spaces shall be provided as specified herein. **11B-221.2.1** *221.2.1*

General Seating
Accessible wheelchair spaces shall be provided in accordance with Table 36A. **11B-221.2.1.1** *221.2.1.1*

Table 36A
Number of Wheelchair Spaces in Assembly Areas

Number of Seats	Minimum Number of Required Wheelchair Spaces
4 to 25	1
26 to 50	2
51 to 150	4
151 to 300	5
301 to 500	6
501 to 5000	6, plus 1 for each 100, or fraction thereof, between 501 through 5000
5001 and over	46, plus 1 for each 200, or fraction thereof, over 5000

Luxury Boxes, Club Boxes, and Suites in Arenas, Stadiums and Grandstands
In each luxury box, club box, and suite within arenas, stadiums, and grandstands, compliant wheelchair spaces shall be provided in accordance with Table 36A. **11B-221.2.1.2** *221.2.1.2*

ADVISORY: Luxury Boxes, Club Boxes, and Suites in Arenas, Stadiums, and Grandstands. The number of wheelchair spaces required in luxury boxes, club boxes, and suites within an arena, stadium or grandstand is to be calculated box by box and suite by suite.

Other Boxes
In boxes other than those specified above, the total number of wheelchair spaces required shall be determined in accordance with Table 36A. Wheelchair spaces shall be located in not less than 20 percent (20%) of all boxes provided. **11B-221.2.1.3** *221.2.1.3*

ADVISORY: Other Boxes. The provision for seating in "other boxes" includes box seating provided in facilities such as performing arts auditoria where tiered boxes are designed for spatial and acoustical purposes. The number of wheelchair spaces required in boxes covered by this provision is calculated based on the total number of seats provided in these other boxes. The resulting number of wheelchair spaces must be located in no fewer than 20% of the boxes covered by this section. For example, a concert hall has 20 boxes, each of which contains 10 seats, totaling 200 seats. In this example, 5 wheelchair spaces would be required, and they must be placed in at least 4 of the boxes. Additionally, because the wheelchair spaces must also meet the dispersion requirements, the boxes containing these wheelchair spaces cannot all be located in one area unless an exception to the dispersion requirements applies.

Team or Player Seating
At least one accessible wheelchair space with a companion seat shall be provided in team or player seating areas serving areas of sport activity. **11B-221.2.1.4** *221.2.1.4*

> **EXCEPTION:** Wheelchair spaces shall not be required in team or player seating areas serving those bowling lanes that are not required to be on an accessible route.

Stadium-Style Movie Theaters

In stadium-style movie theaters, the total number of wheelchair spaces required shall be determined in accordance with Table 36A. The required wheelchair spaces shall be located on risers or cross-aisles in the stadium section that satisfy at least one of the following criteria:

1. Located within the rear 60 percent of the seats provided in the theater; or

2. Located within the area of the theater in which the vertical viewing angles (as measured to the top of the screen) are from the 40th to the 100th percentile of vertical viewing angles for all seats as ranked from the seats in the first row (1st percentile) to seats in the back row (100th percentile). **11B-221.2.1.5**

Specialty Seating Areas

In assembly areas, wheelchair spaces shall be provided in each specialty seating area that provides spectators with distinct services or amenities that generally are not available to other spectators. The number of wheelchair spaces provided in specialty seating areas shall be included in, rather than be in addition to, the total number of wheelchair spaces required by Table 36A.

> **EXCEPTION:** In existing buildings and facilities, if it is not readily achievable for wheelchair spaces to be placed in specialty seating areas, those services or amenities shall be provided to individuals with disabilities, and their companions, at other designated accessible locations at no additional cost. **11B-221.2.1.6**

Integration

Wheelchair spaces shall be an integral part of the seating plan.
11B-221.2.2   *221.2.2*

> *ADVISORY: Integration. The requirement that wheelchair spaces be an "integral part of the seating plan" means that wheelchair spaces must be placed within the footprint of the seating area. Wheelchair spaces cannot be segregated from seating areas. For example, it would be unacceptable to place only the wheelchair spaces, or only the wheelchair spaces and their associated companion seats, outside the seating areas defined by risers in an assembly area.*

Lines of Sight and Dispersion

Wheelchair spaces shall provide lines of sight as detailed herein. In providing lines of sight, wheelchair spaces shall be dispersed and shall provide spectators with choices of seating locations and viewing angles that are substantially equivalent to, or better than, the choices of seating locations and viewing angles available to all other spectators. When the required number of wheelchair spaces have been provided in the required locations, further dispersion shall not be required. In stadiums, arenas and grandstands, wheelchair spaces shall be dispersed to all levels that include seating served by an accessible route.
11B-221.2.3   *221.2.3*

> **EXCEPTION:** Wheelchair spaces in team or player seating areas serving areas of sport activity shall not be required to comply with this provision.

> ***ADVISORY:*** ***Lines of Sight and Dispersion.*** *Consistent with the overall intent of the ADA, individuals who use wheelchairs must be provided equal access so that their experience is substantially equivalent to that of other members of the audience. Thus, while individuals who use wheelchairs need not be provided with the best seats in the house, neither may they be relegated to the worst.*

Horizontal Dispersion

Wheelchair spaces shall be dispersed horizontally. **11B-221.2.3.1** *221.2.3.1*

EXCEPTIONS:

1. Horizontal dispersion shall not be required in assembly areas with 300 or fewer seats if the required companion seats and wheelchair spaces are located within the 2nd or 3rd quartile of the total row length. If the row length in the 2nd and 3rd quartile of a row is insufficient to accommodate the required number of companions seats and wheelchair spaces, the additional companion seats and wheelchair spaces shall be permitted to be located in the 1st and 4th quartile of the row.

2. In row seating, two wheelchair spaces shall be permitted to be located side-by-side.

> ***ADVISORY:*** ***Horizontal Dispersion.*** *Horizontal dispersion of wheelchair spaces is the placement of spaces in an assembly facility seating area from side-to-side or, in the case of an arena or stadium, around the field of play or performance area.*

Vertical Dispersion

Wheelchair spaces shall be dispersed vertically at varying distances from the screen, performance area, or playing field. In addition, wheelchair spaces shall be located in each balcony or mezzanine that is located on an accessible route. **11B-221.2.3.1** *221.2.3.2*

EXCEPTIONS:

1. Vertical dispersion shall not be required in assembly areas with 300 or fewer seats if the wheelchair spaces provide viewing angles that are equivalent to, or better than, the average viewing angle provided in the facility.

2. In bleachers, wheelchair spaces shall not be required to be provided in rows other than rows at points of entry to bleacher seating.

> ***ADVISORY:*** ***Vertical Dispersion.*** *When wheelchair spaces are dispersed vertically in an assembly facility they are placed at different locations within the seating area from front-to-back so that the distance from the screen, stage, playing field, area of sports activity, or other focal point is varied among wheelchair spaces.*
>
> *Points of entry to bleacher seating may include, but are not limited to, cross aisles, concourses, vomitories, and entrance ramps and stairs. Vertical, center,*

or side aisles adjoining bleacher seating that are stepped or tiered are not considered entry points.

Vertical Dispersion Exception 2. *Designing spectator seating for accessibility can be more complicated when folding bleachers are utilized. The lower rows of bleacher seats in a bank of bleachers often are omitted to allow for wheelchair positions with companion seating provided on the end of the adjacent row or on portable chairs.*

Temporary Structures

Wheelchair spaces shall not be located on, or be obstructed by, temporary platforms or other movable structures. **11B-221.2.4** *221.2.4*

> **EXCEPTION:** When an entire seating section is placed on temporary platforms or other movable structures in an area where fixed seating is not provided, in order to increase seating for an event, wheelchair spaces may be placed in that section.

Removable Chairs

When required wheelchair spaces are not occupied by persons eligible for those spaces, individual, removable seats may be placed in those spaces. **11B-221.2.5** *221.2.5*

ADVISORY: ***Removable Chairs.*** *Readily removable seats should be designed to facilitate easy, timely and frequent removal and installation. If mechanically fastened to the floor the release mechanisms need to be easily operated by untrained individuals without special tools or knowledge. Seats which have been removed need to be stored so as not to create obstructions.*

Companion Seats

At least one companion seat as specified herein shall be provided for each required wheelchair space. **11B-221.3** *221.3*

ADVISORY: ***Companion Seats.*** *The following advisory language clarifies the operational requirements of 28 CFR Part 36, Section 36.302(f)(4)(i).*

People purchasing a ticket for an accessible seat may purchase up to three additional seats for their companions in the same row and these seats must be contiguous with the accessible seat. If contiguous seats have already been sold and are not available, the venue must offer other seats as close as possible to the accessible seat. If those seats are in a different price category, the venue is not required to modify the price and may charge the same price as it charges others for those seats. When designing, best practice is to locate wheelchair spaces in rows where seating for a minimum of four is provided. Where two wheelchair spaces are provided adjacent to one another, one can be used as a companion seat.

Where a venue limits ticket sales to fewer than four tickets, those limits also apply to tickets for accessible seats. Similarly, when a venue allows the purchase of more than four tickets, that policy also applies to tickets for accessible seats, but only three companion seats must be contiguous with the accessible seat.

Many venues offer a group sales rate for groups of a pre-determined size. If a group includes one or more individuals who need accessible seating, the entire group should be seated together in an area that includes accessible seating. If it

> *is not possible to seat the entire group together and the group must be split, the tickets should be allocated so that the individuals with disabilities are not isolated from others in their group.*
>
> *Additional information regarding wheelchair accessible seating in venues that sell tickets for assigned seats is available on the US Department of Justice website at http://www.ada.gov/ticketing_2010.htm.*

Designated Aisle Seats

At least 5 percent (5%) of the total number of aisle seats provided shall comply as specified herein and shall be the aisle seats located closest to accessible routes. Accessible signage notifying patrons of the availability of such seats shall be posted in the ticket office.   **11B-221.4**   *221.4*

> **EXCEPTION:** Team or player seating areas serving areas of sport activity shall not be required to comply with this provision.

> ***ADVISORY: Designated Aisle Seats.*** *When selecting which aisle seats will meet the requirements, those aisle seats which are closest to, not necessarily on, accessible routes must be selected first. For example, an assembly area has two aisles (A and B) serving seating areas with an accessible route connecting to the top and bottom of aisle A only. The aisle seats chosen to meet the requirements must be those at the top and bottom of Aisle A, working towards the middle. Only when all seats on Aisle A would not meet the five percent (5%) minimum would seats on Aisle B be designated.*

Semi-Ambulant Seats

At least one percent (1%) of the total number of seats, and no fewer than two (2), shall be semi-ambulant seats that comply as detailed herein.   **11B-221.6**

ACCESSIBLE SEATING IN ASSEMBLY AREAS

WHEELCHAIR SPACES

Floor or Ground Surface

_____A.   The floor or ground surface of wheelchair spaces complies with Section 7, "FLOOR OR GROUND SURFACES".   **11B-802.1.1**   *802.1.1*

_____B.   The floor or ground surfaces at the wheelchair space(s) are sloped not steeper than 1:48 maximum (2.08%).   **11B-802.1.1**   *802.1.1*

Width

_____C.   A single wheelchair space is 36 inches wide minimum. **11B-802.1.2** *802.1.2* **Fig. CD-36A**

_____D.   Where two adjacent wheelchair spaces are provided, each wheelchair space is

---

33 inches wide minimum.   **11B-802.1.2**   *802.1.2*   **Fig. CD-36A**

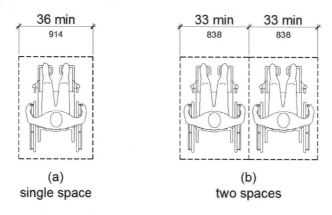

(a)
single space

(b)
two spaces

Fig. CD-36A
Width of Wheelchair Spaces
© ICC Reproduced with Permission

Depth

_____E.   Where a wheelchair space can be entered from the front or rear, the wheelchair space is 48 inches deep minimum. **11B-802.1.3**   *802.1.3*   **Fig. CD-36B**

_____F.   Where a wheelchair space can be entered only from the side, the wheelchair space is 60 inches deep minimum. **11B-802.1.3**   *802.1.3*   **Fig. CD-36B**

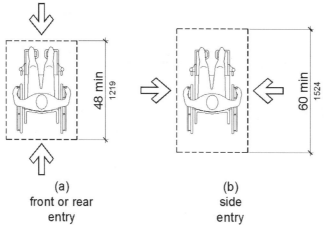

(a)
front or rear
entry

(b)
side
entry

Fig. CD-36B
Depth of Wheelchair Spaces
© ICC Reproduced with Permission

> *do that. A rear or front entry wheelchair space, as depicted in Fig. CD-36B(a) can have a shorter length (48 inches) of maneuvering area. Consideration should be given to the alignment of the wheelchair space and the adjacent companion seat which may require the rear tires of the wheelchair to project behind the back of the companion seat to achieve shoulder-to-shoulder alignment.*

Approach

_____G.　　The wheelchair space(s) adjoin an accessible route.　**11B-802.1.4**　*802.1.4*

_____H.　　The accessible route(s) do not overlap the wheelchair space(s). **11B-802.1.4** *802.1.4*

> **ADVISORY: Approach.** *Because accessible routes serving wheelchair spaces are not permitted to overlap the clear floor space at wheelchair spaces, access to any wheelchair space cannot be through another wheelchair space.*

Overlap

_____I.　　Wheelchair spaces do not overlap circulation paths.　**11B-802.1.5**
　　　　802.1.5

> **ADVISORY: Overlap.** *The term "circulation paths" as used here means aisle width required by applicable building or life safety codes for the specific assembly occupancy. Where the circulation path provided is wider than the required aisle width, the wheelchair space may intrude into that portion of the circulation path that is provided in excess of the required aisle width.*

Lines of Sight
Lines of sight to the screen, performance area, or playing field for spectators in wheelchair spaces shall comply as detailed.　**11B-802.2**　*802.2*

Lines of Sight Over Seated Spectators
Where spectators are expected to remain seated during events, spectators in wheelchair spaces shall be afforded lines of sight as follows:　**11B-802.2.1** *802.2.1*

Lines of Sight Over Heads

_____J.　　Where spectators are provided lines of sight over the heads of spectators seated in the first row in front of their seats, spectators seated in wheelchair spaces are afforded lines of sight over the heads of seated spectators in the first row in front of wheelchair spaces.　**11B-802.2.1.1** *802.2.1.1*　　**Fig. CD-36C**

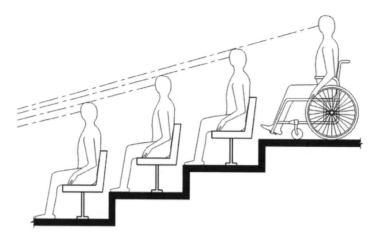

Fig. CD-36C
Lines of Sight Over the Heads of Seated Spectators
© ICC Reproduced with Permission

Lines of Sight Between Heads

_____K.   Where spectators are provided lines of sight over the shoulders and between the heads of spectators seated in the first row in front of their seats, spectators seated in wheelchair spaces are afforded lines of sight over the shoulders and between the heads of seated spectators in the first row in front of wheelchair spaces.   **11B-802.2.1.2**   *802.2.1.2*
Fig. CD-36D

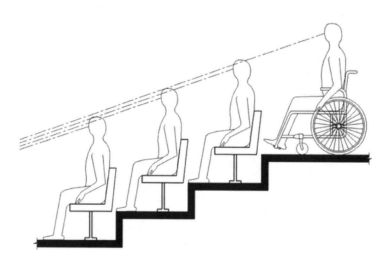

Fig. CD-36D
Lines of Sight Between the Heads of Seated Spectators
© ICC Reproduced with Permission

Lines of Sight Over Standing Spectators
Where spectators are expected to stand during events, spectators in wheelchair spaces shall be afforded lines of sight as follows: **11B-802.2.2** *802.2.2*

Lines of Sight Over Heads

_____L.   Where spectators are provided lines of sight over the heads of spectators standing in the first row in front of their seats, spectators seated in wheelchair spaces are afforded lines of sight over the heads of standing spectators in the first row in front of wheelchair spaces. **11B-802.2.2.1**   *802.2.2.1*   **Fig. CD-36E**

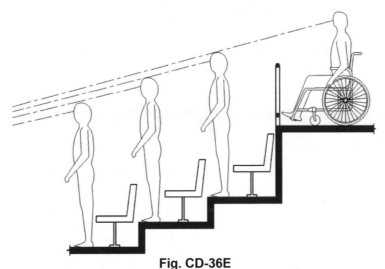

Fig. CD-36E
Lines of Sight Over the Heads of Standing Spectators
© ICC Reproduced with Permission

Lines of Sight Between Heads

_____M.   Where spectators are provided lines of sight over the shoulders and between the heads of spectators standing in the first row in front of their seats, spectators seated in wheelchair spaces are afforded lines of sight over the shoulders and between the heads of standing spectators in the first row in front of wheelchair spaces.   **11B-802.2.2.2**   *802.2.2.2* **Fig. CD-36F**

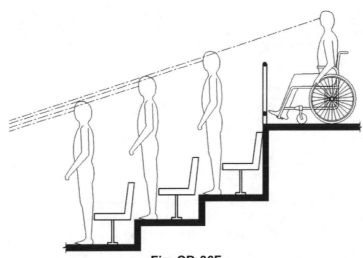

Fig. CD-36F
Lines of Sight Between the Heads of Standing Spectators
© ICC Reproduced with Permission

COMPANION SEATS

Alignment

_____N.   In row seating, companion seats are located to provide shoulder alignment with adjacent wheelchair spaces.   **11B-802.3.1**   *802.3.1*

_____O.   The shoulder alignment point of the wheelchair space is measured 36 Inches from the front of the wheelchair space.   **11B-802.3.1**   *802.3.1*

_____P.   The floor surface of the companion seat is at the same elevation as the floor surface of the wheelchair space.   **11B-802.3.1**   *802.3.1*

Type

_____Q.   Companion seats are equivalent in size, quality, comfort, and amenities to the seating in the immediate area. Companion seats are permitted to be moveable. **11B-802.3.2**   *802.3.2*

DESIGNATED AISLE SEATS

Armrests

_____R.   Where armrests are provided on the seating in the immediate area, folding or retractable armrests are provided on the aisle side of the seat.   **11B-802.4.1** *802.4.1*   **Fig. CD-36G**

Identification

_____S.   Each designated aisle seat is identified with a sign or marker with the International Symbol of Accessibility.   **11B-802.4.2**   *802.4.2*   **Fig. CD-36G**

_____T.   The International Symbol of Accessibility complies with the applicable provisions detailed in Section 30, ""SIGNS".   **11B-802.4.2**

> **ADVISORY: *Identification.*** *Seats with folding or retractable armrests are intended for use by individuals who have difficulty walking. Consider identifying such seats with signs that contrast (light-on-dark or dark-on-light) and that are also photo luminescent.*

_____U.   Signage notifying patrons of the availability of such seats is posted at the ticket office.   **11B-802.4.2**   **Fig. CD-36G**

_____V.   The notification sign complies with the requirements for accessible "Visual Characters" listed in Section 30, "SIGNS".   **11B-802.4.2**

> **ADVISORY: *Identification.*** *Signage notifying patrons of the availability of aisle seats shall be posted at the ticket office. If there is no ticket office, the functional equivalent would be in the lobby or at the entrance to the assembly area in a conspicuous location.*

SEMI-AMBULANT SEATS

_____W.   Semi-ambulant seats provide at least 24 inches clear leg space between the front of the seat to the nearest obstruction or to the back of the seat immediately in front.   **11B-802.5   Fig. CD-36G**

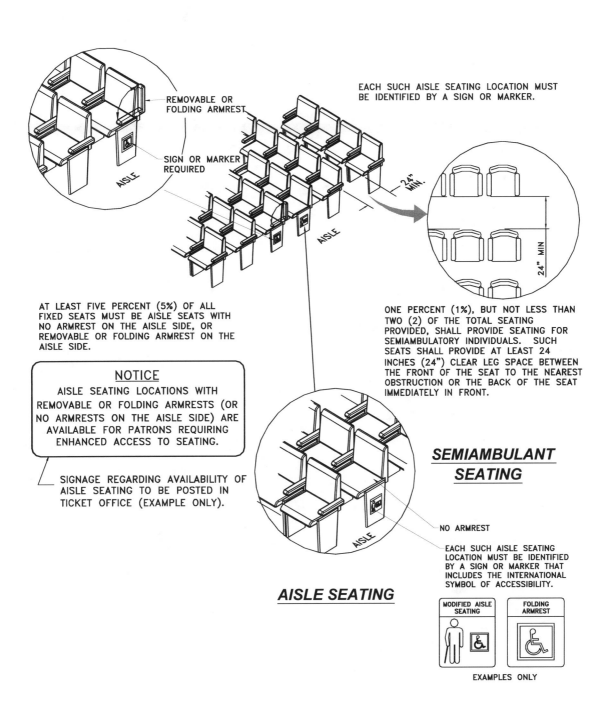

EACH SUCH AISLE SEATING LOCATION MUST BE IDENTIFIED BY A SIGN OR MARKER.

REMOVABLE OR FOLDING ARMREST

SIGN OR MARKER REQUIRED

AISLE

AISLE

24" MIN.

24" MIN

AT LEAST FIVE PERCENT (5%) OF ALL FIXED SEATS MUST BE AISLE SEATS WITH NO ARMREST ON THE AISLE SIDE, OR REMOVABLE OR FOLDING ARMREST ON THE AISLE SIDE.

ONE PERCENT (1%), BUT NOT LESS THAN TWO (2) OF THE TOTAL SEATING PROVIDED, SHALL PROVIDE SEATING FOR SEMIAMBULATORY INDIVIDUALS. SUCH SEATS SHALL PROVIDE AT LEAST 24 INCHES (24") CLEAR LEG SPACE BETWEEN THE FRONT OF THE SEAT TO THE NEAREST OBSTRUCTION OR THE BACK OF THE SEAT IMMEDIATELY IN FRONT.

NOTICE
AISLE SEATING LOCATIONS WITH REMOVABLE OR FOLDING ARMRESTS (OR NO ARMRESTS ON THE AISLE SIDE) ARE AVAILABLE FOR PATRONS REQUIRING ENHANCED ACCESS TO SEATING.

SIGNAGE REGARDING AVAILABILITY OF AISLE SEATING TO BE POSTED IN TICKET OFFICE (EXAMPLE ONLY).

SEMIAMBULANT SEATING

AISLE

NO ARMREST

EACH SUCH AISLE SEATING LOCATION MUST BE IDENTIFIED BY A SIGN OR MARKER THAT INCLUDES THE INTERNATIONAL SYMBOL OF ACCESSIBILITY.

AISLE SEATING

MODIFIED AISLE SEATING

FOLDING ARMREST

EXAMPLES ONLY

Fig. CD-36G
Designated Aisle Seats and Semi-Ambulant Seats

ACCESSIBLE SEATING REQUIREMENTS
IN ASSEMBLY AREAS

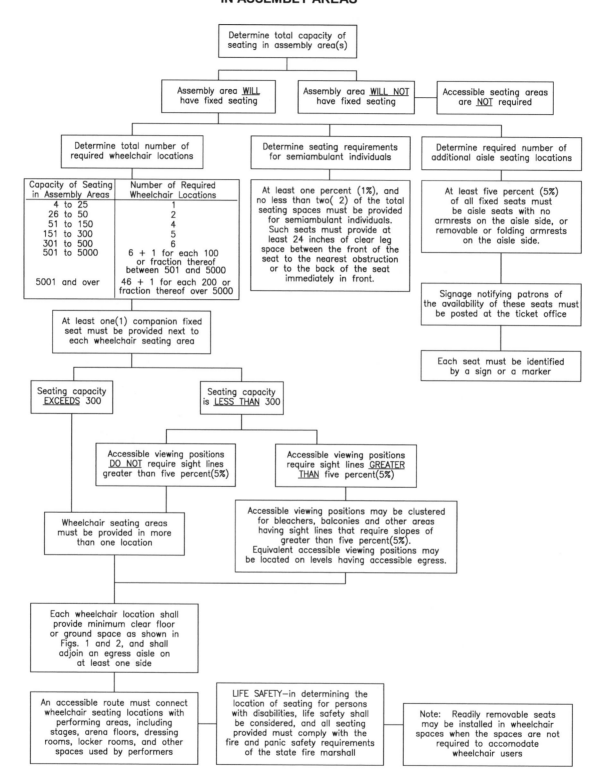

Determine total capacity of seating in assembly area(s)

Assembly area **WILL** have fixed seating

Assembly area **WILL NOT** have fixed seating

Accessible seating areas are **NOT** required

Determine total number of required wheelchair locations

Determine seating requirements for semiambulant individuals

Determine required number of additional aisle seating locations

Capacity of Seating in Assembly Areas	Number of Required Wheelchair Locations
4 to 25	1
26 to 50	2
51 to 150	4
151 to 300	5
301 to 500	6
501 to 5000	6 + 1 for each 100 or fraction thereof between 501 and 5000
5001 and over	46 + 1 for each 200 or fraction thereof over 5000

At least one percent (1%), and no less than two(2) of the total seating spaces must be provided for semiambulant individuals. Such seats must provide at least 24 inches of clear leg space between the front of the seat to the nearest obstruction or to the back of the seat immediately in front.

At least five percent (5%) of all fixed seats must be aisle seats with no armrests on the aisle side, or removable or folding armrests on the aisle side.

Signage notifying patrons of the availability of these seats must be posted at the ticket office

Each seat must be identified by a sign or a marker

At least one(1) companion fixed seat must be provided next to each wheelchair seating area

Seating capacity **EXCEEDS** 300

Seating capacity is **LESS THAN** 300

Accessible viewing positions **DO NOT** require sight lines greater than five percent(5%)

Accessible viewing positions require sight lines **GREATER THAN** five percent(5%)

Wheelchair seating areas must be provided in more than one location

Accessible viewing positions may be clustered for bleachers, balconies and other areas having sight lines that require slopes of greater than five percent(5%). Equivalent accessible viewing positions may be located on levels having accessible egress.

Each wheelchair location shall provide minimum clear floor or ground space as shown in Figs. 1 and 2, and shall adjoin an egress aisle on at least one side

An accessible route must connect wheelchair seating locations with performing areas, including stages, arena floors, dressing rooms, locker rooms, and other spaces used by performers

LIFE SAFETY—in determining the location of seating for persons with disabilities, life safety shall be considered, and all seating provided must comply with the fire and panic safety requirements of the state fire marshall

Note: Readily removable seats may be installed in wheelchair spaces when the spaces are not required to accomodate wheelchair users

37. DRESSING, FITTING AND LOCKER ROOMS

Dressing, Fitting, and Locker Rooms   11B-222, 11B-803   *222, 803*

General
Where dressing rooms, fitting rooms, or locker rooms are provided, at least 5 percent, but no fewer than one, of each type of use in each cluster provided shall comply as detailed. **11B-222.1**   *222.1*

> **EXCEPTION:** In alterations, where it is technically infeasible to provide rooms in accordance with this Section, one room for each sex on each level shall comply as detailed. Where only unisex rooms are provided, unisex rooms shall be permitted.

> *ADVISORY: General. A "cluster" is a group of rooms proximate to one another. Generally, rooms in a cluster are within sight of, or adjacent to, one another. Different styles of design provide users varying levels of privacy and convenience. Some designs include private changing facilities that are close to core areas of the facility, while other designs use space more economically and provide only group dressing facilities. Regardless of the type of facility, dressing, fitting, and locker rooms should provide people with disabilities rooms that are equally private and convenient to those provided others. For example, in a physician's office, if people without disabilities must traverse the full length of the office suite in clothing other than their street clothes, it is acceptable for people with disabilities to be asked to do the same.*

Coat Hooks and Shelves
Where coat hooks or shelves are provided in dressing, fitting or locker rooms without individual compartments, at least one of each type of coat hook or shelf shall comply as detailed. Where coat hooks or shelves are provided in individual compartments, at least one of each type of coat hook or shelf that complies as detailed shall be provided in individual compartments in dressing, fitting, or locker rooms that are required to comply. **11B-222.2**   *222.2*

Mirrors
Where mirrors are provided in dressing, fitting or locker rooms without individual compartments, at least one of each type of mirror shall comply as detailed. Where mirrors are provided in individual compartments at least one of each type of mirror that complies as detailed shall be provided in individual compartments in dressing, fitting or locker rooms that are required to comply.   **11B-222.3**

General
Dressing, fitting, and locker rooms shall comply as detailed.   **11B-803.1**   *803.1*

> *ADVISORY: General. Partitions and doors should be designed to ensure people using accessible dressing and fitting rooms privacy equivalent to that afforded other users of the facility. Section 47, titled "BENCHES", requires dressing room bench seats to be installed so that they are at the same height as a typical wheelchair seat, 17 inches (430 mm) to 19 inches (485 mm). However, wheelchair seats can be lower than dressing room benches for people of short stature or children using wheelchairs.*

Turning Space

_____A.   Turning space that complies with Section 9, "TURNING SPACE", shall be provided within the room.   **11B-803.2**   *803.2*

Door Swing

_____B.   Doors shall not swing into the room unless a turning space complying with Section 9, "TURNING SPACE", is provided beyond the arc of the door swing. **11B-803.3**   *803.3*

Benches

_____C.   A bench complying with Section 47, "BENCHES", shall be provided within the room. **11B-803.4**   *803.4*

Coat Hooks and Shelves

_____D.   Coat hooks provided within the room shall be located within one of the reach ranges specified in Section 12, "REACH RANGES".   **11B-803.5**   *803.5*

_____D1.   Shelves shall be 40 inches (*1016* mm) minimum and 48 inches (*1219* mm) maximum above the finish floor or ground. **11B-803.5**   *803.5*

_____D2.   Coat hooks shall not be located above the bench or other seating in the room.   **11B-803.5**   *803.5*

Mirrors

_____E.   Mirrors shall be installed with the bottom edge of the reflecting surface 20 inches (508 mm) maximum above the finish floor or ground.   **11B-803.6**   *803.6*

_____E1.   Mirrors shall be full length with a reflective surface 18 inches (457 mm) wide minimum by 54 inches (1372 mm) high minimum and shall be mounted in a position affording a view to a person on the bench as well as to a person in a standing position.   **11B-803.6**   *803.6*

38. <u>MEDICAL CARE AND LONG-TERM CARE FACILITIES</u>

<u>Passenger Loading Zone</u>

At least one accessible passenger loading zone shall be provided at an accessible entrance to licensed medical care and licensed long-term care facilities where the period of stay exceeds twenty-four hours. **11B-209.3**   *209.3*

<u>Entrances</u>

_____A.   Weather protection by a canopy or roof overhang shall be provided at a minimum of one accessible entrance to licensed medical care and licensed long-term care facilities where the period of stay may exceed twenty-four hours. **11B-206.4.10** *206.4.10*   **Fig. CD-38A**

_____A1.   The area of weather protection shall include the accessible passenger loading zone and the accessible route from the passenger loading zone to the accessible entrance it serves.   **11B-206.4.10**   *206.4.10* **Fig. CD-38A**

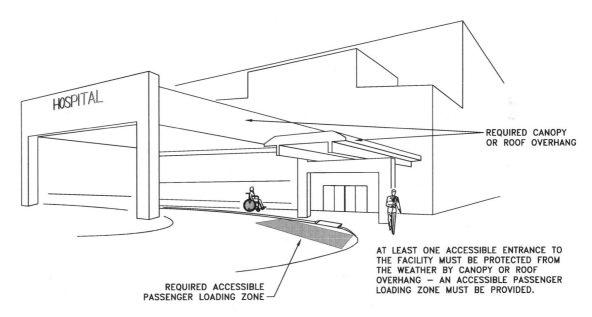

Fig. CD-38A
Weather Protection and Passenger Loading Zone at Entrance

<u>Parking For Hospital Outpatient Facilities</u>

Ten percent of patient and visitor parking spaces provided to serve hospital outpatient facilities, and free-standing buildings providing outpatient clinical services of a hospital shall comply.   **11B-208.2.1**   *208.2.1*

***ADVISORY. Hospital Outpatient Facilities.** The term "outpatient facility" is not defined but is intended to cover facilities or units that are located in hospitals and that provide regular and continuing medical treatment without an overnight stay. Doctor's offices,* *independent clinics, or other facilities not located in hospitals are not considered hospital outpatient facilities for purposes of these requirements.*

The higher percentages required for hospital outpatient facilities or rehabilitation facilities specializing in treating conditions that affect mobility and outpatient physical therapy facilities are intended primarily for visitor and patient parking. If there are separate lots for visitors or patients and employees, the 10% or 20% requirement shall be applied to the visitor/patient lot while accessible parking could be provided in the employee parking lot according to the general scoping requirements in Table 21A. If a lot serves both visitors or patients and employees, 10% or 20% of the spaces intended for use by visitors or patients must be accessible.

At hospitals or other facilities where parking does not specifically serve an outpatient unit, only a portion of the lot would need to comply with the 10% scoping requirement. A local zoning code that requires a minimum number of parking spaces according to occupancy type and square footage may be an appropriate guide in assessing the number of spaces in the lot that "belong" to the outpatient unit. These spaces would be held to the 10% requirement while the rest of the lot would be subject to general scoping requirement in the Table. Those accessible spaces required for the outpatient unit should be located at the accessible entrance serving the unit. This method may also be used in applying the 20% requirement to hospitals or other facilities where only a portion of the unit provides specialized rehabilitation or physical therapy treatment or services for persons with mobility impairments.

Parking For Rehabilitation Facilities and Outpatient Physical Therapy Facilities

Twenty percent of patient and visitor parking spaces provided to serve rehabilitation facilities specializing in treating conditions that affect mobility and outpatient physical therapy facilities shall comply. **11B-208.2.2**   *208.2.2*

*ADVISORY. **Rehabilitation Facilities and Outpatient Physical Therapy Facilities.** Conditions that affect mobility include conditions requiring the use of a brace, cane, crutch, prosthetic device, wheelchair, or powered mobility aid; arthritic, neurological, or orthopedic conditions that severely limit one's ability to walk; respiratory diseases and other conditions which may require the use of portable oxygen; and cardiac conditions that impose significant functional limitations.*

Medical Care and Long-Term Care Facilities   **11B-223**   *223*

General

In licensed medical care facilities and licensed long-term care facilities where the period of stay exceeds twenty-four hours, patient bedrooms or resident sleeping rooms shall be provided in accordance with the requirements contained herein.   **11B-223.1**   *223.1*

> **EXCEPTION:** Toilet rooms that are part of critical or intensive care patient sleeping rooms shall not be required to comply with items A through J of Section 26, "TOILET AND BATHING FACILITIES".

*ADVISORY: **General.** Because medical facilities frequently reconfigure spaces to reflect changes in medical specialties, this section does not include a provision for dispersion of accessible patient bedrooms or resident sleeping rooms. The lack of a design requirement does not mean that covered entities are not required to provide services to people with disabilities where accessible patient bedrooms or resident sleeping rooms are not dispersed in specialty areas. Locate accessible patient bedrooms or resident sleeping rooms near core*

> *areas that are less likely to change over time. While dispersion is not required, the flexibility it provides can be a critical factor in ensuring cost effective compliance with applicable civil rights laws, including titles II and III of the ADA and Section 504 of the Rehabilitation Act of 1973, as amended. Additionally, all types of features and amenities should be dispersed among accessible patient bedrooms or resident sleeping rooms to ensure equal access to and a variety of choices for all patients and residents.*

Alterations

Where patient bedrooms or resident sleeping rooms are altered or added, the requirements listed below for "Facilities Not Specializing in Treating Conditions That Affect Mobility", "Facilities Specializing in Treating Conditions That Affect Mobility", "On Call Rooms", and "Long-Term Care Facilities" shall apply only to the patient bedrooms or resident sleeping rooms being altered or added until the number of patient bedrooms or resident sleeping rooms complies with the minimum number required for new construction. **11B-223.1.1** *223.1.1*

> **ADVISORY: Alterations.** *In alterations and additions, the minimum required number is based on the total number of patient bedrooms or resident sleeping rooms altered or added instead of on the total number of patient bedrooms or resident sleeping rooms provided in a facility. As a facility is altered over time, every effort should be made to disperse accessible patient bedrooms or resident sleeping rooms among patient care areas such as pediatrics, cardiac care, maternity, and other units. In this way, people with disabilities can have access to the full-range of services provided by a medical care facility.*

Area Alterations

Patient bedrooms or resident sleeping rooms added or altered as part of a planned renovation of an entire wing, a department, or other discrete area of an existing medical facility shall comply with items B through E, below, until the number of patient bedrooms or resident sleeping rooms provided within the area of renovation complies with the minimum number required for new construction by the percentage requirements listed below for "Facilities Not Specializing in Treating Conditions That Affect Mobility", "Facilities Specializing in Treating Conditions That Affect Mobility", "On-Call Rooms", and "Long-Term Care Facilities". **11B-223.1.1.1**

Individual Alterations

Patient bedrooms or resident sleeping rooms added or altered individually, and not as part of an alteration of an entire area, shall comply with items B through E, below, until either: a) the number of patient bedrooms or resident sleeping rooms provided in the department or area containing the individually altered or added patient bedrooms or resident sleeping rooms complies with the minimum number required if the percentage requirements listed below for "Facilities Not Specializing in Treating Conditions That Affect Mobility", "Facilities Specializing in Treating Conditions That Affect Mobility", "On-Call Rooms", and "Long-Term Care Facilities", were applied to that department or area; or b) the overall number of patient bedrooms or resident sleeping rooms in the facility complies with the minimum number required for new construction by the percentage requirements listed below for "Facilities Not Specializing in Treating Conditions That Affect Mobility", "Facilities Specializing in Treating Conditions That Affect Mobility", "On-Call Rooms", and "Long-Term Care Facilities". **11B-223.1.1.2**

Toilet and Bathing Facilities

Toilet/bathing rooms which are part of patient bedrooms added or altered and required to be accessible shall comply with items B through E, below. **11B-223.1.1.3**

Hospitals, Rehabilitation Facilities, Psychiatric Facilities and Detoxification Facilities Hospitals, rehabilitation facilities, psychiatric facilities and detoxification facilities shall comply with the requirements listed below for "Facilities Not Specializing in Treating Conditions That Affect Mobility", "Facilities Specializing in Treating Conditions That Affect Mobility", "On-Call Rooms", and "Long-Term Care Facilities". All public use and common use areas shall be accessible in compliance as detailed. **11B-223.2** *223.2*

Facilities Not Specializing in Treating Conditions That Affect Mobility
In facilities not specializing in treating conditions that affect mobility, including hospitals, psychiatric and detoxification facilities, at least 10 percent, but no fewer than one, of the patient bedrooms or resident sleeping rooms shall provide mobility features complying with items B through K, below. Accessible patient bedrooms or resident sleeping rooms shall be dispersed in a manner that is proportionate by type of medical specialty. **11B-223.2.1**

Facilities Specializing in Treating Conditions That Affect Mobility
In facilities specializing in treating conditions that affect mobility, 100 percent of the patient bedrooms shall provide mobility features complying with items B through K, below. **11B-223.2.2** *223.2.2*

> *ADVISORY: Facilities Specializing in Treating Conditions That Affect Mobility. Conditions that affect mobility include conditions requiring the use or assistance of a brace, cane, crutch, prosthetic device, wheelchair, or powered mobility aid; arthritic, neurological, or orthopedic conditions that severely limit one's ability to walk; respiratory diseases and other conditions which may require the use of portable oxygen; and cardiac conditions that impose significant functional limitations. Facilities that may provide treatment for, but that do not specialize in treatment of such conditions, such as general rehabilitation hospitals, are not subject to this requirement but are subject to Section 11B-223.2.1.*

On-Call Rooms
Where physician or staff on-call sleeping rooms are provided, at least 10 percent, but no fewer than one, of the on-call rooms shall provide mobility features complying with items D through H, and items K through O, of Section 39, "TRANSIENT LODGING FACILITIES". **11B-223.2.3**

Long-Term Care Facilities
In licensed long-term care facilities, including skilled nursing facilities, intermediate care facilities and nursing homes, at least 50 percent, but no fewer than one, of each type of patient bedroom or resident sleeping room shall provide mobility features as detailed, below. **11B-223.3**

Professional Offices of Health Care Providers
Professional offices of health care providers shall comply as detailed, below.
11B-223.4

Medical Care and Long-Term Care Facilities **11B-805** *805*

General
Medical care facilities and long-term care facilities shall comply as detailed, below. All common use spaces and public use spaces in medical care facilities and long-term care facilities shall comply with these provisions. **11B-805.1** *805.1*

Patient Bedrooms and Resident Sleeping Rooms

Patient bedrooms and resident sleeping rooms required to provide mobility features shall comply as detailed.　**11B-805.2**

Hand Washing Fixtures

_____B.　Hand washing fixtures shall comply with Section 26F, "LAVATORIES AND SINKS".　**11B-805.2.1**

Beds

_____C.　A 36 inch (914 mm) minimum wide clear space shall be provided along the full length of each side of the beds.　**11B-805.2.2**　*805.3*　**Fig. CD-38B**

Turning Space

_____D.　Turning space complying with Section 9, "TURNING SPACE", shall be provided within the room.　**11B-805.2.3**　*805.2*　**Fig. CD-38B**

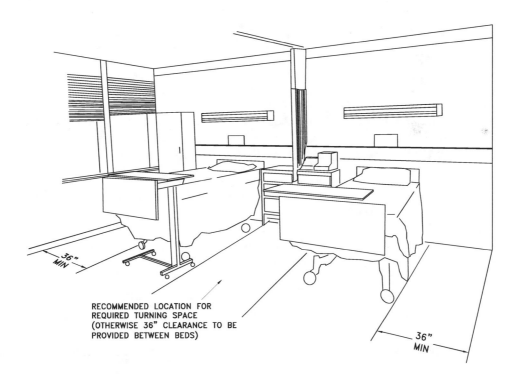

RECOMMENDED LOCATION FOR
REQUIRED TURNING SPACE
(OTHERWISE 36" CLEARANCE TO BE
PROVIDED BETWEEN BEDS)

36"
MIN

36"
MIN

- A 36" MINIMUM CLEAR PATH OF TRAVEL MUST BE PROVIDED THROUGHOUT ALL PORTIONS OF THE BEDROOMS TO BED(S), BATHROOM, PHONE, ETC.
- EACH BEDROOM MUST HAVE A 36" MINIMUM CLEAR FLOOR SPACE ALONG EACH SIDE OF BED.
- A CLEAR SPACE MUST BE PROVIDED IN EACH BEDROOM SUFFICIENT TO INSCRIBE A MINIMUM 60" DIAMETER CIRCLE OR A T-SHAPED SPACE IS PROVIDED THAT ALLOWS FOR 180 DEGREE TURN.

Fig. CD-38B
Clear Space and Turning Space in
Patient Bedrooms and Resident Sleeping Rooms

Toilet and Bathing Rooms

_____E.   Toilet and bathing rooms that are provided as part of patient bedrooms and resident sleeping rooms complying with the requirements listed under "Facilities Not Specializing in Treating Conditions That Affect Mobility", "Facilities Specializing in Treating Conditions That Affect Mobility", "On-Call Rooms", and "Long-Term Care Facilities", above, shall comply with items A through J of Section 26, "TOILET AND BATHING FACILITIES". Where provided, one water closet, one lavatory, and one bathtub or shower shall comply with the applicable requirements of Sections 26 through 26H for accessible toilet and bathing facilities. **11B-805.2.4** *805.4*

Waiting Rooms
Waiting rooms shall comply as detailed. **11B-805.3**

Wheelchair Spaces

_____F.   Where seating is provided in waiting rooms, at least 5 percent of the seating shall be wheelchair spaces complying with items A through I of Section 36, "ASSEMBLY AREAS". **11B-805.3.1**

> **EXCEPTION:** In waiting rooms serving facilities specializing in treating conditions that affect mobility, 10 percent of the seating shall be wheelchair spaces complying with items A through I of Section 36, "ASSEMBLY AREAS".

Examination, Diagnostic and Treatment Rooms
Examination, diagnostic and treatment rooms shall comply as detailed, below. **11B-805.4**

Beds, Exam Tables, Procedure Tables, Gurneys and Lounge Chairs

_____G.   A 36 inch (914 mm) minimum wide clear space shall be provided along the full length of each side of beds, exam tables, procedure tables, gurneys and lounge chairs. **11B-805.4.1**

> **EXCEPTION:** General exam rooms in non-emergency settings may provide clear space on only one side of beds, gurneys and exam tables.

Equipment

_____H.   Clear floor space shall be provided as required for specific equipment that complies with Section 7, "FLOOR OR GROUND SURFACES". Changes in level are not permitted. **11B-805.4.2**

> **EXCEPTION**: Slopes not steeper than 1:48 shall be permitted.

Turning Space

_____I.   Turning space complying with Section 9, "TURNING SPACE", shall be provided within the room. **11B-805.4.3**

Patient Change Areas
Areas where patients change or are prepared for a procedure shall comply as detailed, below. **11B-805.5**

Hand Washing Fixtures, Lavatories and Sinks

_____J.   All hand washing fixtures, lavatories and sinks shall comply with Section 26F, "LAVATORIES AND SINKS". **11B-805.6**

> **EXCEPTION:** Scrub sinks, as defined in California Plumbing Code Section 221.0, shall not be required to comply with this provision.

Built-In Cabinets and Work Surfaces

_____K.   Built-in cabinets, counters and work surfaces shall be accessible, including: patient wardrobes, nurse's stations, administrative centers, reception desks, medicine preparation areas, laboratory work stations, equipment consoles, clean and soiled utility cabinets, and storage areas; and shall comply with Section 45, "STORAGE", and Section 46, "DINING SURFACES AND WORK SURFACES". **11B-805.7**

> **EXCEPTION:**
>
> 1.   Built-in wardrobes in patient bedrooms and resident sleeping rooms not required to be accessible are not required to comply with the provisions of this chapter.
>
> 2.   Clinical laboratory work stations provided in a laboratory area that are in addition to the minimum number required to be accessible (5 percent of the work stations provided, but no fewer than one), are not required to comply with the provisions of Section 46, "DINING SURFACES AND WORK SURFACES".

39. <u>TRANSIENT LODGING GUEST ROOMS</u>

<u>Multi-Story Buidings and Facilities</u>
At least one accessible route shall connect each story and mezzanine in multi-story buildings and facilities. **11B-206.2.3** *206.2.3*

> **EXCEPTION:** Within multi-story transient lodging guest rooms with mobility features, an accessible route shall not be required to connect stories provided that the living and dining areas, exterior spaces and sleeping areas are on an accessible route and sleeping accommodations for two persons minimum are provided on a story served by an accessible route. **11B-206.2.3, Exception 5**, *206.2.3, Exception 5*

> ### <u>Platform Lifts in Guest Rooms and Residential Dwelling Units</u>
> Platform lifts shall be permitted to connect levels within transient lodging guest rooms required to provide mobility features complying with Section 39, "TRANSIENT LODGING GUEST ROOMS", or residential dwelling units required to provide mobility features complying with Section 43, "RESIDENTIAL FACILITIES" and adaptable features complying with Chapter 11A, Division IV. **11B-206.7.6** *206.7.6*

<u>Transient Lodging Guest Rooms</u>   **11B-224**   *224*

<u>General</u>
Hotels, motels, inns, dormitories, resorts and similar transient lodging facilities shall provide guest rooms in accordance with these provisions. **11B-224.1** *224.1*

> *ADVISORY: General. Certain facilities used for transient lodging, including time shares, dormitories, and town homes may be covered by both these requirements and the Fair Housing Amendments Act. The Fair Housing Amendments Act requires that certain residential structures having four or more multi-family dwelling units, regardless of whether they are privately owned or federally assisted, include certain features of accessible and adaptable design according to guidelines established by the U.S. Department of Housing and Urban Development (HUD). This law and the appropriate regulations should be consulted before proceeding with the design and construction of residential housing.*

<u>Alterations</u>
Where guest rooms are altered or added, the requirements of contained herein shall apply only to the guest rooms being altered or added until the number of guest rooms complies with the minimum number required for new construction. **11B-224.1.1** *224.1.1*

> *ADVISORY: Alterations. In alterations and additions, the minimum required number of accessible guest rooms is based on the total number of guest rooms altered or added instead of the total number of guest rooms provided in a facility. Typically, each alteration of a facility is limited to a particular portion of the facility. When accessible guest rooms are added as a result of subsequent alterations, compliance with the requirement for "Dispersion" is more likely to be achieved if all of the accessible guest rooms are not provided in the same area of the facility.*

<u>Guest Room Doors and Doorways</u>
Entrances, doors, and doorways providing user passage into and within guest rooms that are not required to provide compliant mobility features shall comply with the Clear Width requirements for

door openings contained in items C through F of Section 14, "DOORS, DOORWAYS AND GATES". Bathrooms doors shall be either sliding or hung to swing in the direction of egress from the bathroom. **11B-206.5.3, 11B-224.1.2** *206.5.3, 224.1.2*

> **EXCEPTION:** Shower and sauna doors in guest rooms that are not required to provide compliant mobility shall not be required to comply with this provision.

> ***ADVISORY: Guest Room Doors and Doorways.*** *Because of the social interaction that often occurs in lodging facilities, an accessible clear opening width is required for doors and doorways to and within all guest rooms, including those not required to be accessible. This applies to all doors, including bathroom doors, that allow full user passage. Other requirements for doors and doorways contained in Sections 14 and 14A of this publication do not apply to guest rooms not required to provide mobility features.*

Range of Accommodations
Accessible guest rooms or suites shall be dispersed among the various classes of sleeping accommodations to provide a range of options applicable to room sizes, costs, and amenities provided. **11B-224.1.3**

Housing at a Place of Education
Housing at a place of education subject to this section shall comply with the requirements for transient lodging guest rooms contained herein. For the purposes of the application of this section, the term "sleeping room" is interchangeable with "guest room" as used in the transient lodging standards. **11B-224.1.4**

> **EXCEPTIONS:**
>
> 1. Kitchens within housing units containing accessible sleeping rooms with mobility features (including suites and clustered sleeping rooms) or on floors containing accessible sleeping rooms with mobility features shall provide all rooms served by an accessible route with a turning spaces that complies with Section 9, "TURNING SPACE", and kitchen work surfaces that comply with the requirements under "Kitchen Work Surfaces", items E through H, in Section 40, "KITCHENS, KITCHENETTES, WET BARS AND SINKS".
>
> 2. Multi-bedroom housing units containing accessible sleeping rooms with mobility features shall have an accessible route throughout the unit in compliance with items A through A2 of Section 43, "RESIDENTIAL FACILITIES".
>
> 3. Residential dwelling units that are provided by or on behalf of a place of education, which are leased on a year round basis exclusively to graduate students or faculty, and do not contain any public use or common use areas available for educational programming, are not subject to the transient lodging standards and shall comply with the requirements in Section 43, "RESIDENTIAL FACILITIES".

Social Service Center Establishments
Group homes, halfway houses, shelters, or similar social service center establishments that provide either temporary sleeping accommodations or residential dwelling units subject to this section shall comply with the requirements of this section. **11B-224.1.5**

More than Twenty-Five Bed Sleeping Rooms

In sleeping rooms with more than twenty-five beds, a minimum of 5 percent of the beds shall have clear floor space complying with items D through E2, below.   **11B-224.1.5.1**

More than Fifty Bed Facilities

Facilities with more than fifty beds that provide common use bathing facilities, shall provide at least one roll-in shower with a seat that complies with Section 26H, "SHOWER COMPARTMENTS".   When separate shower facilities are provided for men and women, at least one roll-in shower shall be provided for each group.   **11B-224.1.5.2**

Guest Room Toilet and Bathing Rooms

Where toilet and bathing rooms are provided in guest rooms that are not required to provide compliant mobility features as detailed below, toilet and bathing fixtures shall only be required to provide all toilet and bathing fixtures in a location that allows a person using a wheelchair measuring 30 inches by 48 inches (762 mm by 1219 mm) to touch the wheelchair to any lavatory, urinal, water closet, tub, sauna, shower stall and any other similar sanitary installation, if provided.   **11B-224.1.6, 11B-603.6**

Guest Rooms with Mobility Features

In transient lodging facilities, guest rooms with mobility features complying with items A through J, below, shall be provided in accordance with Table 39A, as follows.
11B-224.2   *224.2*

Fifty or Less Guest Room Facilities

Facilities that are subject to the same permit application on a common site that each have 50 or fewer guest rooms may be combined for the purposes of determining the required number of accessible rooms and type of accessible bathing facility.   **11B-224.2.1**

More than Fifty Guest Room Facilities

Facilities with more than 50 guest rooms shall be treated separately for the purposes of determining the required number of accessible rooms and type of accessible bathing facility.   **11B-224.2.2**

Table 39A
Guest Rooms with Mobility Features

Total Number of Guest Rooms Provided	Minimum Number of Required Rooms Without Roll-in Showers	Minimum Number of Required Rooms With Roll-in Showers	Total Number of Required Rooms
1 to 25	*0*	*1*	1
26 to 50	*1*	*1*	2
51 to 75	3	1	4
76 to 100	4	1	5
101 to 150	5	2	7
151 to 200	6	2	8
201 to 300	7	3	10
301 to 400	8	4	12
401 to 500	9	4	13
501 to 1000	2 percent of total	1 percent of total	3 percent of total
1001 and over	20, plus 1 for each 100, or fraction thereof, over 1000	10, plus 1 for each 100, or fraction thereof, over 1000	30, plus 2 for each 100, or fraction thereof, over 1000

> **ADVISORY: Guest Rooms with Mobility Features.** *In hotels, motels or other transient lodging facilities, Table 39A indicates the required total number of mobility accessible rooms, including both rooms without roll-in showers and with roll-in showers. All of these rooms are required to comply with the technical requirements of items A through J, below. Not more than 10 percent of guest rooms required to provide mobility features by Table 39A may be used to satisfy the minimum number of guest rooms required to provide communication features by Table 39B.*
>
> *For example, a new 275-room hotel must have a total of ten mobility accessible guest rooms; a minimum of three of the ten must have roll-in showers per Table 39A. In addition to those ten mobility accessible rooms, 17 guest rooms are required to provide communication features per Table 39B. Only 10 percent of the rooms required to provide mobility features, or one mobility accessible room, may also be used to satisfy the requirements of Table 39B. Sixteen additional communication accessible rooms must be provided.*

Beds
In guest rooms having more than 25 beds, 5 percent minimum of the beds shall have clear floor space complying with items D through E2, below. **11B-224.3** *224.3*

Guest Rooms with Communication Features
In transient lodging facilities, guest rooms with communication features complying with items K through O shall be provided in accordance with Table 39B. **11B-224.4** *224.4*

Table 39B
Guest Rooms with Communication Features

Total Number of Guest Rooms Provided	Minimum Number of Required Guest Rooms With Communication Features
1	1
2 to 25	2
26 to 50	4
51 to 75	7
76 to 100	9
101 to 150	12
151 to 200	14
201 to 300	17
301 to 400	20
401 to 500	22
501 to 1000	5 percent of total
1001 and over	50, plus 3 for each 100 over 1000

Dispersion
Guest rooms required to provide mobility features complying with items A through J, below, and guest rooms required to provide communication features complying with items K through O shall be dispersed among the various classes of guest rooms, and shall provide choices of types of guest rooms, number of beds, and other amenities comparable to the choices provided to other guests. Where the minimum number of guest rooms required to comply is not sufficient to allow for complete dispersion, guest rooms shall be dispersed in the following priority: guest room type, number of beds, and amenities. At least one guest room required to provide compliant mobility features shall also provide compliant communication features. Not more than 10 percent of guest

rooms required to provide compliant mobility features shall be used to satisfy the minimum number of guest rooms required to provide compliant communication features. **11B-224.5**   *224.5*

ADVISORY: Dispersion. Factors to be considered in providing an equivalent range of options may include, but are not limited to, room size, bed size, cost, view, bathroom fixtures such as hot tubs and spas, smoking and nonsmoking, and the number of rooms provided.

Storage

_____A.   Fixed or built-in storage facilities within guest rooms required to provide mobility features shall comply with Section 45, "STORAGE".   **11B-224.6**

Transient Lodging Guest Rooms   **11B-806**   *806*

General
Transient lodging guest rooms shall comply as detailed. Guest rooms required to provide mobility features shall comply with items A through J. Guest rooms required to provide communication features shall comply with items K through O.   **11B-806.1**   *806.1*

Guest Rooms with Mobility Features
Guest rooms required to provide mobility features shall comply with items A through J. **11B-806.2** *806.2*

ADVISORY: Guest Rooms with Mobility Features. The requirements in items A through J do not include requirements that are common to all accessible spaces. For example, closets in guest rooms must comply with the applicable provisions for storage specified in scoping.

Living and Dining Areas

_____B.   Living and dining areas shall be accessible.   **11B-806.2.1**   *806.2.1*

Exterior Spaces

_____C.   Exterior spaces, including patios, terraces and balconies, that serve the gues room shall be accessible.   **11B-806.2.2**   *806.2.2*

Sleeping Areas

_____D.   At least one sleeping area shall provide a 36 inch (914 mm) by 48 inch (1219 mm) minimum clear space on both sides of a bed.   **11B-806.2.3**   *806.2.3* **Fig. CD-39A**

_____D1.   The clear space shall be positioned for parallel approach to the side of the bed.   **11B-806.2.3**   *806.2.3*   **Fig. CD-39A**

EXCEPTION: Where a single clear floor space complying with Section 10, "CLEAR FLOOR OR GROUND SPACE", positioned for parallel approach is provided between two beds, a clear floor or ground space shall not be required on both sides of a bed. **Fig. CD-39B**

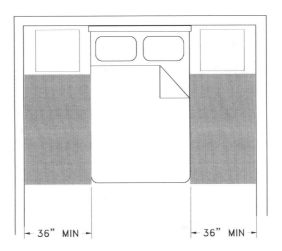

Fig. CD-39A
Clear Space on Both Sides of a Bed

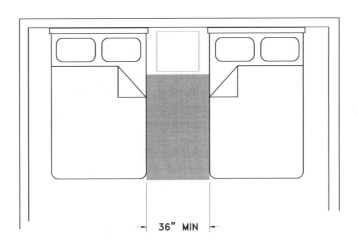

Fig. CD-39B
Clear Space Between Two Beds

Personal Lift Device Floor Space

_____E.   There shall be a clear space under the bed for the use of a personal lift device.
11B-806.2.3.1   Fig. CD-39C

_____E1.   The clear space shall extend under the bed parallel to the long side and shall be adjacent to an accessible route.
11B-806.2.3.1   Fig. CD-39C

_____E2.   The clear space shall extend to points horizontally 30 inches (762 mm), vertically 7 inches (178 mm) and not more than 12 inches (305 mm) from the head and foot end of the bed. **11B-806.2.3.1   Fig. CD-39C**

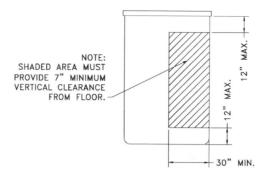

NOTE:
SHADED AREA MUST
PROVIDE 7" MINIMUM
VERTICAL CLEARANCE
FROM FLOOR.

12" MAX.

12" MAX.

30" MIN.

Fig. CD-39C
Personal Lift Device Floor Space

Toilet and Bathing Facilities

_____F.   At least one bathroom that is provided as part of a guest room shall comply with items A through J of Section 26, "TOILET AND BATHING FACILITIES".

_____G.   No fewer than one water closet, one lavatory, and one bathtub or shower shall comply with applicable requirements of Sections 26 through 26H for toilet and bathing facilities. In addition, required roll-in shower compartments shall comply with the requirements for an accessible "Standard Roll-In Type Shower Compartment", or an "Alternate Roll-In Type Shower Compartment", contained in Section 26H, "SHOWER COMPARTMENTS". Toilet and bathing fixtures required to comply with Sections 26 through 26H shall be permitted to be located in more than one toilet or bathing area, provided that travel between fixtures does not require travel between other parts of the guest room.   **11B-806.2.4** *806.2.4*

Vanity Counter Top Space

_____H.   If vanity counter top space is provided in non-accessible guest toilet or bathing rooms, comparable vanity counter top space, in terms of size and proximity to the lavatory, shall also be provided in accessible guest toilet or bathing rooms. **11B-806.2.4.1**   *806.2.4.1*

ADVISORY: *Vanity Counter Top Space. This provision is intended to ensure that accessible guest rooms are provided with comparable vanity counter top space.*

Kitchens, Kitchenettes and Wet Bars

_____I.   Kitchens, kitchenettes and wet bars shall comply with Section 40, "KITCHENS, KITCHENETTES, WET BARS AND SINKS". **11B-806.2.5**   *806.2.5*

Turning Space

_____J.   Turning space complying with Section 9, "TURNING SPACE", shall be provided within the guest room.   **11B-806.2.6**   *806.2.6*

Guest Rooms with Communication Features
Guest rooms required to provide communication features shall comply as detailed, below.
11B-806.3   *806.3*

> **ADVISORY: Guest Rooms with Communication Features.** *In guest rooms required to have accessible communication features, consider ensuring compatibility with adaptive equipment used by people with hearing impairments. To ensure communication within the facility, as well as on commercial lines, provide telephone interface jacks that are compatible with both digital and analog signal use. If an audio headphone jack is provided on a speaker phone, a cutoff switch can be included in the jack so that insertion of the jack cuts off the speaker. If a telephone-like handset is used, the external speakers can be turned off when the handset is removed from the cradle. For headset or external amplification system compatibility, a standard subminiature jack installed in the telephone will provide the most flexibility.*

Alarms

_____K.   Where emergency warning systems are provided, alarms complying with Section 29, "FIRE ALARM SYSTEMS", shall be provided.   **11B-806.3.1**   *806.3.1*

Notification Devices

_____L.   Visible notification devices shall be provided to alert room occupants of incoming telephone calls and a door knock or bell.   **11B-806.3.2**   *806.3.2*

_____M.   Notification devices shall not be connected to visible alarm signal appliances. **11B-806.3.2**   *806.3.2*

_____N.   Telephones shall have volume controls compatible with the telephone system and shall comply with item A of Section 31, "TELEPHONES".   **11B-806.3.2** *806.3.2*

_____O.   Telephones shall be served by an electrical outlet complying with Section 13, "OPERABLE PARTS", located within 48 inches (*1219* mm) of the telephone to facilitate the use of a TTY.   **11B-806.3.2**   *806.3.2*

40. KITCHENS, KITCHENETTES, WET BARS AND SINKS

General
Where provided, kitchens, kitchenettes, wet bars and sinks shall comply as detailed. **11B-212.1** *212.1*

Kitchens, Kitchenettes, and Wet Bars
Kitchens, kitchenettes and wet bars shall comply as detailed. **11B-212.2, 11B-804** *212.2, 804*

Sinks
Where sinks are provided, at least 5 percent, but no fewer than one, of each type provided in each accessible room or space shall comply as detailed.
11B-212.3, 11B-606 *212.3, 606*

> **EXCEPTION:** 1. Mop or service sinks shall not be required to comply.
>
> 2, Scrub sinks, as defined in California Plumbing Code Section 221.0, shall not be required to comply.

Kitchens and Kitchenettes   **11B-804**   *804*

General
Kitchens and kitchenettes shall comply as detailed. **11B-804.1**   *804.1*

Clearance

_____A. Where a pass through kitchen is provided, clearances shall comply with items C and C1, below.

_____B. Where a U-shaped kitchen is provided, clearances shall comply with item D, below. **11B-804.2**   *804.2*

> **EXCEPTION:** Spaces that do not provide a cooktop or conventional range shall not be required to comply with these clearances.

> **ADVISORY: Clearance.** *Clearances are measured from the furthest projecting face of all opposing base cabinets, counter tops, appliances, or walls, excluding hardware.*

Pass through Kitchen

_____C. In pass through kitchens where counters, appliances or cabinets are on two opposing sides, or where counters, appliances or cabinets are opposite a parallel wall, clearance between all opposing base cabinets, counter tops, appliances, or walls within kitchen work areas shall be 40 inches (*1016* mm) minimum. **11B-804.2.1**   *804.2.1* **Fig. CD-40A**

_____C1. Pass through kitchens shall have two entries.   **11B-804.2.1**   *804.2.1* **Fig. CD-40A**

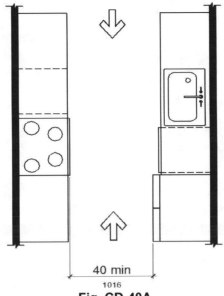

Fig. CD-40A
Pass Through Kitchens
© ICC Reproduced with Permission

U-Shaped Kitchen

_____D.	In U-shaped kitchens enclosed on three contiguous sides, clearance between all opposing base cabinets, counter tops, appliances, or walls within kitchen work areas shall be 60 inches (*1524* mm) minimum **11B-804.2.2** *804.2.2* **Fig. CD-40B**

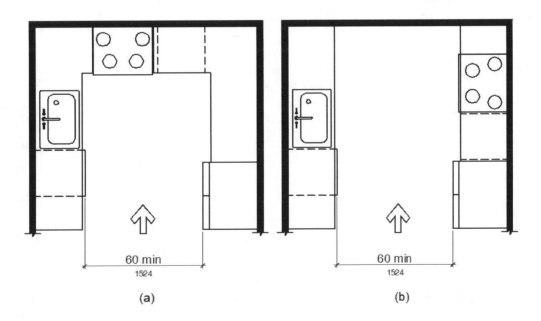

(a)				(b)

Fig. CD-40B
U-Shaped Kitchens
© ICC Reproduced with Permission

Kitchen Work Surface

_____E.   In residential dwelling units required to comply with Section 43, "RESIDENTIAL FACILITIES", at least one 30 inches (762 mm) wide minimum section of counter shall provide a kitchen work surface that complies as detailed, below. **11B-804.3** *804.3*

Clear Floor or Ground Space

_____F.   A clear floor space complying with Section 10, "CLEAR FLOOR OR GROUND SPACE", positioned for a forward approach shall be provided. **11B-804.3.1** *804.3.1*

_____F1.   The clear floor or ground space shall be centered on the kitchen work surface and shall provide knee and toe clearance complying with Section 11, "KNEE AND TOE CLEARANCE". **11B-804.3.1** *804.3.1*

> **EXCEPTION:** Cabinetry shall be permitted under the kitchen work surface provided that all of the following conditions are met:
>
> (a)   the cabinetry can be removed without removal or replacement of the kitchen work surface;
>
> (b)   the finish floor extends under the cabinetry; and
>
> (c)   the walls behind and surrounding the cabinetry are finished.

Height

_____G.   The kitchen work surface shall be 34 inches (864 mm) maximum above the finish floor or ground. **11B-804.3.2** *804.3.2*

> **EXCEPTION:** A counter that is adjustable to provide a kitchen work surface at variable heights, 29 inches (737 mm) minimum and 36 inches (914 mm) maximum, shall be permitted.

Exposed Surfaces

_____H.   There shall be no sharp or abrasive surfaces under the work surface counters. **11B-804.3.3** *804.3.3*

Sinks

_____I.   Sinks shall comply with Section 26F, "LAVATORIES AND SINKS". **11B-804.4** *804.4*

Storage

_____J.   At least 50 percent of shelf space in storage facilities shall comply with Section 45, "STORAGE". **11B-804.5** *804.5*

Appliances

Where provided, kitchen appliances shall comply as detailed. **11B-804.6** *804.6*

Clear Floor or Ground Space

_____K.   A clear floor or ground space complying with Section 10, "CLEAR FLOOR OR GROUND SPACE", shall be provided at each kitchen appliance. Clear floor or ground spaces shall be permitted to overlap.   **11B-804.6.1**   *804.6.1*

Operable Parts

_____L.   All appliance controls shall comply with Section 13, "OPERABLE PARTS".   **11B-804.6.2**   *804.6.2*

EXCEPTIONS:

1.   Appliance doors and door latching devices shall not be required to comply with Section 13, "OPERABLE PARTS".

2.   Bottom-hinged appliance doors, when in the open position, shall not be required to place their operable parts within accessible reach ranges.

Dishwasher

_____M.   Clear floor or ground space shall be positioned adjacent to the dishwasher door. the dishwasher door, in the open position, shall not obstruct the clear floor or ground space for the dishwasher or the sink.   **11B-804.6.3**   *804.6.3*

Range or Cooktop

_____N.   Where a forward approach is provided, the clear floor or ground space shall provide knee and toe clearance complying with Section 11, "KNEE AND TOE CLEARANCE".   **11B-804.6.4**   *804.6.4*

_____N1.   Where knee and toe space is provided, the underside of the range or cooktop shall be insulated or otherwise configured to prevent burns, abrasions, or electrical shock.   **11B-804.6.4**   *804.6.4*

_____N2.   The location of controls shall not require reaching across burners **11B-804.6.4**   *804.6.4*

Oven

_____O.   Ovens shall comply as detailed.   **11B-804.6.5**   *804.6.5*

Side-Hinged Door Ovens

_____O1.   Side-hinged door ovens shall have the work surface required by items E through H, above, positioned adjacent to the latch side of the oven door. **11B-804.6.5.1**   *804.6.5.1*

Bottom-Hinged Door Ovens

_____O2.   Bottom-hinged door ovens shall have the work surface required by items E through H, above, positioned adjacent to one side of the door. **11B-804.6.5.2**   *804.6.5.2*

Controls

_____O3.   Ovens shall have controls on front panels. **11B-804.6.5.** *804.6.5.3*

Refrigerator/Freezer

_____P.   Combination refrigerators and freezers shall have at least 50 percent of the freezer space 54 inches (*1372* mm) maximum above the finish floor or ground. **11B-804.6.6** *804.6.6*

_____Q.   The clear floor or ground space shall be positioned for a parallel approach to the space dedicated to a refrigerator/freezer with the centerline of the clear floor or ground space offset 24 inches (610 mm) maximum from the centerline of the dedicated space. **11B-804.6.6** *804.6.6*

41. <u>JUDICIAL FACILITIES</u>

Entrance for Inmates or Detainees
Where entrances used only by inmates or detainees and security personnel are provided at judicial facilities, detention facilities, or correctional facilities, at least one such entrance shall comply with the requirements of Section 14, "DOORS, DOORWAYS AND GATES", and Section 14A, "MANEUVERING CLEARANCES AT DOORWAYS AND GATES". **11B-206.4.9** *206.4.9*

Platform Lifts
Platform lifts shall be permitted to provide an accessible route to: jury boxes and witness stands; raised courtroom stations including, judges' benches, clerks' stations, bailiffs' stations, deputy clerks' stations, and court reporters' stations; and to depressed areas such as the well of the court. **11B-206.7.4** *206.7.4*

Judicial Facilities   **11B-231**   *231*

General
Judicial facilities shall comply as detailed. **11B-231.1** *231.1*

Courtrooms
Each courtroom shall comply as detailed. **11B-231.2** *231.2*

Holding Cells
Where provided, central holding cells and court-floor holding cells shall comply as detailed, below. **11B-231.3** *231.3*

Central Holding Cells
Where separate central holding cells are provided for adult male, juvenile male, adult female, or juvenile female, one of each type shall comply with items A through D1, below. Where central holding cells are provided and are not separated by age or sex, at least one cell complying with items A through D1, below, shall be provided. **11B-231.3.1** *231.3.1*

Court-Floor Holding Cells

Where separate court-floor holding cells are provided for adult male, juvenile male, adult female, or juvenile female, each courtroom shall be served by one cell of each type complying with items A through D1, below. Where court-floor holding cells are provided and are not separated by age or sex, courtrooms shall be served by at least one cell complying with items A through D1, below. Cells may serve more than one courtroom. **11B-231.3.2** *2321.3.2*

Visiting Areas

Visiting areas shall comply as detailed, below. **11B-231.4** *231.4*

Cubicles and Counters

At least 5 percent, but no fewer than one, of cubicles shall comply with Section *46,* "DINING SURFACES AND WORK SURFACES", on both the visitor and detainee sides. Where counters are provided, at least one shall comply with the requirements for a forward approach per items H through H2 of Section 48, "SALES AND SERVICE", on both the visitor and detainee sides. **11B-231.4.1** *231.4.1*

> **EXCEPTION:** The detainee side of cubicles or counters at non-contact visiting areas not serving holding cells required to comply with these provisions shall not be required to comply with the above provision for cubicles and counters.

Partitions

Where solid partitions or security glazing separate visitors from detainees at least one of each type of cubicle or counter partition shall comply with items I through I2 of Section 48, "SALES AND SERVICE". **11B-231.4.2** *231.4.2*

Holding Cells and Housing Cells   **11B-807**   *807*

General

Holding cells and housing cells shall comply as detailed. **11B-807.1** *807.1*

Cells with Mobility Features

Cells required to provide mobility features shall comply as detailed, below. **11B-807.2** *807.2*

Turning Space

_____A.   Turning space complying with Section 9, "TURNING SPACE", shall be provided within the cell. **11B-807.2.1** *807.2.1*

Benches

_____B.   Where benches are provided, at least one bench shall comply with Section 47, "BENCHES". **11B-807.2.2** *807.2.2*

Beds

_____C.   Where beds are provided, clear floor space complying with Section 10, "CLEAR FLOOR OR GROUND SPACE" shall be provided on at least one side of the bed. **11B-807.2.3** *807.2.3*

_____C1.   The clear floor space shall be positioned for parallel approach to the side of the bed. **11B-807.2.3** *807.2.3*

Toilet and Bathing Facilities

_____D.   Toilet facilities or bathing facilities that are provided as part of a cell shall comply with Section 26, "TOILET AND BATHING FACILITIES". **11B-807.2.4**   *807.2.4*

      _____D1.   Where provided, no fewer than one water closet, one lavatory, and one bathtub or shower shall comply with the applicable requirements of Sections 26 through 26H for toilet and bathing facilities.   **11B-807.2.4**   *807.2.4*

*ADVISORY: **Toilet and Bathing Facilities.** In holding cells, housing cells, or rooms required to be accessible, these requirements do not require a separate toilet room.*

Courtrooms   **11B-808**   *808*

General
Courtrooms shall comply as detailed, below.   **11B-808.1**   *808.1*

Turning Space

_____E.   Where provided, areas that are raised or depressed and accessed by ramps or platform lifts with entry ramps shall provide unobstructed turning space complying with Section 9, "TURNING SPACE".   **11B-808.2**   *808.2*

Clear Floor Space

_____F.   Each jury box and witness stand shall have, within its defined area, clear floor space complying with Section 10, "CLEAR FLOOR OR GROUND SPACE". **11B-808.3**   *808.3*

> **EXCEPTION:** In alterations, wheelchair spaces are not required to be located within the defined area of raised jury boxes or witness stands and shall be permitted to be located outside these spaces where ramp or platform lift access poses a hazard by restricting or projecting into a means of egress required by the appropriate administrative authority.

Judges' Benches and Courtroom Stations

_____G.   Judges' benches, clerks' stations, bailiffs' stations, deputy clerks' stations, court reporters' stations and litigants' and counsel stations shall comply with Section 46, "DINING SURFACES AND WORK SURFACES".   **11B-808.4**   *808.4*

42. <u>DETENTION FACILITIES AND CORRECTIONAL FACILITIES</u>

Entrance for Inmates or Detainees

Where entrances used only by inmates or detainees and security personnel are provided at judicial facilities, detention facilities, or correctional facilities, at least one such entrance shall comply with the requirements of Section 14, "DOORS, DOORWAYS AND GATES", and Section 14A, "MANEUVERING CLEARANCES AT DOORWAYS AND GATES". **11B-206.4.9** *206.4.9*

Detention and Correctional Facilities

In detention and correctional facilities, common use areas that are used only by inmates and detainees and security personnel and that do not serve holding cells or housing cells required to be accessible, shall not be required to comply with these requirements or to be on an accessible route. **11B-203.7** *203.7*

Raised Area

Areas raised primarily for purposes of security, life safety, or fire safety, including but not limited to, observation or lookout galleries, prison guard towers, fire towers, or life guard stands shall not be required to comply with these requirements or to be on an accessible route. **11B-203.3** *203.3*

Multi-Story Buidings and Facilities

At least one accessible route shall connect each story and mezzanine in multi-story buildings and facilities. **11B-206.2.3** *206.2.3*

> **EXCEPTION:** In detention and correctional facilities, an accessible route shall not be required to connect stories where cells with mobility features, all common use areas serving cells with mobility features, and all public use areas are on an accessible route. **11B-206.2.3, Exception 3**, *206.2.3, Exception 3*

Stairways

In detention and correctional facilities, stairways that are not located in public use areas shall not be required to comply. **11B-210.1** *210.1*

Drinking Fountains

In detention or correctional facilities, drinking fountains only serving holding or housing cells not required to have mobility features shall not be required to be accessible. **11B-211.1** *211.1*

General

Buildings, facilities, or portions thereof, in which people are detained for penal or correction purposes, or in which the liberty of the inmates is restricted for security reasons shall comply as detailed. **11B-232** *232.1*

> *ADVISORY: General. Detention facilities include, but are not limited to, jails, detention centers, and holding cells in police stations. Correctional facilities include, but are not limited to, prisons, reformatories, and correctional centers.*

General Holding Cells and General Housing Cells

General holding cells and general housing cells shall be provided as detailed. **11B-232.2** *232.2*

> **ADVISORY:** *General Holding Cells and General Housing Cells. Accessible cells or rooms should be dispersed among different levels of security, housing categories, and holding classifications (e.g., male/female and adult/juvenile) to facilitate access. Many detention and correctional facilities are designed so that certain areas (e.g., "shift" areas) can be adapted to serve as different types of housing according to need. For example, a shift area serving as a medium-security housing unit might be redesignated for a period of time as a high-security housing unit to meet capacity needs. Placement of accessible cells or rooms in shift areas may allow additional flexibility in meeting requirements for dispersion of accessible cells or rooms.*

Cells with Mobility Features
At least 3 percent, but no fewer than one, of the total number of cells in a facility shall provide mobility features that comply as detailed.   **11B-232.2.1, 11B-807.2** *232.2.1, 807.2*

Beds
In cells having more than 25 beds, at least 5 percent of the beds shall have clear floor space that complies as detailed.   **11B-232.2.1.1, 11B-807.2.3** *232.2.1.1, 807.2.3*

Dispersion
Cells with mobility features shall be provided in each classification level.
11B-232.2.1.2 *232.2.1.2*

Substitute Cells
When alterations are made to specific cells, detention and correctional facility operators may satisfy their obligation to provide the required number of cells with mobility features by providing the required mobility features in substitute cells (cells other than those where alterations are original planned), provided that each substitute cell meets the following conditions: **11B-232.2.1.3** *232.2.1.3*

1. Located within the same prison site.

2. Integrated with the other cells to the maximum extent feasible.

3. Has equal physical access as the altered cells to areas used by inmates or detainees for visitation, dining, recreation, educational programs, medical services, work programs, religious services, and participation in other programs that the facility offers to inmates or detainees.

    #### Technically Infeasible
    Where it is technically infeasible to locate a substitute cell within the same prison site as detailed above, a substitute cell shall be provided at another prison site within the correctional system.   **11B-232.2.1.4** *232.2.1.4*

Cells with Communication Features
At least 2 percent, but no fewer than one, of the total number of general holding cells and general housing cells equipped with audible emergency alarm systems and permanently installed telephones within the cell shall provide communication features comply as detailed.   **11B-232.2.2** *232.2.2*

Special Holding Cells and Special Housing Cells
Where special holding cells or special housing cells are provided, at least one cell serving each purpose shall provide mobility features complying as detailed. Cells subject to this requirement include, but are not limited to, those used for purposes of orientation, protective custody,

administrative or disciplinary detention or segregation, detoxification, and medical isolation. **11B-232.3** *232.3*

> **EXCEPTION: Reserved.**

Medical Care Facilities

Patient bedrooms or cells required to comply with the standards for accessible Medical Care and Long-Term Care Facilities shall be provided in addition to any accessible medical isolation cells that are provided. **11B-232.4** *232.4*

VISITING AREAS

Visiting areas shall comply as detailed below. **11B-232.5** *232.5*

Cubicles and Counters

At least 5 percent, but no fewer than one, of cubicles shall comply with Section 46, 'DINING SURFACES AND WORK SURFACES" on both the visitor and detainee sides. Where counters are provided, at least one shall provide a compliant forward approach on both the visitor and detainee or inmate sides. **11B-232.5.1** *232.5.1*

> **EXCEPTION:** The inmate or detainee side of cubicles or counters at non-contact visiting areas not serving holding cells with mobility features shall not be required to comply.

Partitions and Security Glazing

Where solid partitions or security glazing separate visitors from detainees or inmates at least one of each type of cubicle or counter partition shall comply as detailed.
11B-232.5.2, 11B-904.6 *232.5.2, 904.6*

_____A.   Where solid partitions or security glazing separate visitors from detainees, a method to facilitate voice communication shall be provided. Telephone Handset devices, if provided, shall comply as detailed, below. **11B-904.6** *904.6*

> *Advisory. Security Glazing. Accessible assistive listening devices can facilitate voice communication at counters where there is security glazing which promotes distortion in audible information. Where assistive listening devices are installed, the International Symbol of Access for Hearing Loss should be placed to identify those facilities which are so equipped (see Section on "SIGNS"). Other voice communication methods include, but are not limited to, grilles, slats, talk-through baffles, intercoms, or telephone handset devices.*

HOLDING CELLS AND HOUSING CELLS

Cells with Mobility Features

Cells required to provide mobility features shall comply as follows. **11B-807.2** *807.2*

Turning Space

_____B.   An accessible turning space that complies with Section 9, "TURNING SPACE" is provided within the cell.   **11B-807.2.1** *807.2.1*

Benches

_____C.   Where benches are provided, at least one complies with Section 47, "BENCHES". **11B-807.2.2** *807.2.2*

_____D.   Where beds are provided, clear floor space that complies with Section 10, "CLEAR FLOOR OR GROUND SPACE" shall be provided on at least one side of the bed. **11B-807.2.3** *807.2.3*

     _____D1.   The clear floor space shall be positioned for parallel approach to the side of the bed. **11B-807.2.3** *807.2.3*

Toilet and Bathing Facilities

_____E.   Toilet facilities or bathing facilities that are provided as part of a cell shall comply with Section 26, "TOILET AND BATHING ROOMS". Where provided, no fewer than one water closet, one lavatory and one bathtub or shower will comply with the applicable requirements in Sections 26 through 28. **11B-807.2.4** *807.2.4*

> **ADVISORY: Toilet and Bathing Facilities.** *In holding cells, or rooms required to be accessible, these requirements do not require a separate toilet room.*

Cells With Communication Features
Cells required to provide communication features shall comply as detailed. **11B-807.3** *807.3*

Alarms

_____F.   Where audible emergency alarm systems are provided to serve the occupants of cells, visible alarms complying with Section 29, "FIRE ALARM SYSTEMS", shall be provided. **11B-807.3.1** *807.3.1*

     EXCEPTION: Visible alarms shall not be required where inmates or detainees are not allowed independent means of egress

Telephones

_____G.   Telephones, where provided within cells, shall have compliant volume controls as detailed. **11B-807.3.2** *807.3.2*

Volume Control Telephones

_____H.   Telephones required to have volume controls shall be equipped with a receive volume control that provides a gain adjustable up to 20 dB minimum. For incremental volume control, provide at least one intermediate step of 12 dB of gain minimum. An automatic reset shall be provided. **11B-704.3** *704.3*

_____I.   Volume control telephones shall be equipped with a receiver that generates a magnetic field in the area of the receiver cap. **11B-704.3** *704.*

_____J.   Public telephones with volume control shall be hearing aid compatible. **11B-704.3** *704.*

> **ADVISORY: Volume Control Telephones.** *Amplifiers on pay phones are located in the base or the handset or are built into the telephone. Most are operated by pressing a button or key. If the microphone in the handset is not being used, a mute button that temporarily turns off the microphone can also reduce the amount of background noise which the person hears in the earpiece. If a volume adjust is provided that allows the*

> *user to set the level anywhere from the base volume to the upper requirement of 20 dB, there is no need to specify a lower limit. If a stepped volume control is provided, one of the intermediate levels must provide 12 dB of gain. Consider compatibility issues when matching an amplified handset with a phone or phone system. Amplified handsets that can be switched with pay telephone handsets are available. Portable and in-line amplifiers can be used with some phones but are not practical at most public phones covered by these requirements.*

43. RESIDENTIAL FACILITIES

Alterations
Where existing elements or spaces are altered, each altered element or space shall comply with applicable requirements as detailed, including path of travel requirements. **11B-202.3**   *202.3*

> **EXCEPTION:** Residential dwelling units not required to be accessible in compliance with this code shall not be required to comply. **Exception 3**

Multi-Story Buidings and Facilities
At least one accessible route shall connect each story and mezzanine in multi-story buildings and facilities. **11B-206.2.3**   *206.2.3*

> **EXCEPTION:** In residential facilities, an accessible route shall not be required to connect stories where residential dwelling units with required mobility features and also adaptable features complying with Chapter 11A, Division IV, and all common use areas serving residential dwelling units with required mobility features and also adaptable features complying with Chapter 11A, Division IV, and public use areas serving residential dwelling units are on an accessible route. **11B-206.2.3, Exception 4**,   *206.2.3, Exception 4*

Residential Facilities Exception
In public housing residential facilities, common use areas that do not serve residential dwelling units required to provide mobility features, or adaptable features complying with Chapter 11A, Division IV, shall not be required to comply with these requirements or to be on an accessible route. **11B-203.8**   *203.8*

Platform Lifts for Residential Dwelling Units
Platform lifts shall be permitted to connect levels within residential dwelling units required to provide mobility features complying with Section 43, "RESIDENTIAL FACILITIES" and adaptable features complying with Chapter 11A, Division IV. **11B-206.7.6**   *206.7.6*

Parking for Residential Facilities
Parking spaces provided to serve residential facilities shall comply as follows:
11B-208.2.3   *208.2.3*

Parking for Residents
Where at least one parking space is provided for each residential dwelling unit, at least one parking space that complies with Section 21, "PARKING SPACES", shall be provided for each residential dwelling unit required to provide mobility features (see Section 43 "RESIDENTIAL FACILITIES"). **11B-208.2.3.1**   *208.2.3.1*

Additional Parking Spaces for Residents

Where the total number of parking spaces provided for each residential dwelling unit exceeds one parking space per residential dwelling unit, 2 percent, but no fewer than one space, of all the parking spaces not covered by "Parking for Residents", above, shall comply with Section 21, "PARKING SPACES".
11B-208.2.3.2   *208.2.3.2*

Parking for Guests, Employees, and Other Non-Residents

Where parking spaces are provided for persons other than residents, parking shall be provided in accordance with Table 21A of Section 21, "PARKING SPACES".
11B-208.2.3.3   *208.2.3.3*

Requests for Accessible Parking Spaces

When assigned parking is provided, designated accessible parking for the adaptable residential dwelling units shall be provided on requests of residents with disabilities one the same terms and with the full range of choices (e.g. off-street parking, carport or garage) that are available to other residents.
11B-208.2.3.4

Residential Facilities

In residential facilities containing residential units required to provide mobility features per "Residential Dwelling Units" and adaptable features complying with Chapter 11A, Division IV Parking spaces provided for Residential Facilities shall be located on the shortest accessible route to the residential dwelling unit entrance they serve. Spaces provided in accordance with Residential Facilities shall be dispersed throughout all types of parking provided for the residential dwelling units. **11B-208.3.2**   *208.3.2*

> **EXCEPTION:** Parking spaces provided in accordance with "Residential Facilities" shall not be required to be dispersed throughout all types of parking if substantially equivalent or greater accessibility is provided in terms of distance from an accessible entrance, parking fee, and user convenience.

> ***ADVISORY. Residential Facilities Exception.** Factors that could affect "user convenience" include, but are not limited to, protection from the weather, security, lighting, and comparative maintenance of the alternative parking site.*

Private Garages Accessory to Residential Dwelling Units

Private garages accessory to residential dwelling units shall comply. Private garages include individual garages and multiple individual garages grouped together.
11B-208.3.3

Detached private garages accessory to residential dwelling units, shall be accessible and comply as detailed herein. **11B-208.3.3.1**

Attached private garages directly serving a single residential dwelling unit shall provide at least one of the following options: **11B-208.3.3.2**

1. A door leading directly from the residential dwelling unit which immediately enters the garage.

2. An accessible route from the residential dwelling unit to an exterior door entering the garage.

---

3.   An accessible route from the residential dwelling unit's primary entry door to the vehicular entrance of the garage.

Residential Dwelling Unit Primary Entrance

In residential dwelling units, at least one primary entrance shall comply with the requirements of Section 14, "DOORS, DOORWAYS AND GATES", and Section 14A, "MANEUVERING CLEARANCES AT DOORS AND GATES". The primary entrance to a residential dwelling unit shall not be to a bedroom. **11B-206.4.6** *206.4.6*

Doors, Doorways and Gates for Residential Dwelling Units

In residential dwelling units required to provide mobility features, all doors and doorways providing user passage shall comply with Section 14, "DOORS, DOORWAYS AND GATES, and Section 14A, "MANEUVERING CLEARANCES AT DOORS AND GATES".   **11B-206.5.4** *206.5.4*

Residential Facilities   **11B-233**   *233*

General

Public housing facilities with residential dwelling units available for public use with shall comply with this section. See Definition, below. **11B-233.1**   *233.1*

DEFINITION OF "PUBLIC HOUSING"

> **PUBLIC HOUSING.** Housing facilities owned and/or operated by, for or on behalf of a public entity including but not limited to the following:
>
> 1.   Publically owned and/or operated one- or two- family dwelling units or congregate residences;
>
> 2.   Publically owned and/or operated buildings or complexes with three or more residential dwellings units;
>
> 3.   Publically owned and/or operated housing provided by entities subject to regulations issued by the United State Department of Housing and Urban Development under Section 504 of the Rehabilitation Act of 1973 as amended;
>
> 4.   Publically owned and/or operated homeless shelters, group homes and similar social service establishments;
>
> 5.   Publically owned and/or operated transient lodging, such as hotels, motels, hostels and other facilities providing accommodations of a short term nature of not more than 30 days duration;
>
> 6.   Housing at a place of education owned or operated by a public entity, such as housing on or serving a public school, public college or public university campus;
>
> 7.   Privately owned housing made available for public use as housing.

ADVISORY: General. This section addresses long-term living (non-transient), public housing dwelling units.

In public housing projects, the accessibility requirements of this section are triggered when there are one or more units. In contrast, in privately funded housing projects, the

> *accessibility requirements of Chapter 11A are triggered when there are three or more apartments or four or more condominiums within a single building.*

> **ADVISORY: General. This** *section outlines the requirements for residential facilities subject to the Americans with Disabilities Act of 1990. The facilities covered by this section, as well as other facilities not covered by this section, may still be subject to other Federal laws such as the Fair Housing Act and Section 504 of the Rehabilitation Act of 1973, as amended. For example, the Fair Housing Act requires that certain residential structures having four or more multi-family dwelling units, regardless of whether they are privately owned or federally assisted, include certain features of accessible and adaptable design according to guidelines established by the U.S. Department of Housing and Urban Development (HUD). These laws and the appropriate regulations should be consulted before proceeding with the design and construction of residential facilities.*
>
> *Residential facilities containing residential dwelling units provided by entities subject to HUD's Section 504 regulations and residential dwelling units covered by this section must comply with the technical and scoping requirements in Divisions 1 through 10 included this chapter. This section is not a stand-alone section; this section only addresses the minimum number of residential dwelling units within a facility required to comply with Division 8. However, residential facilities must also comply with the requirements of this section. For example, it is required that all doors and doorways providing user passage in residential dwelling units providing mobility features to comply with the requirements for accessible doors; another section permits platform lifts to be used to connect levels within residential dwelling units providing mobility features. General scoping for accessible parking specifies the required number of accessible parking spaces for each residential dwelling unit providing mobility features. Mail boxes are required to be within reach ranges when they serve residential dwelling units providing mobility features, and play areas and swimming pools have specific obligations.*

Residential Dwelling Units Provided by Entities Subject to HUD Section 504 Regulations
Where public housing facilities with residential dwelling units are provided by entities subject to regulations issued by the Department of Housing and Urban Development (HUD) under Section 504 of the Rehabilitation Act of 1973, as amended, such entities shall provide residential dwelling units with compliant mobility features in a number required by the applicable HUD regulations. Residential dwelling units required to provide compliant mobility features shall be on an accessible route as required. In addition, such entities shall provide residential dwelling units with compliant communication features as detailed in this section in a number required by the applicable HUD regulations. Entities subject to this section shall not be required to comply with the requirements for residential dwelling units provided by entities NOT subject to HUD Section 504 regulations.
11B-233.2 *233.2*

> **ADVISORY: Residential Dwelling Units Provided by Entities Subject to HUD Section 504 Regulations.** *The provision above requires that entities subject to HUD's regulations implementing Section 504 of the Rehabilitation Act of 1973, as amended, provide residential dwelling units containing mobility features and residential dwelling units containing communication features complying with these regulations in a number specified in HUD's Section 504 regulations. Further, the residential dwelling units provided must be dispersed according to HUD's Section 504 criteria. In addition, this provision defers to HUD the specification of criteria by which the technical requirements of this chapter will apply to alterations of existing facilities subject to HUD's Section 504 regulations.*

Residential Dwelling Units Provided by Entities Not Subject to HUD Section 504 Regulations

Public housing facilities with residential dwelling units provided by entities not subject to regulations issued by the Department of Housing and Urban Development (HUD) under Section 504 of the Rehabilitation Act of 1973, as amended, shall comply as detailed. **11B-233.3**   *233.3*

Minimum Number: New Construction

Newly constructed public housing facilities with residential dwelling units shall comply as detailed. **11B-233.3.1**   *233.3.1*

> **EXCEPTION:** Where facilities contain 15 or fewer residential dwelling units, the requirements for residential dwelling units with mobility features and residential dwelling units with communication features shall apply to the total number of residential dwelling units that are constructed under a single contract, or are developed as a whole, whether or not located on a common site.

Residential Dwelling Units with Mobility Features

In public housing facilities with residential dwelling units, at least 5 percent, but no fewer than one unit, of the total number of residential dwelling units shall provide compliant mobility features as detailed and shall be on an accessible route as required.   **11B-233.3.1.1**   *233.3.1.1*

Residential Dwelling Units with Adaptable Features

In public housing facilities with residential dwelling units, adaptable residential dwelling units complying with Chapter 11A, Division IV – Dwelling Unit Features shall be provided as required, below. Adaptable residential dwelling units shall be on an accessible route as required by Section 2, "ACCESSIBLE ROUTES". **11B-233.3.1.2**   *233.3.1.2*

> **EXCEPTION:** The number of required adaptable residential dwelling units shall be reduced by the number of units required by "Residential Units with Mobility Features", above.

Elevator Buildings

Residential dwelling units on floors served by an elevator shall be adaptable. **11B-233.3.1.2.1**

Non-Elevator Buildings

Ground floor residential dwelling units in non-elevator buildings shall be adaptable.   **11B-233.3.1.2.2**

Ground Floors Above Grade

Where the first floor in a building containing residential dwelling units is a floor above grade, all units on that floor shall be adaptable. **11B-233.3.1.2.3**

Multi-Story Residential Dwelling Units

In elevator buildings, public housing facilities with multi-story residential dwelling units shall comply with the following: **11B-233.3.1.2.4.**

> **EXCEPTION:** In non-elevator buildings, a minimum of 10 percent but not less than one of the ground floor multi-story residential dwelling units shall comply with the requirements for "Residential Dwelling Units with

Adaptable Features", above, calculated using the total number of multi-story residential dwelling units in buildings on a site.

1. The primary entry of the multi-story residential dwelling unit shall be on an accessible route. In buildings with elevators the primary entry shall be on the floor served by the elevator.

2. At least one powder room or bathroom shall be located on the primary entry level.

3. Rooms or spaces located on the primary entry level shall be served by an accessible route and comply with Chapter 11A, Division IV – Dwelling Unit Features.

Public Housing Facility Site Impracticality

The number of adaptable residential dwelling units required in non-elevator building public housing facilities shall be determined in accordance with Chapter 11A, Section 1150A.1. The remaining ground floor residential dwelling units shall comply with the following requirements: **11B-233.3.1.2.5**

1. Grab bar reinforcement complying with items E through O of Section 26H, "COMPARTMENT SHOWERS".

2. Doors complying with Section 14, "DOORS, DOORWAYS AND GATES", and Section 14A, "MANEUVERING CLEARANCES AT DOORS, DOORWAYS AND GATES".

3. Communication features complying with items I through I5, below.

4. Electrical receptacles and switches complying with Section 12, "REACH RANGES".

5. Toilet and bathing facilities complying with items C through D, below.

6. Kitchen sink removable cabinets complying with Section 26F, "LAVATORIES AND SINKS", Item A, Exception 3.

Residential Dwelling Units with Communication Features

In public housing facilities with residential dwelling units, at least 2 percent, but no fewer than one unit, of the total number of residential dwelling units shall provide communication features complying with items E through J, below.   **11B-233.3.1.3** *233.3.1.3*

Residential Dwelling Units for Sale

Residential dwelling units designed and constructed or altered by public entities that will be offered for sale to individuals shall provide accessible features to the extent required by this chapter.

> **EXCEPTION:** Existing residential dwellings or residential dwelling units acquired by public entities that will be offered for resale to individuals without additions or alterations shall not be required to comply with this chapter.

ADVISORY: Residential Dwelling Units for Sale. A public entity that conducts a program to build housing for purchase by individual home buyers must provide access according to the requirements of this chapter and the ADA regulations and a program receiving Federal financial assistance must comply with the applicable Section 504 regulation.

Additions
Where an addition to an existing public housing facility results in an increase in the number of residential dwelling units, the requirements listed above under "Minimum Number: New Construction" shall apply only to the residential dwelling units that are added until the total number of residential dwelling units complies with the minimum number required by "Residential Dwelling Units with Mobility Features", and "Residential Dwelling Units with Adaptable Features". Residential dwelling units required to comply with these provisions shall be on an accessible route as required by Section 2, "ACCESSIBLE ROUTES". **11B-233.3.3**   *233.3.3*

Alterations
Alterations to a public housing facility shall comply as detailed, below.
11B-233.3.4   *233.3.4*

> **EXCEPTION:** Where compliance with items A through D, below, is technically infeasible, or where it is technically infeasible to provide an accessible route to a residential dwelling unit, the entity shall be permitted to alter or construct a comparable residential dwelling unit to comply with items A through D provided that the minimum number of residential dwelling units required by "Residential Units with Mobility Features", "Residential Units with Adaptable Features", and "Residential Units with Communication Features", as applicable, is satisfied.

ADVISORY: Alterations Exception. A substituted dwelling unit must be comparable to the dwelling unit that is not made accessible. Factors to be considered in comparing one dwelling unit to another should include the number of bedrooms; amenities provided within the dwelling unit; types of common spaces provided within the facility; and location with respect to community resources and services, such as public transportation and civic, recreational, and mercantile facilities.

Alterations to Vacated Buildings
Where a building is vacated for the purposes of alteration for use as public housing, and the altered building contains more than 15 residential dwelling units, at least 5 percent of the residential dwelling units shall comply with items A through D, below, and shall be on an accessible route as required by Section 2, "ACCESSIBLE ROUTES". Residential dwelling units with adaptable features shall be provided in compliance with "Residential Dwelling Units with Adaptable Features", above. In addition, at least 2 percent of the residential dwelling units shall comply with items E through J, below. **11B-233.3.4.1**   *233.3.4.1*

> **EXCEPTION:** Where any portion of a building's exterior is preserved, but the interior of the building is removed, including all structural portions of floors and ceilings and a new building intended for use as public housing is constructed behind the existing exterior, the building is considered a new building for determining the application of this chapter.

> **ADVISORY: *Alterations to Vacated Buildings.*** *This provision is intended to apply where a building is vacated with the intent to alter the building. Buildings that are vacated solely for pest control or asbestos removal are not subject to the requirements to provide residential dwelling units with mobility features or communication features.*

Alterations to Individual Residential Dwelling Units

In public housing facilities with individual residential dwelling units, where a bathroom or a kitchen is substantially altered, and at least one other room is altered, the requirements of "Residential Dwelling Units with Mobility Features", above, shall apply to the altered residential dwelling units until the total number of residential dwelling units complies with the minimum number required by "Residential Units with Mobility Features", "Residential Units with Adaptable Features", and "Residential Units with Communication Features". Residential dwelling units required to comply with "Residential Units with Mobility Features", and "Residential Units with Adaptable Features", shall be on an accessible route as required by Section 2, "ACCESSIBLE ROUTES". **11B-233.3.4.2**   *233.3.4.2*

> **EXCEPTION:** Where public housing facilities contain 15 or fewer residential dwelling units, the requirements of "Residential Units with Mobility Features", "Residential Units with Adaptable Features", and "Residential Units with Communication Features", shall apply to the total number of residential dwelling units that are altered under a single contract, or are developed as a whole, whether or not located on a common site.

> **ADVISORY: *Alterations to Individual Residential Dwelling Units.*** *The provision above uses the terms "substantially altered" and "altered." A substantial alteration to a kitchen or bathroom includes, but is not limited to, alterations that are changes to or rearrangements in the plan configuration, or replacement of cabinetry. Substantial alterations do not include normal maintenance or appliance and fixture replacement, unless such maintenance or replacement requires changes to or rearrangements in the plan configuration, or replacement of cabinetry. The term "alteration" is defined in Chapter 2, Section 202.*

Dispersion

In public housing facilities, residential dwelling units required to provide mobility features complying with items A through D, below, and residential dwelling units required to provide communication features complying with items E through J, below, and adaptable features complying with Chapter 11A, Division IV shall be dispersed among the various types of residential dwelling units in the facility and shall provide choices of residential dwelling units comparable to, and integrated with, those available to other residents.   **11B-233.3.5**   *233.3.5*

> **EXCEPTION:** In public housing facilities where multi-story residential dwelling units are one of the types of residential dwelling units provided, one-story residential dwelling units shall be permitted as a substitute for multi-story residential dwelling units where equivalent spaces and amenities are provided in the one-story residential dwelling unit.

Graduate Student and Faculty Housing at a Place of Education

Residential dwelling units that are provided by or on behalf of a place of education, which are leased on a year round basis exclusively to graduate students or faculty, and do not contain any public use or common use areas available for educational programming, are not subject to Section 39, "TRANSIENT LODGING GUEST ROOMS", and shall comply with the requirements for "Residential Facilities", above, and the requirements for "Residential Dwelling Units", below. **11B-233.3.6**

Residential Dwelling Units   **11B-809**   *809*

General
When located within public housing facilities, residential dwelling units shall comply as detailed, below. Residential dwelling units required to provide mobility features shall comply with items A through D. below. Residential dwelling units required to provide communication features shall comply with items E through J, below.   **11B-809.1**   *809.1*

Accessible Routes

_____A.   Accessible routes complying with Section 2, "ACCESSIBLE ROUTES", shall be provided within residential dwelling units in accordance with items A through A2.   **11B-809.2**   *809.2*

> **EXCEPTION:** Accessible routes shall not be required to or within unfinished attics or unfinished basements.

Location

_____A1.   At least one accessible route shall connect all spaces and elements which are a part of the residential dwelling unit. Where only one accessible route is provided, it shall not pass through bathrooms, closets, or similar spaces.   **11B-809.2.1**   *809.2.1*

Turning Space

_____A2.   All rooms served by an accessible route shall provide a turning space complying with Section 9, "TURNING SPACE".   **11B-809.2.2**   *809.2.2*

> **EXCEPTION:** Turning space shall not be required in exterior spaces 30 inches (762 mm) maximum in depth or width.

> ***ADVISORY: Turning Space.*** *It is generally acceptable to use required clearances to provide wheelchair turning space. For example, in kitchens, Section 40, "KITCHENS, KITCHENETTES, WET BARS AND SINKS", requires at least one work surface with clear floor space complying with Section 10, "CLEAR FLOOR OR GROUND SPACE", to be centered beneath. If designers elect to provide clear floor space that is at least 36 inches (915 mm) wide, as opposed to the required 30 inches (760 mm) wide, that clearance can be part of a T-turn, thereby maximizing efficient use of the kitchen area. However, the overlap of turning space must be limited to one segment of the T-turn so that back-up maneuvering is not restricted. It would, therefore, be unacceptable to use both the clearances under the work surface and the sink as part of a T-turn. See Section 9, "TURNING SPACE", regarding T-turns.*

Kitchen

_____B.   Where a kitchen is provided, it shall comply with Section 40, "KITCHENS, KITCHENETTES, WET BARS AND SINKS".   **11B-809.3**   *809.3*

Toilet Facilities and Bathing Facilities

_____C.   At least one bathroom shall comply with Section 26, "TOILET FACILITIES AND BATHING FACILITIES".   **11B-809.4**   *809.4*

_____C1.   No fewer than one of each type of fixture provided within the bathroom shall comply with applicable requirements of Sections 26 through 26F for toilet and bathing facilities.   **11B-809.4**   *809.4*

_____C2.   Toilet and bathing fixtures required to comply with Sections 26 through 26F for toilet and bathing facilities shall be located in the same bathroom or toilet and bathing area, such that travel between fixtures does not require travel between other parts of the residential dwelling unit.   **11B-809.4**   *809.4*

> **ADVISORY: Toilet Facilities and Bathing Facilities.** *In an effort to promote space efficiency, vanity counter top space in accessible residential dwelling units is often omitted. This omission does not promote equal access or equal enjoyment of the unit. Where comparable units have vanity counter tops, accessible units should also have vanity counter tops located as close as possible to the lavatory for convenient access to toiletries.*

Subsequent Bathrooms

_____D.   In residential dwelling units with more than one bathroom, when a bathtub is installed in the first bathroom in compliance with Section 26G, "BATHTUBS" and a shower compartment is provided in a subsequent bathroom, at least one shower compartment shall comply with Section 26H, "SHOWER COMPARTMENTS".   **11B-809.4.1**

Residential Dwelling Units with Communication Features
Residential dwelling units required to provide communication features shall comply with items E through J, below.   **11B-809.5**   *809.5*

Building Fire Alarm System

_____E.   Where a building fire alarm system is provided, the system wiring shall be extended to a point within the residential dwelling unit in the vicinity of the residential dwelling unit smoke detection system.   **11B-809.5.1**   *809.5.1*

Alarm Appliances

_____E1.   Where alarm appliances are provided within a residential dwelling unit as part of the building fire alarm system, they shall comply with CBC Chapter 9, Section 907.5.2.3.4.   **11B-809.5.1.1**   *809.5.1.1*

Activation

_____E2.   All visible alarm appliances provided within the residential dwelling unit for building fire alarm notification shall be activated upon activation of the building fire alarm in the portion of the building containing the residential dwelling unit.   **11B-809.5.1.2**   *809.5.1.2*

Residential Dwelling Unit Smoke Detection System

_____F.   Residential dwelling unit smoke detection systems shall comply with CBC Chapter 9, Section 907.2.11.   **11B-809.5.2**   *809.5.2*

Activation

_____F1.   All visible alarm appliances provided within the residential dwelling unit for smoke detection notification shall be activated upon smoke detection. **11B-809.5.2.1**   *809.5.2.1*

Interconnection

_____G.   The same visible alarm appliances shall be permitted to provide notification of residential dwelling unit smoke detection and building fire alarm activation. **11B-809.5.3**   *809.5.3*

Prohibited Use

_____H.   Visible alarm appliances used to indicate residential dwelling unit smoke detection or building fire alarm activation shall not be used for any other purpose within the residential dwelling unit.   **11B-809.5.4**   *809.5.4*

Residential Dwelling Unit Primary Entrance

_____I.   Communication features shall be provided at the residential dwelling unit primary entrance complying with *Section 11B*-809.5.5.   **11B-809.5.5**   *809.5.5*

Notification

_____I1.   A hard-wired electric doorbell shall be provided.   **11B-809.5.5.1** *809.5.5.1*

_____I2.   A button or switch shall be provided outside the residential dwelling unit primary entrance.   **11B-809.5.5.1**   *809.5.5.1*

_____I3.   Activation of the button or switch shall initiate an audible tone and visible signal within the residential dwelling unit.   **11B-809.5.5.1**   *809.5.5.1*

_____I4.   Where visible doorbell signals are located in sleeping areas, they shall have controls to deactivate the signal.   **11B-809.5.5.1**   *809.5.5.1*

Identification

_____I5.   A means for visually identifying a visitor without opening the residential welling unit entry door shall be provided and shall allow for a minimum 180 degree range of view.   **11B-809.5.5.2**   *809.5.5.2*

> **ADVISORY: Identification.** *In doors, peepholes that include prisms clarify the image and should offer a wide-angle view of the hallway or exterior for both standing persons and wheelchair users. Such peepholes can be placed at a standard height and permit a view from several feet from the door.*

Site, Building, or Floor Entrance

_____J.   Where a system, including a closed-circuit system, permitting voice Communication between a visitor and the occupant of the residential dwelling unit is provided, the system shall comply with items C and D of Section 35, "TWO-WAY COMMUNICATIONS".   **11B-809.5.6**   *809.5.6*

44. <u>TRANSPORTATION FACILITIES</u>

<u>Entrances to Transportation Facilities</u>
In addition to the requirements for Parking Structures Entrances, Entrances from Tunnels or Elevated Walkways, Tenant Spaces, Residential Dwelling Unit Primary Entrance, Restricted Entrances. Service Entrances and Entrances for Inmates or Detainees, transportation facilities shall provide entrances as detailed.   **11B-206.4.4**   *206.4.4*

<u>Location</u>
In transportation facilities, where different entrances serve different transportation fixed routes or groups of fixed routes, entrances serving each fixed route or group of fixed routes shall comply with the requirements of Section 14, "DOORS, DOORWAYS AND GATES", and Section 14A, "MANEUVERING CLEARANCES AT DOORWAYS AND GATES".   **11B-206.4.4.1**   *206.4.4.1*

> **EXCEPTION:**   Entrances to key stations or existing intercity rail stations retrofitted in accordance with 49 CFR 37.49 or 49 CFR 37.51 shall not be required to comply.

<u>Direct Connections</u>
Direct connections to the other facilities shall provide an accessible route which complies with the requirements of Section 14, "DOORS, DOORWAYS AND GATES", and Section 14A, "MANEUVERING CLEARANCES AT DOORWAYS AND GATES" from the point of connection to boarding platforms and all transportation system elements required to be accessible. Any elements provided to facilitate future direct connections shall be on an accessible route connecting boarding platforms and all transportation system elements required to be accessible. **11B-206.4.4.2**   *206.4.4.2*

> **EXCEPTION:** In key stations and existing intercity rail stations, existing direct connections shall not be required to comply.

<u>Key Stations and Intercity Rail Stations</u>
Key stations and existing intercity rail stations required by Subpart C of 49 CFR part 37 to be altered, shall have entrances that comply with the requirements of Section 14, "DOORS, DOORWAYS AND GATES", and Section 14A, "MANEUVERING CLEARANCES AT DOORWAYS AND GATES".   **11B-206.4.4.3**   *206.4.4.3*

<u>Transportation Facilities</u>   **11B-218**   *218*

<u>General</u>
Transportation facilities shall comply as detailed.   **11B-218.1**   *218*

<u>Bus Loading Zones</u>
In bus loading zones restricted to use by designated or specified public transportation vehicles, each bus bay, bus stop, or other area designated for lift or ramp deployment shall comply as detailed. **11B-209.2.2.**   *209.2.2*

<u>On-Street Bus Stops</u>
On-street bus stops shall comply as detailed. **11B-209.2.3.**   *209.2.3*

New and Altered Fixed Guideway Stations
New and altered stations in rapid rail, light rail, commuter rail, intercity rail, high speed rail, and other fixed guideway systems shall comply with items I through R, below.   **11B-218.2**   *218.2*

Key Stations and Existing Intercity Rail Stations
Key stations and existing intercity rail stations shall comply with items I through R, below. **11B-218.3**   *218.3*

Bus Shelters
Where provided, bus shelters shall comply with items G and G1, below.   **11B-218.4**   *218.4*

Other Transportation Facilities
In other transportation facilities, public address systems shall comply with item O, below, and clocks shall comply with items P through P2, below.   **11B-218.5**   *218.5*

Transportation Facilities   **11B-810**   *810*

General
Transportation facilities shall comply as detailed.

Vehicle Boarding

_____A.   Stations shall not be designed or constructed so as to require persons with disabilities to board or alight from a vehicle at a location other than one used by the general public.   **11B-810.1.1**

Baggage Systems

_____B.   Baggage check-in and retrieval systems shall be on an accessible route complying with Section 2, "ACCESSIBLE ROUTES", and shall have space immediately adjacent complying with Section 10, "CLEAR FLOOR OR GROUND SPACE".   **11B-810.1.2**

Bus Boarding and Alighting Areas
Bus boarding and alighting areas shall comply as detailed, below.   **11B-810.2**   *810.2*

> **ADVISORY: Bus Boarding and Alighting Areas.** *At bus stops where a shelter is provided, the bus stop pad can be located either within or outside of the shelter.*

Surface

_____C.   Bus stop boarding and alighting areas shall have a firm, stable surface. **11B-810.2.1**   *810.2.1*

Dimensions

_____D.   Bus stop boarding and alighting areas shall provide a clear length of 96 inches (*2438* mm) minimum, measured perpendicular to the curb or vehicle roadway edge, and a clear width of 60 inches (*1524* mm) minimum, measured parallel to the vehicle roadway.   **11B-810.2.2**   *810.2.2*   **Fig. CD-44A**

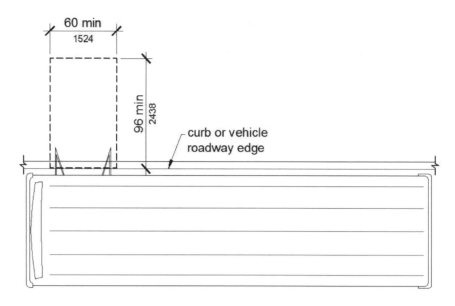

Fig. CD-44A
Dimensions of Bus Boarding and Alighting Areas
© ICC Reproduced with Permission

Connection

_____ E.   Bus stop boarding and alighting areas shall be connected to streets, sidewalks, or pedestrian paths by an accessible route complying with Section 2, "ACCESSIBLE ROUTES". **11B-810.2.3** *810.2.3*

_____ E1.   Newly constructed bus stop pads shall provide a square curb transition between the pad and roadway elevations or detectable warnings complying with item M of Section 32, "DETECTABLE WARNINGS". **11B-810.2.3** *810.2.3*

Slope

_____ F.   Parallel to the roadway, the slope of the bus stop boarding and alighting area shall be the same as the roadway, to the maximum extent practicable. **11B-810.2.4** *810.2.4*

_____ F1.   Perpendicular to the roadway, the slope of the bus stop boarding and alighting area shall not be steeper than 1:48. **11B-810.2.4** *810.2.4*

Bus Shelters
Bus shelters shall comply as detailed. **810.3** *810.2.3*

_____ G.   Bus shelters shall provide a minimum clear floor or ground space complying with Section 10, "CLEAR FLOOR OR GROUND SPACE" entirely within the shelter. **11B-810.3** *810.2.3* **Fig. CD-44B**

_____ G1.   Bus shelters shall be connected by an accessible route complying with Section 2, "ACCESSIBLE ROUTES" to a boarding and alighting area complying with items C through F1, above. **11B-810.3** *810.2.3* **Fig. CD-44B**

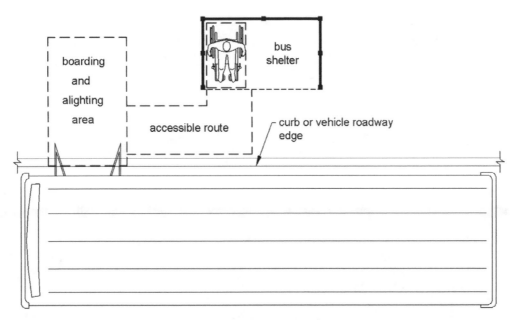

Fig. CD-44B
Bus Shelters
© ICC Reproduced with Permission

Bus Signs

Bus signs shall comply as detailed.   **810.4**   *810.4*

_____H.   Bus route identification signs shall comply with items A through D, and items G and H of Section 30D, "VISUAL CHARACTERS".   **11B-810.4**   *810.4*

_____H1.   In addition, to the maximum extent practicable, bus route identification signs shall comply with item E of Section 30D, "VISUAL CHARACTERS". **11B-810.4**   *810.4*

> **EXCEPTION:** Bus schedules, timetables and maps that are posted at the bus stop or bus bay shall not be required to comply.

Rail Platforms   Rail platforms shall comply as detailed.   **11B-810.5**   *810.5*

Slope

_____I.   Rail platforms shall not exceed a slope of 1:48 in all directions. **11B-810.5.1**   *810.5.1*

> **EXCEPTION:** Where platforms serve vehicles operating on existing track or track laid in existing roadway, the slope of the platform parallel to the track shall be permitted to be equal to the slope (grade) of the roadway or existing track.

Detectable Warnings

_____J.   Platform boarding edges not protected by platform screens or guards shall have detectable warnings complying with Section 32, "DETECTABLE WARNINGS", along the full length of the public use area of the platform.   **11B-810.5.2** *810.5.2*

Platform and Vehicle Floor Coordination

_____K.   Station platforms shall be positioned to coordinate with vehicles in accordance with the applicable requirements of 36 CFR Part 1192. Low-level platforms shall be 8 inches (*203* mm) minimum above top of rail.   **11B-810.5.3**   *810.5.3*

> **EXCEPTION:** Where vehicles are boarded from sidewalks or street-level, low-level platforms shall be permitted to be less than 8 inches (*203* mm).

> **ADVISORY: Platform and Vehicle Floor Coordination.** *The height and position of a platform must be coordinated with the floor of the vehicles it serves to minimize the vertical and horizontal gaps, in accordance with the ADA Accessibility Guidelines for Transportation Vehicles (36 CFR Part 1192). The vehicle guidelines, divided by bus, van, light rail, rapid rail, commuter rail, intercity rail, are available at* **www.access-board.gov**. *The preferred alignment is a high platform, level with the vehicle floor. In some cases, the vehicle guidelines permit use of a low platform in conjunction with a lift or ramp. Most such low platforms must have a minimum height of eight inches above the top of the rail. Some vehicles are designed to be boarded from a street or the sidewalk along the street and the exception permits such boarding areas to be less than eight inches high.*

Rail Station Signs
Rail station signs shall comply as detailed.   **11B-810.6**   *810.6*

> **EXCEPTION.** Signs shall not be required to comply with items L through M2, below, where audible signs are remotely transmitted to hand-held receivers, or are user- or proximity-actuated.

> **ADVISORY: Rail Station Signs Exception.** *Emerging technologies such as an audible sign systems using infrared transmitters and receivers may provide greater accessibility in the transit environment than traditional Braille and raised letter signs. The transmitters are placed on or next to print signs and transmit their information to an infrared receiver that is held by a person. By scanning an area, the person will hear the sign. This means that signs can be placed well out of reach of Braille readers, even on parapet walls and on walls beyond barriers. Additionally, such signs can be used to provide wayfinding information that cannot be efficiently conveyed on Braille signs.*

Entrances

_____L.   Where signs identify a station or its entrance, at least one sign at each entrance shall comply with Section 30A, "RAISED CHARACTERS", and shall be placed in uniform locations to the maximum extent practicable.

_____L1.   Where signs identify a station that has no defined entrance, at least one sign shall comply with Section 30A, "RAISED CHARACTERS", and shall be placed in a central location.   **11B-810.6.1**   *810.6.1*

Routes and Destinations

_____M.   Lists of stations, routes and destinations served by the station which are located on boarding areas, platforms, or mezzanines shall comply with Section 30A, "RAISED CHARACTERS". **11B-810.6.2**   *810.6.2*

      _____M1.   At least one tactile sign identifying the specific station and complying with Section 30A, "RAISED CHARACTERS", shall be provided on each platform or boarding area. **11B-810.6.2**   *810.6.2*

      _____M2.   Signs covered by this requirement shall, to the maximum extent practicable, be placed in uniform locations within the system. **11B-810.6.2**   *810.6.2*

> **EXCEPTION:** Where sign space is limited, characters shall not be required to exceed 3 inches (*76* mm).

> **ADVISORY: Routes and Destinations.** *Route maps are not required to comply with the informational sign requirements in this chapter.*

Station Names

_____N.   Stations covered by this section shall have identification signs complying with items A through D of Section 30A, "RAISED CHARACTERS".   **11B-810.6.3** *810.6.3*

      _____N1.   Signs shall be clearly visible and within the sight lines of standing and sitting passengers from within the vehicle on both sides when not obstructed by another vehicle. **11B-810.6.3**   *810.6.3*

> **ADVISORY: Station Names.** *It is also important to place signs at intervals in the station where passengers in the vehicle will be able to see a sign when the vehicle is either stopped at the station or about to come to a stop in the station. The number of signs necessary may be directly related to the size of the lettering displayed on the sign.*

Public Address Systems

_____O.   Where public address systems convey audible information to the public, the same or equivalent information shall be provided in a visual format. **11B-810.7** *810.7*

Clocks

_____P.   Where clocks are provided for use by the public, the clock face shall be uncluttered so that its elements are clearly visible. **11B-810.8**   *810.8*

      _____P1.   Hands, numerals and digits shall contrast with the background, either light-on-dark or dark-on-light. **11B-810.8**   *810.8*

      _____P2.   Where clocks are installed overhead, numerals and digits shall comply with items A through D of Section 30A, "RAISED CHARACTERS". **11B-810.8**   *810.8*

Escalators

_____Q.   Where provided, escalators shall comply with Sections 6.1.3.5.6 and 6.1.3.6.5 of ASME A17.1 and shall have a clear width of 32 inches (*813* mm) minimum. **11B-810.9**   *810.9*

> **EXCEPTION:** Existing escalators in key stations shall not be required to comply with this provision.

Track Crossings

_____R.   Where a circulation path serving boarding platforms crosses tracks, it shall comply with Section 2, "ACCESSIBLE ROUTES".   **11B-810.10**   *810.10*

> **EXCEPTION:** Openings for wheel flanges shall be permitted to be 2½ inches (64 mm) maximum.   **Fig. CD-44C**

Fig. CD-44C (Exception)
Track Crossings
© ICC Reproduced with Permission

45. STORAGE

Storage   **11B-225**   *225*

General
Storage facilities shall comply with *Section 11B*-225.   **11B-225.1**   *225.1*

Storage
Where storage is provided in accessible spaces, at least one of each type shall comply as detailed.   **11B-225.2**   *225.2*

> ***ADVISORY: Storage.*** *Types of storage include, but are not limited to, closets, cabinets, shelves, clothes rods, hooks, and drawers. Where provided, at least one of each type of storage must be within the reach ranges specified in Section 11B-308; however, it is permissible to install additional storage outside the reach ranges.*

Lockers
Where lockers are provided, at least 5 percent, but no fewer than one of each type, shall comply as detailed, below.   **11B-225.2.1**   *225.2.1*

> ***ADVISORY: Lockers.*** *Different types of lockers may include full-size and half-size lockers, as well as those specifically designed for storage of various sports equipment.*

Self-Service Shelving

_____A.   Self-service shelves shall be located on a compliant accessible route. Self-service shelving shall not be required to comply with Section 12, "REACH RANGES". **11B-225.2.2**   *225.2.2*

> **ADVISORY: Self-Service Shelving.** *Self-service shelves include, but are not limited to, library, store, or post office shelves.*

Library Book Stacks

_____B.   Book stacks available for public use shall be 54 inches (1372 mm) maximum above the finish floor.   **11B-225.2.3**   **Fig. CD-45A**

EXCEPTIONS:

1.   Book stacks available for public use may be higher than 54 inches (1372 mm) maximum above the finish floor when an attendant is available to assist persons with disabilities.

2.   Book stacks restricted to employee use are not required to comply with these requirements.

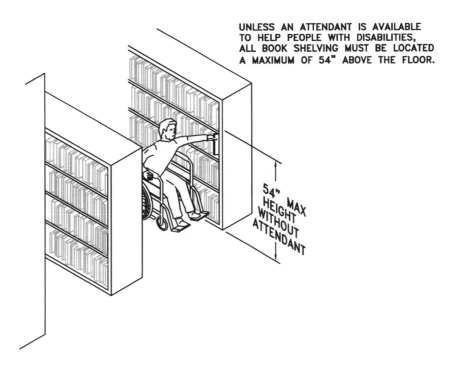

UNLESS AN ATTENDANT IS AVAILABLE TO HELP PEOPLE WITH DISABILITIES, ALL BOOK SHELVING MUST BE LOCATED A MAXIMUM OF 54" ABOVE THE FLOOR.

54" MAX HEIGHT WITHOUT ATTENDANT

Fig. CD-45A
Library Book Stacks

Self-Service Storage Facilities

Self-service storage facilities shall provide individual self-service storage spaces complying with these requirements in accordance with Table 45A.   **11B-225.3**   *225.3*

Table 45A
Self-Service Storage Facilities

Total Spaces in Facility	Minimum Number of Spaces Required to be Accessible
1 to 200	5 percent, but no fewer than 1
201 and over	10, plus 2 percent of total number of units over 200

ADVISORY: Self-Service Storage Facilities. Although there are no technical requirements that are unique to self-service storage facilities, elements and spaces provided in facilities containing self-service storage spaces required to comply with these requirements must comply with this chapter where applicable. For example: the number of storage spaces required to comply with these requirements must provide Accessible Routes complying with Section 2, "ACCESSIBLE ROUTES"; Accessible Means of Egress complying with Section 4, "ACCESSIBLE MEANS OF EGRESS"; Parking Spaces complying with Section 21, "
PARKING SPACES"; and, where provided, other public use or common use elements and facilities such as toilet rooms, drinking fountains, and telephones must comply with the applicable requirements of this chapter.

Dispersion

Individual self-service storage spaces shall be dispersed throughout the various classes of spaces provided. Where more classes of spaces are provided than the number required to be accessible, the number of spaces shall not be required to exceed that required by Table 45A. Self-service storage spaces complying with Table 45A shall not be required to be dispersed among buildings in a multi-building facility.   **11B-225.3.1**   *225.3.1*

Storage   **11B-811**   *811*

General

Storage shall comply as detailed.   **11B-811.1**   *811.1*

Clear Floor or Ground Space

_____C.   A clear floor or ground space complying with Section 10, "CLEAR FLOOR OR GROUND SPACE", shall be provided.   **11B-811.2**   *911.2*

Height

_____D.   Storage elements shall comply with at least one of the reach ranges specified in Section 12, "REACH RANGES".   **11B-811.3**   *811.3*

Operable Parts

_____E.   Operable parts shall comply with Section 13, "OPERABLE PARTS".   **11B-811.4**   *811.4*

46. <u>DINING SURFACES AND WORK SURFACES</u>

<u>Accessible Route in Restaurants, Cafeterias, Banquet Facilities and Bars</u>

In restaurants, cafeterias, banquet facilities, bars, and similar facilities, an accessible route shall be provided to all functional areas, including raised or sunken dining areas, and outdoor dining areas.   **11B-206.2.5**   *206.2.5*

EXCEPTIONS:

1.   In alterations of buildings or facilities not required to provide an accessible route between stories, an accessible route shall not be required to a mezzanine dining area where the mezzanine contains less than 25 percent of the total combined area for seating and dining and where the same décor and services are provided in the accessible area.

2.   **<u>Reserved.</u>**

3.   In sports facilities, tiered dining areas providing seating required to comply with the requirements for accessible Assembly Areas shall be required to have accessible routes serving at least 25 percent of the dining area provided that accessible routes serve seating required to comply with the requirements for accessible Assembly Areas, and each tier is provided with the same services.

<u>Dining Surfaces and Work Surfaces</u>   **11B-226**   *226*

<u>General</u>

Where dining surfaces are provided for the consumption of food or drink, at least 5 percent of the seating spaces and standing spaces at the dining surfaces shall comply with items B through G, below. In addition, where work surfaces are provided for use by other than employees, at least 5 percent shall comply with items B through G, below.   **11B-226.1**   *226.1*

EXCEPTIONS:

1.   Sales counters and service counters shall not be required to comply with E through G, below. See Section 48, "SALES AND SERVICE".

2.   Check writing surfaces provided at check-out aisles not required to comply with items A through D of Section 48, "SALES AND SERVICE", shall not be required to comply with items B through G, below.

> **ADVISORY: General.** *In facilities covered by the ADA, this requirement does not apply to work surfaces used only by employees. However, the ADA and, where applicable, Section 504 of the Rehabilitation Act of 1973, as amended, provide that employees are entitled to "reasonable accommodations." With respect to work surfaces, this means that employers may need to procure or adjust work stations such as desks, laboratory and work benches, fume hoods, reception counters, teller windows, study carrels, commercial kitchen counters, and conference tables to accommodate the individual needs of employees with disabilities on an "as needed" basis. Consider work surfaces that are flexible and permit installation at variable heights and clearances.*

Dispersion

Dining surfaces required to comply with items B through G, below, shall be dispersed throughout the space or facility containing dining surfaces for each type of seating in a functional area. Work surfaces required to comply with Section 46, "DINING SURFACES AND WORK SURFACES", shall be dispersed throughout the space or facility containing work surfaces. **11B-226.2** *226.2*

Dining Surfaces Exceeding 34 Inches in Height

_____A. Where food or drink is served for consumption at a counter exceeding 34 inches (864 mm) in height, a portion of the main counter 60 inches (1525 mm) minimum in length shall be provided in compliance with Section 11B-902.3. **11B-226.3**

Dining Surfaces and Work Surfaces **11B-902** *902*

General

Dining surfaces and work surfaces shall comply with items B through G, below. **11B-902.1** *902.1*

> **EXCEPTION:** Dining surfaces and work surfaces for children's use shall be permitted to comply with items E through G, below.

> **ADVISORY: General.** *Dining surfaces include, but are not limited to, bars, tables, lunch counters, and booths. Examples of work surfaces include writing surfaces, study carrels, student laboratory stations, baby changing and other tables or fixtures for personal grooming, coupon counters, and where covered by the ABA scoping provisions, employee work stations.*

Clear Floor or Ground Space

_____B. A clear floor space complying with Section 10, "CLEAR FLOOR OR GROUND SPACE", positioned for a forward approach shall be provided. **11B-902.2** *902.2*

_____C. Knee and toe clearance complying with Section 11, "KNEE AND TOE CLEARANCE" shall be provided. **11B-902.2** *902.2*

Height

_____D. The tops of dining surfaces and work surfaces shall be 28 inches (*711* mm) minimum and 34 inches (*864* mm) maximum above the finish floor or ground. **11B-902.3** *902.3*

Dining Surfaces and Work Surfaces for Children's Use

Accessible dining surfaces and work surfaces for children's use shall comply as detailed, below. **11B-902.4** *902.4*

> **EXCEPTION:** Dining surfaces and work surfaces that are used primarily by children 5 years and younger shall not be required to comply these provisions where a clear floor or ground space complying with Section 10, "CLEAR FLOOR OR GROUND SPACE", positioned for a parallel approach is provided.

Clear Floor or Ground Space

_____E. A clear floor space complying with Section 10, "CLEAR FLOOR OR GROUND

SPACE", positioned for forward approach shall be provided.   **11B-902.4.1**
902.4.1

_____F.   Knee and toe clearance complying with Section 11, "KNEE AND TOE
CLEARANCE" shall be provided, except that knee clearance 24 inches (610 mm)
minimum above the finish floor or ground shall be permitted.   **11B-902.4.1**
902.4.1

Height

_____G.   The tops of tables and counters shall be 26 inches (660 mm) minimum and 30
inches (*762* mm) maximum above the finish floor or ground.   **11B-902.4.2**
902.4.2

47. BENCHES

Benches   **11B-903**   *903*

General
Benches shall comply as detailed.   **11B-903.1**   *903.1*

Clear Floor or Ground Space

_____A.   Clear floor or ground space complying with Section 10, "CLEAR FLOOR OR GROUND
SPACE", shall be provided and shall be positioned at the end of the bench seat and
parallel to the short axis of the bench.   **11B-903.2**   *903.2*

Size

_____B.   Benches shall have seats that are 48 inches (*1219* mm) long minimum and 20 inches
(*508* mm) deep minimum and 24 inches (610 mm) deep maximum.   **11B-903.3**
903.3

Back Support

_____C.   The bench shall provide for back support or shall be affixed to a wall along its long
dimension.

_____C1.   Back support shall be 48 inches (*1219* mm) long minimum and shall extend from
a point 2 inches (51 mm) maximum above the seat surface to a point 18 inches
(*457* mm) minimum above the seat surface.   **11B-903.3**   *903.3*   **Fig. CD-47A**

_____C2.   Back support shall be 2½ inches (64 mm) maximum from the rear edge of the
seat measured horizontally.   **11B-903.4**   *903.4*   **Fig. CD-47A**

> **ADVISORY: Back Support.** *To assist in transferring to the bench, consider providing
> grab bars on a wall adjacent to the bench, but not on the seat back. If provided, grab bars
> cannot obstruct transfer to the bench.*

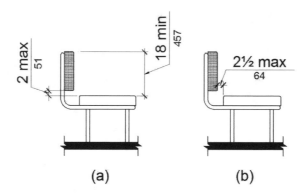

Fig. CD-47A
Bench Back Support
© ICC Reproduced with Permission

Height

_____D.　　The top of the bench seat surface shall be 17 inches (*432* mm) minimum and 19 inches (*483* mm) maximum above the finish floor or ground.　**11B-903.5**　*903.5*

Structural Strength

_____E.　　Benches shall be affixed to the wall or floor. Allowable stresses shall not be exceeded for materials used when a vertical or horizontal force of 250 pounds (1112 N) is applied at any point on the seat, fastener, mounting device, or supporting structure.　**11B-903.6** *903.6*

Wet Locations

_____F.　　Where installed in wet locations, the surface of the seat shall be slip resistant and shall not accumulate water.　**11B-903.7**　*903.7*

48. SALES AND SERVICE

Sales and Service　**11B-227**　*227*

General
Where provided, check-out aisles, sales counters, service counters, food service lines, queues, and waiting lines shall comply as detailed.　**11B-227.1**　*227.1*

General
Check-out aisles and sales and service counters shall comply with the applicable requirements as detailed.　**11B-904.1**　*904.1*

Queues and Waiting Lines
Queues and waiting lines servicing required accessible sales and service counters or check-out aisles shall comply with Section 5, "WALKING SURFACES".　**11B-227.5**　*227.5*

CHECK-OUT AISLES

Check-Out Aisles

Where check-out aisles are provided, check-out aisles that comply as detailed shall be provided in accordance with Table 48A. Where check-out aisles serve different functions, check-out aisles that comply as detailed shall be provided in accordance with Table 48A for each function. Where check-out aisles are dispersed throughout the building or facility, check-out aisles that comply as detailed shall be dispersed. When check-out aisles are open for customer use, a minimum of one accessible check-out aisle shall always be available. As check-out aisles are opened and closed based on fluctuating customer levels, the number of accessible check-out aisles available shall comply with Table 48A. When not all check-out aisles are accessible, accessible check-out aisles shall be identified by the International Symbol of Accessibility complying with Section 30F, "SYMBOLS OF ACCESSIBILITY". **11B-227.2** *227.2*

> **EXCEPTION:** In existing buildings, where the selling space is under 5000 square feet (465 m^2) no more than one check-out aisle that complies as detailed shall be required.

Table 48A
Check-Out Aisles

Number of Check-Out Aisles of Each Function	Minimum Number of Check-Out Aisles of Each Function Required to Comply
1 to 4	1
5 to 8	2
9 to 15	3
16 and over	3, plus 20 percent of additional aisles

ADVISORY: Check-Out Aisles. Where check-out aisles are provided, accessible check-out aisles are required. This section provides requirements specific to check-out aisle design and construction. In addition to the requirements of this section, operational procedures are often necessary to ensure the Americans with Disabilities Act accessibility requirements are met. When check-out aisles are open for customer use, business owners should ensure that a minimum of one accessible check-out aisle is always available for use by persons with disabilities. As check-out aisles are opened and closed based on fluctuating customer levels, business owners should ensure that the number of accessible check-out aisles available complies with Table 48A.

Altered Check-Out Aisles

Where check-out aisles are altered, at least one of each check-out aisle serving each function shall comply as detailed until the number of check-out aisles complies as required. **11B-227.2.1** *227.2.1*

Check-Out Aisles

Check-out aisles shall comply as detailed. **11B-904.3** *904.3*

Aisle

_____A.   Aisles shall comply with Section 5, "WALKING SURFACES". **11B-904.3.1** *904.3.1*

Counter

_____B. The counter surface height shall be 38 inches (965 mm) maximum above the finish floor or ground. **11B-904.3.2** *904.3.2* **Fig. CD-48A**

_____B1. The top of the counter edge protection shall be 2 inches (51 mm) maximum above the top of the counter surface on the aisle side of the check-out counter. **11B-904.3.2** *904.3.2* **Fig. CD-48A**

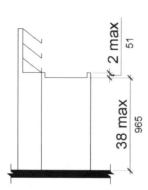

Fig. CD-48A
Check-Out Aisle Counters
© ICC Reproduced with Permission

Check Writing Surfaces

_____C. Where provided, the tops of check writing surfaces shall be 28 inches (710 mm) minimum and 34 inches (865 mm) maximum above the finish floor or ground. **11B-904.3.3, 902.3** *904.3.3, 902.3*

Identification Sign

_____D. When not all check-out aisles are accessible, accessible check-out aisles shall be identified by a sign clearly visible to a person in a wheelchair displaying the International Symbol of Accessibility complying with Section 30F, "SYMBOLS OF ACCESSIBILITY". The sign shall be a minimum of 4 inches by 4 inches (102 mm by 102 mm). **11B-904.3.4** *904.3.4* **Fig. CD-48B**

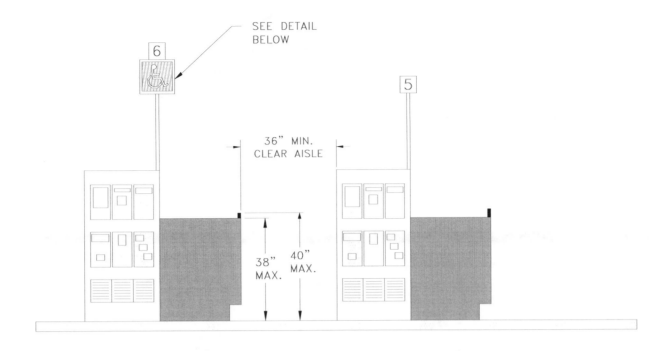

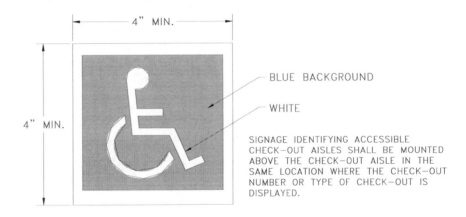

Fig. CD-48B
Identification Sign

SALES AND SERVICE COUNTERS

Counters

Where provided, at least one of each type of sales counter and service counter shall comply as detailed. Where counters are dispersed throughout the building or facility, counters that comply as detailed also shall be dispersed. **11B-227.3** *227.3*

> **ADVISORY: Counters.** *Types of counters that provide different services in the same facility include, but are not limited to, order, pick-up, express, and returns. One continuous counter can be used to provide different types of service. For example, order and pick-up are different services. It would not be acceptable to provide access only to the part of the counter where orders are taken when orders are picked-up at a different location on the same counter. Both the order and pick-up section of the counter must be accessible.*

Approach

_____E.   All portions of counters required to comply shall be located adjacent to a walking surface complying with Section 5, "WALKING SURFACES".   **11B-904.2**   *904.2*

> **ADVISORY: Approach.** *If a cash register is provided at the sales or service counter, locate the accessible counter close to the cash register so that a person using a wheelchair is visible to sales or service personnel and to minimize the reach for a person with a disability.*

Sales and Service Counters

_____F.   Sales counters and service counters shall comply with the requirements for an accessible parallel approach or forward approach as detailed below. The accessible portion of the counter top shall extend the same depth as the sales or service counter top. **11B-904.4** *904.4*

> **EXCEPTION:** In alterations, when the provision of a fully compliant counter would result in a reduction of the number of existing counters at work stations or a reduction of the number of existing mail boxes, the counter shall be permitted to have a portion which is 24 inches (610 mm) long minimum that provides an accessible parallel approach, provided that the required clear floor or ground space is centered on the accessible length of the counter.   **Fig. CD-48C**

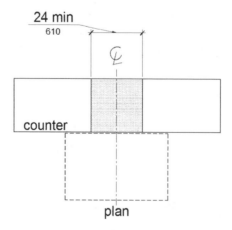

Fig. CD-48C (Exception)
Alteration of Sales and Service Counters
© ICC Reproduced with Permission

Parallel Approach

_____G.   A portion of the counter surface that is 36 inches (*914* mm) long minimum and 3 inches (*864* mm) high maximum above the finish floor shall be provided. **11B-904.4.1**   *904.4.1*

_____G1.   A clear floor or ground space complying with Section 10, "CLEAR FLOOR OR GROUND SPACE", shall be positioned for a parallel approach adjacent to the 36 inch (*914* mm) minimum length of counter. **11B-904.4.1**   *904.4.1*

> **EXCEPTION:** Where the provided counter surface is less than 36 inches (*914* mm) long, the entire counter surface shall be 34 inches (*864* mm) high maximum above the finish floor.

Forward Approach

_____H.   A portion of the counter surface that is *36* inches (*914* mm) long minimum and 34 inches (*864* mm) high maximum shall be provided. **11B-904.4.2**   *904.4.2*

_____H1.   Knee and toe space complying with Section 11, "KNEE AND TOE CLEARANCE", shall be provided under the counter. **11B-904.4.2**   *904.4.2*

_____H2.   A clear floor or ground space complying with Section 10, "CLEAR FLOOR OR GROUND SPACE", shall be positioned for a forward approach to the counter.   **11B-904.4.2**   *904.4.2*

Security Glazing

_____I.   Where counters or teller windows have security glazing to separate personnel from the public, a method to facilitate voice communication shall be provided. **11B-904.6**   *904.6*

_____I1.   Telephone handset devices, if provided, shall comply with the requirements for volume control telephones.   **11B-904.6**   *904.6*

Volume Control Telephones

_____I2.   Public telephones required to have volume controls shall be equipped with a receive volume control that provides a gain adjustable up to 20 dB minimum. For incremental volume control, provide at least one intermediate step of 12 dB of gain minimum. An automatic reset shall be provided. Volume control telephones shall be equipped with a receiver that generates a magnetic field in the area of the receiver cap. Public telephones with volume control shall be hearing aid compatible. **11B-704.3**   *704.3*

> **ADVISORY: Security Glazing.** *Assistive listening devices complying with Section 33, "ASSISTIVE LISTENING SYSTEMS", can facilitate voice communication at counters or teller windows where there is security glazing which promotes distortion in audible information. Where assistive listening devices are installed, place signs with the International Symbol of Access For Hearing Loss, per Section 30F, "SYMBOLS OF ACCESSIBILITY", to identify those facilities which are so equipped. Other voice communication methods include, but are not limited to, grilles, slats, talk-through baffles, intercoms, or telephone handset devices.*

FOOD SERVICE LINES

Food Service Lines
Food service lines shall comply as detailed. Where self-service shelves are provided, at least 50 percent, but no fewer than one, of each type provided shall comply with Section 12, "REACH RANGES". **11B-227.4** *227.4*

Food Service Lines
Counters in food service lines shall comply as detailed. **11B-904.5** *904.5*

Self-Service Shelves and Dispensing Devices

_____J.   Self-service shelves and dispensing devices for tableware, dishware, condiments, food and beverages shall comply with Section 12, "REACH RANGES". **11B-904.5.1** *904.5.1*

Tray Slides

_____K.   The tops of tray slides shall be 28 inches (*711* mm) minimum and 34 inches (864 mm) maximum above the finish floor or ground. **11B-904.5.2** *904.5.2*

49. AMUSEMENT RIDES

Amusement Rides   **11B-234**   *234*

Accessible Routes for Amusement Rides
Amusement rides required to comply shall provide accessible routes as detailed below. Accessible routes serving amusement rides shall comply except as modified. **11B-206.2.9**   *206.2.9*

Load and Unload Areas
Load and unload areas shall be on an accessible route. Where load and unload areas have more than one loading or unloading position, at least one loading and unloading position shall be on an accessible route. **11B-206.2.9.1**   *206.2.9.1*

Wheelchair Spaces, Ride Seats Designed for Transfer, and Transfer Devices
When amusement rides are in the load and unload position, wheelchair spaces, amusement ride seats designed for transfer, and transfer devices shall be on an accessible route. **11B-206.2.9.2**   *206.2.9.2*

Platform Lifts
Platform lifts shall be permitted to provide accessible routes to load and unload areas serving amusement rides. **11B-206.7.7**   *206.7.7*

Water Slides
Water slides shall not be required to comply with these requirements or to be on an accessible route. **11B-203.11**   *203.11*

General
Amusement rides shall comply as detailed. **11B-234.1**   *234.1*

EXCEPTION: Mobile or portable amusement rides shall not be required to comply with these provisions.

ADVISORY: General. *These requirements apply generally to newly designed and constructed amusement rides and attractions. A custom designed and constructed ride is new upon its first use, which is the first time amusement park patrons take the ride. With respect to amusement rides purchased from other entities, new refers to the first permanent installation of the ride, whether it is used off the shelf or modified before it is installed. Where amusement rides are moved after several seasons to another area of the park or to another park, the ride would not be considered newly designed or newly constructed.*

Some amusement rides and attractions that have unique designs and features are not addressed by these requirements. In those situations, these requirements are to be applied to the extent possible. An example of an amusement ride not specifically addressed by these requirements includes "virtual reality" rides where the device does not move through a fixed course within a defined area. An accessible route must be provided to these rides. Where an attraction or ride has unique features for which there are no applicable scoping provisions, then a reasonable number, but at least one, of the features must be located on an accessible route. Where there are appropriate technical provisions, they must be applied to the elements that are covered by the scoping provisions.

ADVISORY: General Exception. *Mobile or temporary rides are those set up for short periods of time such as traveling carnivals, State and county fairs, and festivals. The amusement rides that are covered by this Section are ones that are not regularly assembled and disassembled.*

Load and Unload Areas
Load and unload areas serving amusement rides shall comply as detailed. **11B-234.2** *234.2*

Minimum Number
Amusement rides shall provide at least one wheelchair space complying with items C through L, below, or at least one amusement ride seat designed for transfer complying with items M through P, below, or at least one transfer device complying with items Q through S, below. **11B-234.3** *234.3*

EXCEPTIONS:

1. Amusement rides that are controlled or operated by the rider shall not be required to comply with the requirement for wheelchair space or transfer.

2. Amusement rides designed primarily for children, where children are assisted on and off the ride by an adult, shall not be required to comply with the requirement for wheelchair space or transfer.

3. Amusement rides that do not provide amusement ride seats shall not be required to comply with the requirement for wheelchair space or transfer.

> **ADVISORY: Minimum Number Exceptions 1 through 3.** *Amusement rides controlled or operated by the rider, designed for children, or rides without ride seats are not required to comply with the requirement for wheelchair space or transfer. These rides are not exempt from the other provisions in this Section requiring an accessible route to the load and unload areas and to the ride. The exception does not apply to those rides where patrons may cause the ride to make incidental movements, but where the patron otherwise has no control over the ride.*

> **ADVISORY: Minimum Number Exception 2.** *The exception is limited to those rides designed "primarily" for children, where children are assisted on and off the ride by an adult. This exception is limited to those rides designed for children and not for the occasional adult user. An accessible route to and turning space in the load and unload area will provide access for adults and family members assisting children on and off these rides.*

Existing Amusement Rides
Where existing amusement rides are altered, the alteration shall comply as detailed below.   **11B-234.4**   *234.4*

Load and Unload Areas
Where load and unload areas serving existing amusement rides are newly designed and constructed, the load and unload areas shall provide an accessible turning space complying with Section 9, "TURNING SPACE". **11B-234.4.1** *234.4.1*

Minimum Number
Where the structural or operational characteristics of an amusement ride are altered to the extent that the amusement ride's performance differs from that specified by the manufacturer or the original design, the amusement ride shall comply with the requirement to provide wheelchair space and transfer. **11B-234.4.2** *234.4.2*

> **ADVISORY: Existing Amusement Rides.** *Routine maintenance, painting, and changing of theme boards are examples of activities that do not constitute an alteration subject to this section.*

Amusement Rides   **11B-1002**   *1002*

General
Amusement rides shall comply as detailed.   **11B-1002.1**   *1002.1*

Accessible Routes

_____A.   Accessible routes serving amusement rides shall comply. **11B-1002.2** *1002.2*

EXCEPTIONS:

1.   In load or unload areas and on amusement rides, where providing a ramp run with a maximum running slope of 1:12 (8.33%) is not structurally or operationally feasible, ramp slope shall be permitted to be 1:8 (12.5%) maximum.

2.   In load or unload areas and on amusement rides, handrails

provided along accessible walking surfaces and required on accessible ramps shall not be required to comply with the requirements for accessible handrails where compliance is not structurally or operationally feasible.

ADVISORY: Accessible Routes Exception 1. Steeper slopes are permitted on accessible routes connecting the amusement ride in the load and unload position where it is "structurally or operationally infeasible." In most cases, this will be limited to areas where the accessible route leads directly to the amusement ride and where there are space limitations on the ride, not the queue line. Where possible, the least possible slope should be used on the accessible route that serves the amusement ride.

Load and Unload Areas

_____B.   A turning space complying with Section 9, "TURNING SPACE", shall be provided in load and unload areas. **11B-1002.3**   *1002.3*

Wheelchair Spaces in Amusement Rides
Wheelchair spaces in amusement rides shall comply as detailed.   **11B-1002.4**   *1002.4*

Floor or Ground Surface

_____C.   The floor or ground surface of wheelchair spaces shall be stable and firm.   **11B-1002.4.1**   *1002.4.1*

Slope

_____D.   The floor or ground surface of wheelchair spaces shall have a slope not steeper than 1:48 when in the load and unload position. **11B-1002.4.2**   *1002.4.2*

Gaps

_____E.   Floors of amusement rides with wheelchair spaces and floors of load and unload areas shall be coordinated so that, when amusement rides are at rest in the load and unload position, the vertical difference between the floors shall be within plus or minus 5/8 inches (*15.9* mm) and the horizontal gap shall be 3 inches (*76* mm) maximum under normal passenger load conditions.   **11B-1002.4.3**   *1002.4.3*

EXCEPTION: Where compliance is not operationally or structurally feasible, ramps, bridge plates, or similar devices complying with the applicable requirements of 36 CFR 1192.83(c) shall be provided.

ADVISORY: Gaps Exception. 36 CFR 1192.83(c) ADA Accessibility Guidelines for Transportation Vehicles - Light Rail Vehicles and Systems - Mobility Aid Accessibility is available at www.access-board.gov. It includes provisions for bridge plates and ramps that can be used at gaps between wheelchair spaces and floors of load and unload areas.

Clearances
Clearances for wheelchair spaces shall comply as detailed.   **11B-1002.4.3**   *1002.4.3*

EXCEPTIONS:

1. Where provided, securement devices shall be permitted to overlap required clearances.

2. Wheelchair spaces shall be permitted to be mechanically or manually repositioned.

3. Wheelchair spaces shall not be required to comply with the requirement to provide minimum vertical clearance of 80 inches high, or provide guardrails or other barriers when vertical clearance is less than 80 inches high.

> **ADVISORY: Clearances Exception 3.** *This exception for protruding objects applies to the ride devices, not to circulation areas or accessible routes in the queue lines or the load and unload areas.*

Width and Length

_____F.   Wheelchair spaces shall provide a clear width of 30 inches (*762* mm) minimum and a clear length of 48 inches (*1219* mm) minimum measured to 9 inches (*229* mm) minimum above the floor surface.
11B-1002.4.4.1   *1002.4.4.1*

Side Entry

_____G.   Where wheelchair spaces are entered only from the side, amusement rides shall be designed to permit sufficient maneuvering clearance for individuals using a wheelchair or mobility aid to enter and exit the ride.
11B-1002.4.4.2   *1002.4.4.2*

> **ADVISORY: Side Entry.** *The amount of clear space needed within the ride, and the size and position of the opening are interrelated. A 32 inch (815 mm) clear opening will not provide sufficient width when entered through a turn into an amusement ride. Additional space for maneuvering and a wider door will be needed where a side opening is centered on the ride. For example, where a 42 inch (1065 mm) opening is provided, a minimum clear space of 60 inches (1525 mm) in length and 36 inches (915mm) in depth is needed to ensure adequate space for maneuvering.*

Permitted Protrusions in Wheelchair Spaces

_____H.   Objects are permitted to protrude a distance of 6 inches (*152* mm) maximum along the front of the wheelchair space, where located 9 inches (*229* mm) minimum and 27 inches (*686* mm) maximum above the floor or ground surface of the wheelchair space. Objects are permitted to protrude a distance of 25 inches (*635* mm) maximum along the front of the wheelchair space, where located more than 27 inches (*686* mm) above the floor or ground surface of the wheelchair space.
11B-1002.4.4.3   *1002.4.4.3*   **Fig. CD-49A**

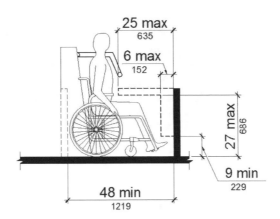

Fig. CD-49A
Protrusions in Wheelchair Spaces in Amusement Rides
© ICC Reproduced with Permission

Ride Entry

_____ I.   Openings providing entry to wheelchair spaces on amusement rides shall be 32 inches (*813* mm) minimum clear. **11B-1002.4.5**   *1002.4.5*

Approach

_____ J.   One side of the wheelchair space shall adjoin an accessible route when in the load and unload position. **11B-1002.4.6**   *1002.4.6*

Companion Seats

_____ K.   Where the interior width of the amusement ride is greater than 53 inches (*1346* mm), seating is provided for more than one rider, and the wheelchair is not required to be centered within the amusement ride, a companion seat shall be provided for each wheelchair space. **11B-1002.4.7**   *1002.4.7*

Shoulder-to-Shoulder Seating

_____ L.   Where an amusement ride provides shoulder-to-shoulder seating, companion seats shall be shoulder-to-shoulder with the adjacent wheelchair space. **11B-1002.4.7.1**   *1002.4.7.1*

> **EXCEPTION:** Where shoulder-to-shoulder companion seating is not operationally or structurally feasible, compliance with this requirement shall be required to the maximum extent practicable.

Amusement Ride Seats Designed for Transfer
Amusement ride seats designed for transfer shall comply with *Section 11B*-1002.5 when positioned for loading and unloading. **11B-1002.5**   *1002.5*

> *ADVISORY: Amusement Ride Seats Designed for Transfer. The proximity of the clear floor or ground space next to an element and the height of the element one is transferring to are both critical for a safe and independent transfer. Providing additional*

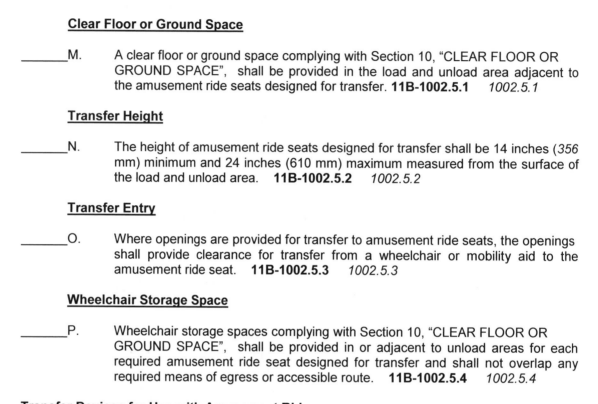

clear floor or ground space both in front of and diagonal to the element will provide flexibility and will increase usability for a more diverse population of individuals with disabilities. Ride seats designed for transfer should involve only one transfer. Where possible, designers are encouraged to locate the ride seat no higher than 17 to 19 inches (430 to 485 mm) above the load and unload surface. Where greater distances are required for transfers, providing gripping surfaces, seat padding, and avoiding sharp objects in the path of transfer will facilitate the transfer.

Clear Floor or Ground Space

_____M.   A clear floor or ground space complying with Section 10, "CLEAR FLOOR OR GROUND SPACE", shall be provided in the load and unload area adjacent to the amusement ride seats designed for transfer. **11B-1002.5.1**   *1002.5.1*

Transfer Height

_____N.   The height of amusement ride seats designed for transfer shall be 14 inches (*356 mm*) minimum and 24 inches (610 mm) maximum measured from the surface of the load and unload area.   **11B-1002.5.2**   *1002.5.2*

Transfer Entry

_____O.   Where openings are provided for transfer to amusement ride seats, the openings shall provide clearance for transfer from a wheelchair or mobility aid to the amusement ride seat.   **11B-1002.5.3**   *1002.5.3*

Wheelchair Storage Space

_____P.   Wheelchair storage spaces complying with Section 10, "CLEAR FLOOR OR GROUND SPACE", shall be provided in or adjacent to unload areas for each required amusement ride seat designed for transfer and shall not overlap any required means of egress or accessible route.   **11B-1002.5.4**   *1002.5.4*

Transfer Devices for Use with Amusement Rides
Transfer devices for use with amusement rides shall comply with as detailed when positioned for loading and unloading.   **11B-1002.6**   *1002.6*

ADVISORY: Transfer Devices for Use with Amusement Rides. Transfer devices for use with amusement rides should permit individuals to make independent transfers to and from their wheelchairs or mobility devices. There are a variety of transfer devices available that could be adapted to provide access onto an amusement ride. Examples of devices that may provide for transfers include, but are not limited to, transfer systems, lifts, mechanized seats, and custom designed systems. Operators and designers have flexibility in developing designs that will facilitate individuals to transfer onto amusement rides. These systems or devices should be designed to be reliable and sturdy.

Designs that limit the number of transfers required from a wheelchair or mobility device to the ride seat are encouraged. When using a transfer device to access an amusement ride, the least number of transfers and the shortest distance is most usable. Where possible, designers are encouraged to locate the transfer device seat no higher than 17 to 19 inches (430 to 485 mm) above the load and unload surface. Where greater distances are required for transfers, providing gripping surfaces, seat padding, and avoiding sharp objects in the path of transfer will facilitate the transfer. Where a series of transfers are required to reach the amusement ride seat, each vertical transfer should not exceed 8 inches (205 mm).

Clear Floor or Ground Space.

_____Q.   A clear floor or ground space complying with Section 10, "CLEAR FLOOR OR GROUND SPACE", shall be provided in the load and unload area adjacent to the transfer device.   **11B-1002.6.1**   *1002.6.1*

Transfer Height

_____R.   The height of transfer device seats shall be 14 inches (*356* mm) minimum and 24 inches (610 mm) maximum measured from the load and unload surface.   **11B-1002.6.2**   *1002.6.2*

Wheelchair Storage Space

_____S.   Wheelchair storage spaces complying with Section 10, "CLEAR FLOOR OR GROUND SPACE", shall be provided in or adjacent to unload areas for each required transfer device and shall not overlap any required means of egress or accessible route.   **11B-1002.6.3**   *1002.6.3*

50. <u>RECREATIONAL BOATING FACILITIES</u>

<u>**Recreational Boating Facilities**</u>   **11B-235**   *235*

<u>General</u>
Recreational boating facilities shall comply as detailed.   **11B-235.1**   *235.1*

<u>Accessible Routes for Recreational Boating Facilities</u>
Boat slips required to comply, and boarding piers at boat launch ramps required to comply, shall be on an accessible route. Accessible routes serving recreational boating facilities shall comply except as modified.   **11B-206.2.10**   *206.2.10*

<u>Platform Lifts</u>
Platform lifts shall be permitted to be used instead of gangways that are part of accessible routes serving recreational boating facilities and fishing piers and platforms.   **11B-206.7.10**   *206.7.10*

<u>Boat Slips</u>
Boat slips complying as detailed shall be provided in accordance with Table 50A. Where the number of boat slips is not identified, each 40 feet (*12192 mm*) of boat slip edge provided along the perimeter of the pier shall be counted as one boat slip for the purpose of this section.
11B-235.2   *235.2*

Table 50A
<u>Boat Slips</u>

Total Number of Boat Slips Provided in Facility	Minimum Number of Required Accessible Boat Slips
1 to 25	1
26 to 50	2
51 to 100	3
101 to 150	4
151 to 300	5
301 to 400	6
401 to 500	7
501 to 600	8
601 to 700	9
701 to 800	10
801 to 900	11
901 to 1000	12
1001 and over	12, plus 1 for every 100, or fraction thereof, over 1000

ADVISORY: Boat Slips. The requirement for boat slips also applies to piers where boat slips are not demarcated. For example, a single pier 25 feet (7620 mm) long and 5 feet (1525 mm) wide (the minimum width specified by Section 11B-1003.3) allows boats to moor on three sides. Because the number of boat slips is not demarcated, the total length of boat slip edge (55 feet, 17 m) must be used to determine the number of boat slips provided (two). This number is based on the specification in that each 40 feet (12 m) of boat slip edge, or fraction thereof, counts as one boat slip. In this example, Table 50A would require one boat slip to be accessible.

Dispersion

Boat slips complying with items B and B1, below shall be dispersed throughout the various types of boat slips provided. Where the minimum number of boat slips required to comply with items B and B1 per Table 50A has been met, no further dispersion shall be required.   **11B-235.2.1** *235.2.1*

> **ADVISORY: Dispersion.** *Types of boat slips are based on the size of the boat slips; whether single berths or double berths, shallow water or deep water, transient or longer-term lease, covered or uncovered; and whether slips are equipped with features such as telephone, water, electricity or cable connections. The term "boat slip" is intended to cover any pier area other than launch ramp boarding piers where recreational boats are moored for purposes of berthing, embarking, or disembarking. For example, a fuel pier may contain boat slips, and this type of short term slip would be included in determining compliance with these provisions.*

Boarding Piers at Boat Launch Ramps

Where boarding piers are provided at boat launch ramps, at least 5 percent, but no fewer than one, of the boarding piers shall comply as detailed.   **11B-235.3**   *235.3*

Recreational Boating Facilities   **11B-1003**   *1003*

General

Recreational boating facilities shall comply as detailed.   **11B-1003.1**   *1003.1*

Accessible Routes

A.   Accessible routes serving recreational boating facilities, including gangways and floating piers, shall comply as required except as modified by the exceptions listed under "Boat Slips", below.   **11B-1003.2**   *1003.2*

Boat Slips

Accessible routes serving boat slips shall be permitted to use the exceptions listed below.   **11B-1003.2.1**   *1003.2.1*

EXCEPTIONS:

1.   Where an existing gangway or series of gangways is replaced or altered, an increase in the length of the gangway shall not be required to comply with the accessible route requirements of this section unless required by the Path of Travel requirements for alterations to existing buildings or facilities contained in Chapter 2.

2.   Gangways shall not be required to comply with the maximum 30 inch rise specified for accessible ramps.

3.   Where the total length of a gangway or series of gangways serving as part of a required accessible route is 80 feet (*12192 mm*) minimum, gangways shall not be required to comply with the accessible ramp maximum running slope limitation of 1:2 (8.33%).

4.   Where facilities contain fewer than 25 boat slips and the total

length of the gangway or series of gangways serving as part of a required accessible route is 30 feet (*9144* mm) minimum, gangways shall not be required to comply with the accessible ramp maximum running slope limitation of 1:2 (8.33%).

5.   Where gangways connect to transition plates, landings shall not be required to comply with the requirements for accessible landings at ramps.

6.   Where gangways and transition plates connect and are required to have handrails, handrail extensions shall not be required. Where handrail extensions are provided on gangways or transition plates, the handrail extensions shall not be required to be parallel with the ground or floor surface.

7.   The cross slope of 1:48 maximum that is required for accessible walkways and ramps, when applied for gangways, transition plates, and floating piers that are part of accessible routes shall be measured in the static position.

8.   Changes in level that are a maximum of ¼ inch vertical or between ¼ inch and ½ inch high maximum that are beveled with a slope not steeper than 1:2 shall be permitted on the surfaces of gangways and boat launch ramps.

> **ADVISORY: Boat Slips Exception 3.** *The following example shows how exception 3 would be applied: A gangway is provided to a floating pier which is required to be on an accessible route. The vertical distance is 10 feet (3050 mm) between the elevation where the gangway departs the landside connection and the elevation of the pier surface at the lowest water level. Exception 3 permits the gangway to be 80 feet (24 m) long. Another design solution would be to have two 40 foot (12 m) plus continuous gangways joined together at a float, where the float (as the water level falls) will stop dropping at an elevation five feet below the landside connection. The length of transition plates would not be included in determining if the gangway(s) meet the requirements of the exception.*

Boarding Piers at Boat Launch Ramps
Accessible routes serving boarding piers at boat launch ramps shall be permitted to use the exceptions listed below.   **11B-1003.2.2**   *1003.2.2*

EXCEPTIONS:

1.   Accessible routes serving floating boarding piers shall be permitted to use Exceptions 1, 2, 5, 6, 7 and 8 in the accessible route requirements for "Boat Slips", above.

2.   Where the total length of the gangway or series of gangways serving as part of a required accessible route is 30 feet (*9144* mm) minimum, gangways shall not be required to comply with the accessible ramp maximum running slope limitation of 1:2 (8.33%).

3.   Where the accessible route serving a floating boarding pier or

skid pier is located within a boat launch ramp, the portion of the accessible route located within the boat launch ramp shall not be required to comply with the requirements for an accessible ramp.

Clearances

Clearances at boat slips and on boarding piers at boat launch ramps shall comply as detailed. **11B-1003.3**   *1003.3*

> *ADVISORY: Clearances.   Although the minimum width of the clear pier space is 60 inches (1525 mm), it is recommended that piers be wider than 60 inches (1525 mm) to improve the safety for persons with disabilities, particularly on floating piers.*

Boat Slip Clearance

_____B.   Boat slips shall provide clear pier space 60 inches (*1524* mm) wide minimum and at least as long as the boat slips.   **11B-1003.3.1**   *1003.3.1*   **Fig. CD-50A**

_____B1.   Each 10 feet (*3048* mm) maximum of linear pier edge serving boat slips shall contain at least one continuous clear opening 60 inches (*1524* mm) wide minimum.   **11B-1003.3.1**   *1003.3.1*   **Fig. CD-50A**

EXCEPTIONS:

1.   Clear pier space shall be permitted to be 36 inches (*914* mm) wide minimum for a length of 24 inches (610 mm) maximum, provided that multiple 36 inch (*914* mm) wide segments are separated by segments that are 60 inches (*1524* mm) wide minimum and 60 inches (*1524* mm) long minimum.   **Fig. CD-50B**

2.   Edge protection shall be permitted at the continuous clear openings, provided that it is 4 inches (*102* mm) high maximum and 2 inches (51 mm) wide maximum. **Fig. CD-50C**

3.   In existing piers, clear pier space shall be permitted to be located perpendicular to the boat slip and shall extend the width of the boat slip, where the facility has at least one boat slip complying with these provisions, and further compliance with these provisions would result in a reduction in the number of boat slips available or result in a reduction of the widths of existing slips.

> *ADVISORY: Boat Slip Clearance Exception 3. Where the conditions in exception 3 are satisfied, existing facilities are only required to have one accessible boat slip with a pier clearance which runs the length of the slip. All other accessible slips are allowed to have the required pier clearance at the head of the slip. Under this exception, at piers with perpendicular boat slips, the width of most "finger piers" will remain unchanged. However, where mooring systems for floating piers are replaced as part of pier alteration projects, an opportunity may exist for increasing accessibility. Piers may be reconfigured to allow an increase in the number of wider finger piers, and serve as accessible boat slips.*

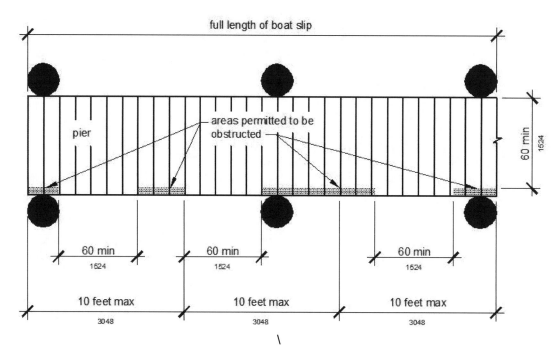

Fig. CD-50A
Boat Slip Clearance
© ICC Reproduced with Permission

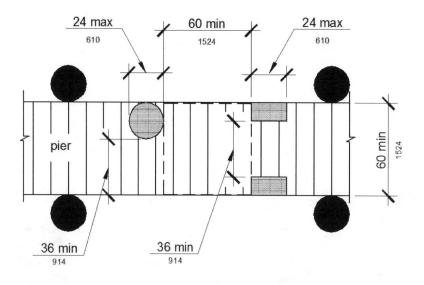

Fig. CD-50B (Exception 1)
Clear Pier Space Reduction at Boat Slips
© ICC Reproduced with Permission

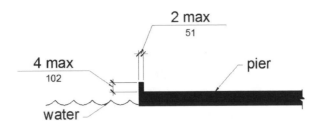

Fig. CD-50C (Exception 2)
Edge Protection at Boat Slips
© ICC Reproduced with Permission

Boarding Pier Clearances

_____C.   Boarding piers at boat launch ramps shall provide clear pier space 60 inches (*1524* mm) wide minimum and shall extend the full length of the boarding pier. **11B-1003.3.2**   *1003.3.2*   **Fig. CD-50D**

_____C1.   Every 10 feet (*3048* mm) maximum of linear pier edge shall contain at least one continuous clear opening 60 inches (*1524* mm) wide minimum. **11B-1003.3.2**   *1003.3.2*   **Fig. CD-50D**

EXCEPTIONS:

1.   The clear pier space shall be permitted to be 36 inches (*914* mm) wide minimum for a length of 24 inches (610 mm) maximum provided that multiple 36 inch (*914* mm) wide segments are separated by segments that are 60 inches (*1524* mm) wide minimum and 60 inches (*1524* mm) long minimum.   **Fig. CD-50E**

2.   Edge protection shall be permitted at the continuous clear openings provided that it is 4 inches (*102* mm) high maximum and 2 inches (*51* mm) wide maximum. **Fig. CD-50F**

ADVISORY: Boarding Pier Clearances. These requirements do not establish a minimum length for accessible boarding piers at boat launch ramps. The accessible boarding pier should have a length at least equal to that of other boarding piers provided at the facility. If no other boarding pier is provided, the pier would have a length equal to what would have been provided if no access requirements applied. The entire length of accessible boarding piers would be required to comply with the same technical provisions that apply to accessible boat slips. For example, at a launch ramp, if a 20 foot (6100 mm) long accessible boarding pier is provided, the entire 20 feet (6100 mm) must comply with the pier clearance requirements in Section 11B-1003.3. Likewise, if a 60 foot (18 m) long accessible boarding pier is provided, the pier clearance requirements detailed would apply to the entire 60 feet (18 m).

The following example applies to a boat launch ramp boarding pier: A chain of floats is provided on a launch ramp to be used as a boarding pier which is required to be accessible. At high water, the entire chain is floating and a transition plate connects the first float to the surface of the launch ramp. As the water level decreases, segments of the chain end up resting on the launch ramp surface, matching the slope of the launch ramp.

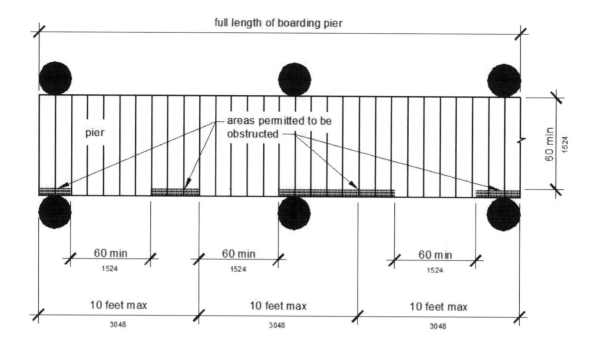

Fig. CD-50D
Boarding Pier Clearance
© ICC Reproduced with Permission

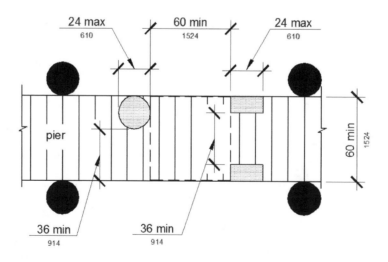

Fig. CD-50E (Exception 1)
Clear Pier Space Reduction at Boarding Piers
© ICC Reproduced with Permission

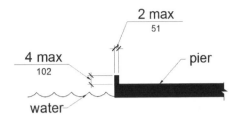

Fig. CD-50F (Exception 2)
Edge Protection at Boarding Piers
© ICC Reproduced with Permission

51. <u>EXERCISE MACHINES AND EQUIPMENT</u>

<u>Exercise Machines and Equipment</u>   **11B-236**   *236*

<u>General</u>
At least one of each type of exercise machine and equipment shall comply as detailed.
11B-236.1   *236.1*

<u>Accessible Route</u>

_____A.   Exercise machines and equipment required to comply shall be on an accessible route.
11B-206.2.13   *206.2.13*

> *ADVISORY: General. Most strength training equipment and machines are considered different types. Where operators provide a biceps curl machine and cable-cross-over machine, both machines are required to meet the provisions in this section, even though an individual may be able to work on their biceps through both types of equipment.*
>
> *Similarly, there are many types of cardiovascular exercise machines, such as stationary bicycles, rowing machines, stair climbers, and treadmills. Each machine provides a cardiovascular exercise and is considered a different type for purposes of these requirements.*

<u>Exercise Machines and Equipment</u>   **11B-1004**   *1004*

<u>Clear Floor Space</u>

_____B.   Exercise machines and equipment shall have a clear floor space complying with Section 10, "CLEAR FLOOR OR GROUND SPACE", positioned for transfer or for use by an individual seated in a wheelchair. Clear floor or ground spaces required at exercise machines and equipment shall be permitted to overlap.   **11B-1004.1**   *1004.1*

> *ADVISORY: Clear Floor Space. One clear floor or ground space is permitted to be shared between two pieces of exercise equipment. To optimize space use, designers should carefully consider layout options such as connecting ends of the row and center aisle spaces. The position of the clear floor space may vary greatly depending on the use of the equipment or machine. For example, to provide access to a shoulder press machine, clear floor space next to the seat would be appropriate to allow for transfer. Clear floor space for a bench press machine designed for use by an individual seated in a wheelchair, however, will most likely be centered on the operating mechanisms.*

52. <u>FISHING PIERS AND PLATFORMS</u>

<u>Fishing Piers and Platforms</u>   **11B-237**   *237*

<u>General</u>
Fishing piers and platforms shall comply as detailed.   **11B-237.1**   *237.1*

<u>Accessible Route</u>
Fishing piers and platforms shall be on an accessible route. Accessible routes serving fishing piers and platforms shall comply except as modified.   **11B-206.2.14**   *206.2.14*

<u>Platform Lifts</u>
Platform lifts shall be permitted to be used instead of gangways that are part of accessible routes serving recreational boating facilities and fishing piers and platforms.   **11B-206.7.10**   *206.7.10*

<u>Fishing Piers and Platforms</u>   **11B-1005**   *1005*

<u>Accessible Routes</u>

_____A.   Accessible routes serving fishing piers and platforms, including gangways and floating piers, shall comply.   **11B-1005.1**   *1005.1*

> **EXCEPTIONS:**
>
> 1.   Accessible routes serving floating fishing piers and platforms shall be permitted to use Exceptions 1, 2, 5, 6, 7 and 8 in item A of Section 50, "RECREATIONAL BOATING FACILITIES".
>
> 2.   Where the total length of the gangway or series of gangways serving as part of a required accessible route is 30 feet (*9144 mm*) minimum, gangways shall not be required to comply with the accessible ramp maximum running slope limitation of 1:2 (8.33%)..

<u>Railings</u>

_____B.   Where provided, railings, guards, or handrails shall comply as detailed.   **11B-1005.2**   *1005.2*

<u>Height</u>

_____B1.   At least 25 percent of the railings, guards, or handrails shall be 34 inches (*864 mm*) maximum above the ground or deck surface.   **11B-1005.2.1**   *1005.2.1*

> **EXCEPTION:** Where a guard complying with CBC Chapter 10, Sections 1013.2 through 1013.4 is provided, the guard shall not be required to comply with this provision.

<u>Dispersion</u>

_____B2.   Railings, guards, or handrails required to comply with these provisions shall be dispersed throughout the fishing pier or platform.   **11B-1005.2.1.1**   *1005.2.1.1*

Edge Protection

_____C.   Where railings, guards, or handrails complying with the requirements for "Railings", above are provided, edge protection complying with the requirements for "Curb or Barrier", or "Extended Ground or Floor Surface", as detailed below shall be provided. **11B-1005.3**   *1005.3*

Curb or Barrier

_____C1.   Curbs or barriers shall extend 2 inches (51 mm) minimum above the surface of the fishing pier or platform.   **11B-1005.3.1**   *1005.3.1*

Extended Ground or Deck Surface

_____C2.   The ground or deck surface shall extend 12 inches (305 mm) minimum beyond the inside face of the railing. Toe clearance shall be provided and shall be 30 inches (*762 mm*) wide minimum and 9 inches (*229 mm*) minimum above the ground or deck surface beyond the railing.   **11B-1005.3.2**   *1005.3.2*
Fig. CD-52A

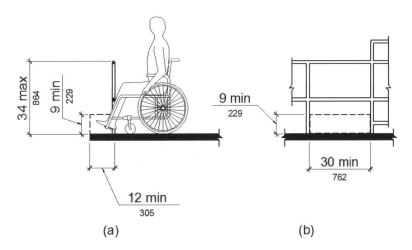

(a)   (b)

Fig. CD-52A
Extended Ground or Deck Surface at Fishing Piers and Platforms
© ICC Reproduced with Permission

Clear Floor or Ground Space

_____D.   At each location where there are railings, guards, or handrails complying with the requirements under "Railings", above, a clear floor or ground space complying with Section 10, "CLEAR FLOOR OR GROUND SPACE", shall be provided. **11B-1005.4** *1005.4*

       _____D1.   Where there are no railings, guards, or handrails, at least one clear floor or ground space complying with Section 10, "CLEAR FLOOR OR GROUND SPACE", shall be provided on the fishing pier or platform. **11B-1005.4** *1005.4*

Turning Space

_____E.   At least one turning space complying with Section 9, "TURNING SPACE", shall be provided on fishing piers and platforms. **11B-1005.5** *1005.5*

53. <u>GOLF FACILITIES</u>

<u>Golf Facilities</u> **11B-238** *238*

<u>General</u>
Golf facilities shall comply as detailed. **11B-238.1** *238.1*

<u>Accessible Route</u>
At least one accessible route shall connect accessible elements and spaces within the boundary of the golf course. In addition, accessible routes serving golf car rental areas; bag drop areas; course weather shelters, course toilet rooms and practice putting greens, practice teeing grounds, and teeing stations at driving ranges shall comply except as modified. **11B-206.2.15** *206.2.15*

> **EXCEPTION:** Golf car passages that comply shall be permitted to be used for all or part of accessible routes required.

<u>Golf Courses</u>
Golf courses shall comply as detailed. **11B-238.2** *238.2*

<u>Teeing Grounds</u>
Where one teeing ground is provided for a hole, the teeing ground shall be designed and constructed so that a golf car can enter and exit the teeing ground. Where two teeing grounds are provided for a hole, the forward teeing ground shall be designed and constructed so that a golf car can enter and exit the teeing ground. Where three or more teeing grounds are provided for a hole, at least two teeing grounds, including the forward teeing ground, shall be designed and constructed so that a golf car can enter and exit each teeing ground.
11B-238.2.1 *238.2.1*

> **EXCEPTION:** In existing golf courses, the forward teeing ground shall not be required to be one of the teeing grounds on a hole designed and constructed so

that a golf car can enter and exit the teeing ground where compliance is not feasible due to terrain.

Putting Greens
Putting greens shall be designed and constructed so that a golf car can enter and exit the putting green.　**11B-238.2.2**　*238.2.2*

Weather Shelters
Where provided, weather shelters shall be designed and constructed so that a golf car can enter and exit the weather shelter and shall comply with item D, below.　**11B-238.2.3**　*238.2.3*

Practice Putting Greens, Practice Teeing Grounds, and Teeing Stations at Driving Ranges
At least 5 percent, but no fewer than one, of practice putting greens, practice teeing grounds, and teeing stations at driving ranges shall be designed and constructed so that a golf car can enter and exit the practice putting greens, practice teeing grounds, and teeing stations at driving ranges.　**11B-238.3**　*238.3*

Golf Facilities　**11B-1006**　*1006*

General
Golf facilities shall comply as detailed.　**11B-1006.1**　*1006.1*

Accessible Routes

A.　Accessible routes serving teeing grounds, practice teeing grounds, putting greens, practice putting greens, teeing stations at driving ranges, course weather shelters, golf car rental areas, bag drop areas, and course toilet rooms shall comply as required and shall be 48 inches (*1219* mm) wide minimum. Where handrails are provided, accessible routes shall be 60 inches (*1524* mm) wide minimum.　**11B-1006.2**　*1006.2*

> **EXCEPTION:** Handrails shall not be required on golf courses. Where handrails are provided on golf courses, the handrails shall not be required to comply with the requirements for accessible handrails.

ADVISORY: Accessible Routes. The 48 inch (1220 mm) minimum width for the accessible route is necessary to ensure passage of a golf car on either the accessible route or the golf car passage. This is important where the accessible route is used to connect the golf car rental area, bag drop areas, practice putting greens, practice teeing grounds, course toilet rooms, and course weather shelters. These are areas outside the boundary of the golf course, but are areas where an individual using an adapted golf car may travel. A golf car passage may not be substituted for other accessible routes to be located outside the boundary of the course. For example, an accessible route connecting an accessible parking space to the entrance of a golf course clubhouse is not covered by this provision.

Providing a golf car passage will permit a person that uses a golf car to practice driving a golf ball from the same position and stance used when playing the game. Additionally, the space required for a person using a golf car to enter and maneuver within the teeing stations required to be accessible should be considered.

Golf Car Passages
Golf car passages shall comply as detailed.　**11B-1006.3**　*1006.3*

Clear Width

_____B.   The clear width of golf car passages shall be 48 inches (*1219* mm) minimum. **11B-1006.3.1**   *1006.3.1*

Barriers

_____C.   Where curbs or other constructed barriers prevent golf cars from entering a fairway, openings 60 inches (*1524* mm) wide minimum shall be provided at intervals not to exceed 75 yards (*69* m).   **11B-1006.3.2**   *1006.3.2*

Weather Shelters

_____D.   A clear floor or ground space 60 inches (*1524* mm) minimum by 96 inches (*2438* mm) minimum shall be provided within weather shelters.   **11B-1006.4**   *1006.4*

54. <u>MINIATURE GOLF FACILITIES</u>

<u>Miniature Golf Facilities</u>   **11B-239**   *239*

General
Miniature golf facilities shall comply with *Section 11B-239.*   **11B-239.1**   *239.1*

Accessible Route
Holes required to comply, including the start of play, shall be on an accessible route. Accessible routes serving miniature golf facilities shall comply except as modified.   **11B-206.2.16** *206.2.16*

Minimum Number
At least 50 percent of holes on miniature golf courses shall comply with the requirements for "Miniature Golf Holes", as detailed below.   **11B-239.2**   *239.2*

> ***ADVISORY: Minimum Number.*** *Where possible, providing access to all holes on a miniature golf course is recommended. If a course is designed with the minimum 50 percent accessible holes, designers or operators are encouraged to select holes which provide for an equivalent experience to the maximum extent possible.*

Miniature Golf Course Configuration
Miniature golf courses shall be configured so that the holes complying with the requirements for "Miniature Golf Holes", below are consecutive. Miniature golf courses shall provide an accessible route from the last hole complying with the requirements for "Miniature Golf Holes" to the course entrance or exit without requiring travel through any other holes on the course.   **11B-239.3** *239.3*

> **EXCEPTION:** One break in the sequence of consecutive holes shall be permitted provided that the last hole on the miniature golf course is the last hole in the sequence.

> **ADVISORY: Miniature Golf Course Configuration.** *Where only the minimum 50 percent of the holes are accessible, an accessible route from the last accessible hole to the course exit or entrance must not require travel back through other holes. In some cases, this may require an additional accessible route. Other options include increasing the number of accessible holes in a way that limits the distance needed to connect the last accessible hole with the course exit or entrance.*

Miniature Golf Facilities   **11B-1007**   *1007*

General
Miniature golf facilities shall comply as detailed.   **11B-1007.1**   *1007.1*

Accessible Routes

_____A.   Accessible routes serving holes on miniature golf courses shall comply as required. Accessible routes located on playing surfaces of miniature golf holes shall be permitted to use the exceptions detailed below.   **11B-1007.2**   *1007.2*

> **EXCEPTIONS:**
>
> 1. Playing surfaces shall not be required to comply with the requirements for accessible carpet in Section 7, "FLOOR OR GROUND SURFACES".
>
> 2. Where accessible routes intersect playing surfaces of holes, a 1 inch (25 mm) maximum curb shall be permitted for a width of 32 inches (*813* mm) minimum.
>
> 3. A slope not steeper than 1:4 (25%) for a 4 inch (*102* mm) maximum rise shall be permitted.
>
> 4. Ramp landing slopes shall be permitted to be 1:20 maximum.
>
> 5. Ramp landing length shall be permitted to be 48 inches (*1219* mm) long minimum.
>
> 6. Ramp landing size specified when ramps change direction between runs at landings (60 inches minimum by 72 inches minimum in the direction of downward travel from the upper ramp run) shall be permitted to be 48 inches (*1219* mm) minimum by 60 inches (*1524* mm) minimum.
>
> 7. Handrails shall not be required on holes. Where handrails are provided on holes, the handrails shall not be required to comply with the requirements for accessible handrails.

Miniature Golf Holes
Miniature golf holes shall comply as detailed.   **11B-1007.3**   *1007.3*

Start of Play

_____B.   A clear floor or ground space 48 inches (*1219* mm) minimum by 60 inches (*1524* mm) minimum with slopes not steeper than 1:48 shall be provided at the start of play. **11B-1007.3.1**   *1007.3.1*

Golf Club Reach Range Area

_____C.   All areas within holes where golf balls rest shall be within 36 inches (*914* mm) maximum of a clear floor or ground space 36 inches (*914* mm) wide minimum and 48 inches (*1219* mm) long minimum having a running slope not steeper than 1:20. The clear floor or ground space shall be served by an accessible route. **11B-1007.3.2**   *1007.3.2*   **Fig. CD-54A**

> **ADVISORY: Golf Club Reach Range Area.** *The golf club reach range applies to all holes required to be accessible. This includes accessible routes provided adjacent to or, where provided, on the playing surface of the hole.*

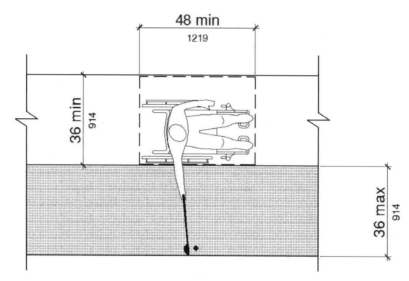

Note: Running Slope of Clear Floor or Ground Space Not Steeper Than 1:20

Fig. CD-54A
Golf Club Reach Range Area
© ICC Reproduced with Permission

55. PLAY AREAS

Accessible Routes for Play Areas
Play areas shall provide accessible routes as required. Accessible routes serving play areas shall comply except as modified.   **11B-206.2.17**   *206.2.17*

Platform Lifts
Platform lifts shall be permitted to provide accessible routes to play components or soft contained play structures.   **11B-206.7.8**   *206.7.8*

Ground Level and Elevated Play Components

At least one accessible route shall be provided within the play area. The accessible route shall connect ground level play components and elevated play components required to comply, including entry and exit points of the play components. **11B-206.2.17.1** *206.2.17.1*

Soft Contained Play Structures
Where three or fewer entry points are provided for soft contained play structures, at least one entry point shall be on an accessible route. Where four or more entry points are provided for soft contained play structures, at least two entry points shall be on an accessible route. **11B-206.2.17.2** *206.2.17.2*

Area of Sport Activity
An accessible route shall be provided to the boundary of each area of sport activity. **11B-206.2.18**

Team or Player Seating
Platform lifts shall be permitted to provide accessible routes to team or player seating areas serving areas of sport activity. **11B-206.7.9** *206.7.9*

Bowling Lanes
Where bowling lanes are provided, at least 5 percent, but no fewer than one of each type of bowling lane, shall be on an accessible route. **11B-206.2.11** *206.2.11*

Court Sports
In court sports, at least one accessible route shall directly connect both sides of the court. **11B-206.2.12** *206.2.12*

Raised Refereeing, Judging, and Scoring Areas
Raised structures used solely for refereeing, judging, or scoring a sport shall not be required to comply with these requirements or to be on an accessible route. A compliant accessible route shall be provided to the ground- or floor-level entry points, where provided, of stairs, ladders, or other means of reaching the raised elements or areas. **11B-203.10** *203.10*

Raised Boxing or Wrestling Rings
Raised boxing or wrestling rings shall not be required to comply with these requirements or to be on an accessible route. A compliant accessible route shall be provided to the ground- or floor-level entry points, where provided, of stairs, ladders or other means of reaching raised elements or areas. **11B-203.13** *2103.13*

Stairs
Stairs that connect play components shall not be required to comply. **11B-210** *210*

Play Areas   **11B-240**   *240*

General
Play areas for children ages 2 and over shall comply as detailed herein. Where separate play areas are provided within a site for specific age groups, each play area shall comply as detailed herein. **11B-240.1** *240.1*

EXCEPTIONS:

1. Play areas located in family child care facilities where the proprietor actually resides shall not be required to comply with these provisions.

2. In existing play areas, where play components are relocated for the purposes of creating safe use zones and the ground surface is not altered or extended for more than one use zone, the play area shall not be required to comply with these provisions.

3. Amusement attractions shall not be required to comply with these provisions.

4. Where play components are altered and the ground surface is not altered, the ground surface shall not be required to comply with items H or I, below, unless required by the Path of Travel requirements for alterations to existing buildings or facilities contained in Chapter 2.

> **ADVISORY: General.** *Play areas may be located on exterior sites or within a building. Where separate play areas are provided within a site for children in specified age groups (e.g., preschool (ages 2 to 5) and school age (ages 5 to 12)), each play area must comply with this section. Where play areas are provided for the same age group on a site but are geographically separated (e.g., one is located next to a picnic area and another is located next to a softball field), they are considered separate play areas and each play area must comply with this section.*

Additions

Where play areas are designed and constructed in phases, the requirements contained herein shall apply to each successive addition so that when the addition is completed, the entire play area complies with all the applicable requirements of this Section.
11B-240.1.1   *240.1.1*

> **ADVISORY: Additions.** *These requirements are to be applied so that when each successive addition is completed, the entire play area complies with all applicable provisions. For example, a play area is built in two phases. In the first phase, there are 10 elevated play components and 10 elevated play components are added in the second phase for a total of 20 elevated play components in the play area. When the first phase was completed, at least 5 elevated play components, including at least 3 different types, were to be provided on an accessible route. When the second phase is completed, at least 10 elevated play components must be located on an accessible route, and at least 7 ground level play components, including 4 different types, must be provided on an accessible route. At the time the second phase is complete, ramps must be used to connect at least 5 of the elevated play components and transfer systems are permitted to be used to connect the rest of the elevated play components required to be located on an accessible route.*

Play Components

Where provided, play components shall comply as detailed.   **11B-240.2**   *240.2*

Ground Level Play Components

Ground level play components shall be provided in the number and types required as detailed herein. Ground level play components that are provided to comply with "Minimum Number and Types", below shall be permitted to satisfy the additional number required by "Additional Number and Types", below if the minimum required types of play components are satisfied. Where two or more required ground level play components are provided, they shall be dispersed throughout the play area and integrated with other play components.
11B-240.2.1   *240.2.1*

---

*ADVISORY: **Ground Level Play Components.** Examples of ground level play components may include spring rockers, swings, diggers, and stand-alone slides. When distinguishing between the different types of ground level play components, consider the general experience provided by the play component. Examples of different types of experiences include, but are not limited to, rocking, swinging, climbing, spinning, and sliding. A spiral slide may provide a slightly different experience from a straight slide, but sliding is the general experience and therefore a spiral slide is not considered a different type of play component from a straight slide.*

Ground level play components accessed by children with disabilities must be integrated into the play area. Designers should consider the optimal layout of ground level play components accessed by children with disabilities to foster interaction and socialization among all children. Grouping all ground level play components accessed by children with disabilities in one location is not considered integrated.

Where a stand-alone slide is provided, an accessible route must connect the base of the stairs at the entry point to the exit point of the slide. A ramp or transfer system to the top of the slide is not required. Where a sand box is provided, an accessible route must connect to the border of the sand box. Accessibility to the sand box would be enhanced by providing a transfer system into the sand or by providing a raised sand table with knee clearance complying with the requirements for accessible "Play Tables" included in this Section.

Ramps are preferred over transfer systems since not all children who use wheelchairs or other mobility devices may be able to use, or may choose not to use, transfer systems. Where ramps connect elevated play components, the maximum rise of any ramp run is limited to 12 inches (305 mm). Where possible, designers and operators are encouraged to provide ramps with a slope less than the 1:12 maximum. Berms or sculpted dirt may be used to provide elevation and may be part of an accessible route to composite play structures.

Platform lifts are permitted as a part of an accessible route. Because lifts must be independently operable, operators should carefully consider the appropriateness of their use in unsupervised settings.

Minimum Number and Types

Where ground level play components are provided, at least one of each type shall be on an accessible route and shall comply with items L through P of "Play Components", below.
11B-240.2.1.1   *240.2.1.1*

Additional Number and Types

Where elevated play components are provided, ground level play components shall be provided in accordance with Table 55A and shall comply with items L through P of "Play Components", below.   **11B-240.2.1.2**   *240.2.1.2*

EXCEPTION: If at least 50 percent of the elevated play components are connected by a ramp and at least 3 of the elevated play components connected by the ramp are different types of play components, the play area shall not be required to comply with the provisions detailed in "Additional Number and Types", above.

Table 55A
Number and Types of Ground Level Play
Components Required to be on Accessible Routes

Number of Elevated Play Components Provided	Minimum Number of Ground Level Play Components Required to be on an Accessible Route	Minimum Number of Different Types of Ground Level Play Components Required to be on an Accessible Route
1	Not applicable	Not applicable
2 to 4	1	1
5 to 7	2	2
8 to 10	3	3
11 to 13	4	3
14 to 16	5	3
17 to 19	6	3
20 to 22	7	4
23 to 25	8	4
26 and over	8, plus 1 for each additional 3, or fraction thereof, over 25	5

ADVISORY: Additional Number and Types. Where a large play area includes two or more composite play structures designed for the same age group, the total number of elevated play components on all the composite play structures must be added to determine the additional number and types of ground level play components that must be provided on an accessible route.

Elevated Play Components
Where elevated play components are provided, at least 50 percent shall be on an accessible route and shall comply with items L through P of "Play Components", below.   **11B-240.2.2** *240.2.2*

ADVISORY: Elevated Play Components. A double or triple slide that is part of a composite play structure is one elevated play component. For purposes of this section, ramps, transfer systems, steps, decks, and roofs are not considered elevated play components. Although socialization and pretend play can occur on these elements, they are not primarily intended for play.

Some play components that are attached to a composite play structure can be approached or exited at the ground level or above grade from a platform or deck. For example, a climber attached to a composite play structure can be approached or exited at the ground level or above grade from a platform or deck on a composite play structure. Play components that are attached to a composite play structure and can be approached from a platform or deck (e.g., climbers and overhead play components) are considered elevated play components. These play components are not considered ground level play components and do not count toward the requirements regarding the number of ground level play components that must be located on an accessible route.

Play Areas   **11B-1008**   *1008*

General
Play areas shall comply as detailed.   **11B-1008.1**   *1008.1*

Accessible Routes

_____A.   Accessible routes serving play areas shall comply with all applicable standards and as detailed herein and shall be permitted to use the exceptions under "Ground Level and Elevated Play Components", "Soft Contained Play Structures", and "Water Play Components", below.   **11B-1008.2**   *1008.2*

_____A1.   Where accessible routes serve ground level play components, the vertical clearance shall be 80 inches high (*2032* mm) minimum.   **11B-1008.2**   *1008.2*

Ground Level and Elevated Play Components
Accessible routes serving ground level play components and elevated play components shall be permitted to use the exceptions as detailed, below.
11B-1008.2.1   *1008.2.1*

EXCEPTIONS:

1.   Transfer systems complying with items J through J6 of "Transfer Systems", below, shall be permitted to connect elevated play components except where 20 or more elevated play components are provided no more than 25 percent of the elevated play components shall be permitted to be connected by transfer systems.

2.   Where transfer systems are provided, an elevated play component shall be permitted to connect to another elevated play component as part of an accessible route.

Soft Contained Play Structures
Accessible routes serving soft contained play structures shall be permitted to use the exception detailed below.   **11B-1008.2.2**   *1008.2.2*

EXCEPTION: Transfer systems complying with items J through J6 of "Transfer Systems", below, shall be permitted to be used as part of an accessible route.

Water Play Components
Accessible routes serving water play components shall be permitted to use the exceptions listed below.   **11B-1008.2.3**   *1008.2.3*

EXCEPTIONS:

1.   Where the surface of the accessible route, clear floor or ground spaces, or turning spaces serving water play components is submerged, compliance with the requirements of Section 7, "FLOOR OR GROUND SURFACES", the maximum 1:20 running slope and maximum 1:48 cross slope requirements of Section 5, "WALKING SURFACES", the maximum 1:12 running slope and maximum 1:48 cross slope requirements of Section 15, "RAMPS", and the requirements of items H and I of the

requirements for "Ground Surfaces", detailed below, shall not be required.

2.   Transfer systems complying with items J through J6 of "Transfer Systems", below, shall be permitted to connect elevated play components in water.

> ***ADVISORY: Water Play Components.*** *Personal wheelchairs and mobility devices may not be appropriate for submerging in water when using play components in water. Some may have batteries, motors, and electrical systems that when submerged in water may cause damage to the personal mobility device or wheelchair or may contaminate the water. Providing an aquatic wheelchair made of non-corrosive materials and designed for access into the water will protect the water from contamination and avoid damage to personal wheelchairs.*

Clear Width

_____B.   Accessible routes connecting play components shall provide a clear width that complies as detailed, below.   **11B-1008.2.4**   *1008.2.4*

Ground Level

_____B1.   At ground level, the clear width of accessible routes shall be 60 inches (*1524* mm) minimum.   **11B-1008.2.4.1**   *1008.2.4.1*

EXCEPTIONS:

1.   In play areas less than 1000 square feet (93 m^2), the clear width of accessible routes shall be permitted to be 44 inches (*1118* mm) minimum, if at least one turning space complying with item B of Section 9, "TURNING SPACE", is provided where the restricted accessible route exceeds 30 feet (*9144* mm) in length.

2.   The clear width of accessible routes shall be permitted to be 36 inches (*914* mm) minimum for a distance of 60 inches (*1524* mm) maximum provided that multiple reduced width segments are separated by segments that are 60 inches (*1524* mm) wide minimum and 60 inches (*1524* mm) long minimum.

Elevated

_____B2.   The clear width of accessible routes connecting elevated play components shall be 36 inches (*914* mm) minimum.   **11B-1008.2.4.2** *1008.2.4.2*

EXCEPTIONS:

1.   The clear width of accessible routes connecting elevated play components shall be permitted to be reduced to 32 inches (*813* mm) minimum for

a distance of 24 inches (610 mm) maximum provided that reduced width segments are separated by segments that are 48 inches (*1219 mm*) long minimum and 36 inches (*914* mm) wide minimum.

2.   The clear width of transfer systems connecting elevated play components shall be permitted to be 24 inches (610 mm) minimum.

Ramps

Within play areas, ramps connecting ground level play components and ramps connecting elevated play components shall comply as detailed.   **11B-1008.2.5** *1008.2.5*

Ground Level

_____C.   Ramp runs connecting ground level play components shall have a running slope not steeper than 1:16.   **11B-1008.2.5.1**   *1008.2.5.1*

Elevated

_____D.   The rise for any ramp run connecting elevated play components shall be 12 inches (305 mm) maximum.   **11B-1008.2.5.2**   *1008.2.5.2*

Handrails

Where required on ramps serving play components, the handrails shall comply with Section 24, "HANDRAILS", except as modified below.   **11B-1008.2.5.3**   *1008.2.5.3*

EXCEPTIONS:

1.   Handrails shall not be required on ramps located within ground level use zones.

2.   Handrail extensions shall not be required.

Handrail Gripping Surfaces

_____E.   Handrail gripping surfaces with a circular cross section shall have an outside diameter of 0.95 inch (24 mm) minimum and 1.55 inches (39 mm) maximum.

_____F.   Where the shape of the gripping surface is non-circular, the handrail shall provide an equivalent gripping surface.
11B-1008.2.5.3.1   *1008.2.5.3.1*

Handrail Height

_____G.   The top of handrail gripping surfaces shall be 20 inches (*508* mm) minimum and 28 inches (*711* mm) maximum above the ramp surface.
11B-1008.2.5.3.2   *1008.2.5.3.2*

Ground Surfaces

Ground surfaces on accessible routes, clear floor or ground spaces, and turning spaces shall comply as detailed, below.   **11B-1008.2.6**   *1008.2.6*

Accessibility

_____H. Ground surfaces shall comply with ASTM F 1951. Ground surfaces shall be inspected and maintained regularly and frequently to ensure continued compliance with ASTM F 1951.   **11B-1008.2.6.1** *1008.2.6.1*

Use Zones

_____I. Ground surfaces located within use zones shall comply with ASTM F 1292.   **11B-1008.2.6.2**   *1008.2.6.2*

> ***ADVISORY: Ground Surfaces.*** *Ground surfaces must be inspected and maintained regularly to ensure continued compliance with the ASTM F 1951 standard. The type of surface material selected and play area use levels will determine the frequency of inspection and maintenance activities.*

Transfer Systems

Where transfer systems are provided to connect to elevated play components, transfer systems shall comply as detailed.   **11B-1008.3**   *1008.3*

> ***ADVISORY: Transfer Systems.*** *Where transfer systems are provided, consideration should be given to the distance between the transfer system and the elevated play components. Moving between a transfer platform and a series of transfer steps requires extensive exertion for some children. Designers should minimize the distance between the points where a child transfers from a wheelchair or mobility device and where the elevated play components are located. Where elevated play components are used to connect to another elevated play component instead of an accessible route, careful consideration should be used in the selection of the play components used for this purpose.*

Transfer Platforms

_____J. Transfer platforms shall be provided where transfer is intended from wheelchairs or other mobility aids. Transfer platforms shall comply as detailed, below. **11B-1008.3.1**   *1008.3.1*

Size

_____J1. Transfer platforms shall have level surfaces 14 inches (*356* mm) deep minimum and 24 inches (610 mm) wide minimum. **11B-1008.3.1.1**   *1008.3.1.1*   **Fig. CD-55A**

Height

_____J2. The height of transfer platforms shall be 11 inches (*279* mm) minimum and 18 inches (*457* mm) maximum measured to the top of the surface from the ground or floor surface.   **11B-1008.3.1.2**   *1008.3.1.2* **Fig. CD-55A**

Transfer Space

_____ J3.　A transfer space that is 30 inches minimum by 48 inches minimum and complies with the requirements of Section 7, "FLOOR OR GROUND SURFACES", shall be provided adjacent to the transfer platform. Changes in level are not permitted.　**11B-1008.3.1.3**　*1008.3.1.3* **Fig. CD-55A**

> **EXCEPTION:** Slopes not steeper than 1:48 shall be permitted.

_____ J4.　The 48 inch (*1219* mm) long minimum dimension of the transfer space shall be centered on and parallel to the 24 inch (610 mm) long minimum side of the transfer platform.　**11B-1008.3.1.3**　*1008.3.1.3*　**Fig. CD-55A**

_____ J5.　The side of the transfer platform serving the transfer space shall be unobstructed.　**11B-1008.3.1.3**　*1008.3.1.3*

Transfer Supports

_____ J6.　At least one means of support for transferring shall be provided. **11B-1008.3.1.4**　*1008.3.1.4*

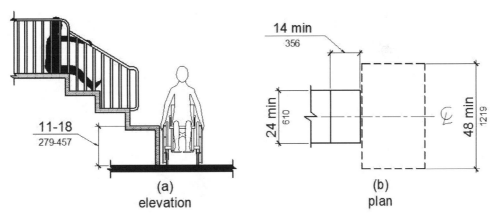

Fig. CD-55A
Transfer Platforms
© ICC Reproduced with Permission

Transfer Steps

_____ K.　Transfer steps shall be provided where movement is intended from transfer platforms to levels with elevated play components required to be on accessible routes. Transfer steps shall comply as detailed, below.　**11B-1008.3.2** *1008.3.2*

Size

_____ K1.　Transfer steps shall have level surfaces 14 inches (*356* mm) deep minimum and 24 inches (610 mm) wide minimum. **11B-1008.3.2.1**　*1008.3.2.1*　**Fig. CD-55B**

Height

_____ K2.   Each transfer step shall be 8 inches (*203* mm) high maximum.
11B-1008.3.2.2   *1008.3.2.2*   **Fig. CD-55B**

Transfer Supports

_____ K3.   At least one means of support for transferring shall be provided.
11B-1008.3.2.3   *1008.3.2.3*

> **ADVISORY: Transfer Supports.** *Transfer supports are required on transfer platforms and transfer steps to assist children when transferring. Some examples of supports include a rope loop, a loop type handle, a slot in the edge of a flat horizontal or vertical member, poles or bars, or D rings on the corner posts.*

Contrasting Stripe

_____ K4.   Striping providing clear visual contrast shall be provided at each transfer
step.   **11B-1008.3.2.4, 504.4.1**

_____ K5.   The stripe shall be a minimum of 2 inches wide to a maximum of 4
inches wide placed parallel to and not more than 1 inch from the nose of
the step or upper approach.   **11B-504.4.1**   **Fig. CD-23A**

_____ K6.   The stripe shall extend the full width of the step or upper approach and
shall be of a material that is at least as slip-resistant as the other treads
of the stair.   A painted stripe shall be acceptable.   Grooves shall not be
used to satisfy this requirement. **11B-504.4.1**

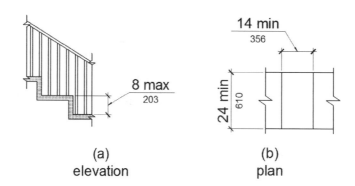

(a)
elevation

(b)
plan

Fig. CD-55B
Transfer Steps
© ICC Reproduced with Permission

Play Components
Ground level play components on accessible routes and elevated play components connected by
ramps shall comply as detailed.   **11B-1008.4**   *1008.4*

Turning Space

_____ L.   At least one turning space complying with Section 9, "TURNING SPACE", shall
be provided on the same level as play components.   **11B-1008.4.1**   *1008.4.1*

_____L1.   Where swings are provided, the turning space shall be located immediately adjacent to the swing.   **11B-1008.4.1**   *1008.4.1*

Clear Floor or Ground Space

_____M.   A clear floor or ground space that is 30 inches minimum by 48 inches minimum and complies with the requirements of Section 7, "FLOOR OR GROUND SURFACES", shall be provided at play components. Changes in level are not permitted.   **11B-1008.4.2**   *1008.4.2*   **Fig. CD-55A**

> **EXCEPTION:**   Slopes not steeper than 1:48 shall be permitted.

> **ADVISORY: Clear Floor or Ground Space.** *Clear floor or ground spaces, turning spaces, and accessible routes are permitted to overlap within play areas. A specific location has not been designated for the clear floor or ground spaces or turning spaces, except swings, because each play component may require that the spaces be placed in a unique location. Where play components include a seat or entry point, designs that provide for an unobstructed transfer from a wheelchair or other mobility device are recommended. This will enhance the ability of children with disabilities to independently use the play component.*
>
> *When designing play components with manipulative or interactive features, consider appropriate reach ranges for children seated in wheelchairs. The following table provides guidance on reach ranges for children seated in wheelchairs. These dimensions apply to either forward or side reaches. The reach ranges are appropriate for use with those play components that children seated in wheelchairs may access and reach. Where transfer systems provide access to elevated play components, the reach ranges are not appropriate.*

Children's Reach Ranges			
Forward or Side Reach	**Ages 3 and 4**	**Ages 5 through 8**	**Ages 9 through 12**
High (maximum)	36 in (915 mm)	40 in (1015 mm)	44 in (1120 mm)
Low (minimum)	20 in (510 mm)	18 in (455 mm)	16 in (405 mm)

Play Tables

_____N.   Where play tables are provided, knee clearance 24 inches (610 mm) high minimum, 17 inches deep (432 mm) minimum, and 30 inches (762 mm) wide minimum shall be provided.   ***11B*-1008.4.3**   *1008.4.3*

_____N1.   The tops of rims, curbs, or other obstructions shall be 31 inches (787 mm) high maximum.   ***11B*-1008.4.3**   *1008.4.3*

> **EXCEPTION:** Play tables designed and constructed primarily for children 5 years and younger shall not be required to provide knee clearance where the clear floor or ground space required by item M, above, is arranged for a parallel approach.

Entry Points and Seats

_____O.   Where play components require transfer to entry points or seats, the entry points or seats shall be 11 inches (279 mm) minimum and 24 inches (610 mm) maximum from the clear floor or ground space.   **11B-1008.4.4**   *1008.4.4*

> **EXCEPTION:** Entry points of slides shall not be required to comply with this provision.

Transfer Supports

_____P.   Where play components require transfer to entry points or seats, at least one means of support for transferring shall be provided.   **11B-1008.4.5**   *1008.4.5*

56. SWIMMING POOLS, WADING POOLS AND SPAS

Swimming Pools, Wading Pools, and Spas   **11B-242**   *242*

General
Swimming pools, wading pools, and spas shall comply as detailed.   **11B-242.1**   *242.1*

Raised Diving Boards and Diving Platforms
Raised diving boards and diving platforms shall not be required to comply with these requirements or to be on an accessible route.   **11B-203.14**   *203.14*

Swimming Pools
At least two accessible means of entry shall be provided for swimming pools. Accessible means of entry shall be swimming pool lifts as detailed below; sloped entries as detailed below; transfer walls as detailed below; transfer systems as detailed below; and pool stairs as detailed below. At least one accessible means of entry provided shall comply with the requirements for a pool lift as detailed or sloped entry as detailed.   **11B-242.2**   *242.2*

> **EXCEPTIONS:**
>
> 1.   Where a swimming pool has less than 300 linear feet (91 m) of swimming pool wall, no more than one accessible means of entry shall be required provided that the accessible means of entry is a swimming pool lift or sloped entry complying with the requirements as detailed.
>
> 2.   Wave action pools, leisure rivers, sand bottom pools, and other pools where user access is limited to one area shall not be required to provide more than one accessible means of entry provided that the accessible means of entry is a swimming pool lift, sloped entry, or a transfer system complying with the requirements as detailed.

3.   Catch pools shall not be required to provide an accessible means of entry provided that the catch pool edge is on an accessible route.

> **ADVISORY: Swimming Pools.** *Where more than one means of access is provided into the water, it is recommended that the means be different. Providing different means of access will better serve the varying needs of people with disabilities in getting into and out of a swimming pool. It is also recommended that where two or more means of access are provided, they not be provided in the same location in the pool. Different locations will provide increased options for entry and exit, especially in larger pools.*

> **ADVISORY: Swimming Pools Exception 1.** *Pool walls at diving areas and areas along pool walls where there is no pool entry because of landscaping or adjacent structures are to be counted when determining the number of accessible means of entry required.*

Wading Pools

At least one accessible means of entry shall be provided for wading pools. Accessible means of entry shall comply with sloped entries as detailed below.   **11B-242.3**   *242.3*

Spas

At least one accessible means of entry shall be provided for spas. Accessible means of entry shall comply with the requirements for swimming pool lifts; transfer walls, or transfer systems as detailed below.   **11B-242.4**   *242.4*

> **EXCEPTION:** Where spas are provided in a cluster, no more than 5 percent, but no fewer than one, spa in each cluster shall be required to comply with this provision.

Swimming Pools, Wading Pools, and Spas   **11B-1009**   *1009*

General

Where provided, pool lifts, sloped entries, transfer walls, transfer systems, and pool stairs shall comply as detailed.   **11B-1009.1**   *1009.1*

Pool Decks

Any mechanism provided to assist persons with disabilities in gaining entry into the pool and in exiting from the pool shall comply as detailed. **CBC 3113B.1, Exception 4**

Pool Lifts

Pool lifts shall comply as detailed.   **11B-1009.2**   *1009.2*

> **ADVISORY: Pool Lifts.** *There are a variety of seats available on pool lifts. Pool lift seats with backs enable a larger population of persons with disabilities to use the lift. Pool lift seats that consist of materials that resist corrosion and provide a firm base to transfer will be usable by a wider range of people with disabilities. Additional options such as armrests, head rests, seat belts, and leg support will enhance accessibility and better accommodate people with a wide range of disabilities.*

Pool Lift Location

_____A.   Pool lifts shall be located where the water level is 36 inches (914 mm) minimum and 48 inches (1219 mm) maximum.   **11B-1009.2.1**   *1009.2.1*

EXCEPTIONS:

1.   Where the entire pool depth is greater than 48 inches (*1219* mm), compliance with this requirement shall not be required.

2.   Where multiple pool lift locations are provided, no more than one pool lift shall be required to be located in an area where the water level is 48 inches (*1219* mm) maximum.

3.   Where the water depth of the entire swimming pool, wading pool or spa is less than 36 inches (*914* mm), pool lifts shall be located where the water level is less than 36 inches (*914* mm).

Seat Location

_____ B.   In the raised position, the centerline of the seat shall be located over the deck and 16 inches (*406* mm) minimum from the edge of the pool. **11B-1009.2.2**   *1009.2.2*   **Fig. CD-56A**

_____ B1.   The deck surface between the centerline of the seat and the pool edge shall have a slope not steeper than 1:48.   **11B-1009.2.2** *1009.2.2*

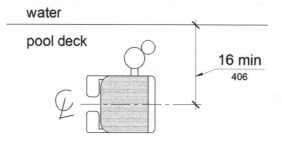

Fig. CD-56A
Pool Lift Seat Location
© ICC Reproduced with Permission

Clear Deck Space

_____ C.   On the side of the seat opposite the water, a clear deck space shall be provided parallel with the seat. **11B-1009.2.3**   *1009.2.3*   **Fig. CD-56B**

_____ C1.   The space shall be 36 inches (*914* mm) wide minimum and shall extend forward 48 inches (*1219* mm) minimum from a line located 12 inches (305 mm) behind the rear edge of the seat.   **11B-1009.2.3** *1009.2.3*   **Fig. CD-56B**

_____ C2.   The clear deck space shall have a slope not steeper than 1:48. **11B-1009.2.3**   *1009.2.3*

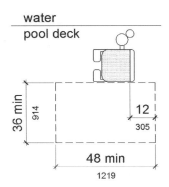

Fig. CD-56B
Clear Deck Space at Pool Lifts
© ICC Reproduced with Permission

Seat

_____D.   The seat shall be rigid and shall have a back support that is at least 12 inches (305 mm) tall.   **11B-1009.2.4**

    _____D1.   The height of the lift seat shall be designed to allow a stop at 17 inches (423 mm) minimum to 19 inches (*483* mm) maximum measured from the deck to the top of the seat surface when in the raised (load) position.   **11B-1009.2.4**   *1009.2.4*
               Fig. CD-56C

    _____D2.   The seat shall have a restraint for the use of the occupant with operable parts complying with Section 13, "OPERABLE PARTS".   **11B-1009.2.4**

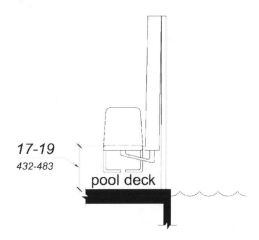

Fig. CD-56C
Pool Lift Seat Height
© ICC Reproduced with Permission

Seat Width

_____E.   The seat shall be 16 inches (*406* mm) wide minimum.   **11B-1009.2.5**
     1009.2.5

Footrests and Armrests

_____F.   Footrests shall be provided and shall move with the seat. **11B-1009.2.6**
1009.2.6

_____F1.   The seat shall have two armrests. **11B-1009.2.6**

_____F2.   The armrest positioned opposite the water shall be removable or
shall fold clear of the seat when the seat is in the raised (load) position.
11B-1009.2.6 *1009.2.6*

> **EXCEPTION:** Footrests shall not be required on pool lifts
> provided in spas.

Operation

_____G.   The lift shall be capable of unassisted operation from both the deck and
water levels. **11B-1009.2.7** *1009.2.7*

_____G1.   Controls and operating mechanisms shall be unobstructed when
the lift is in use and shall comply with Section 13, "CONTROLS AND
OPERATING LEVELS". **11B-1009.2.7** *1009.2.7*

_____G2.   The lift shall be stable and not permit unintended movement
when a person is getting into or out of the seat. **11B-1009.2.7**

ADVISORY: Operation. Pool lifts must be capable of unassisted operation from both the deck and water levels. This will permit a person to call the pool lift when the pool lift is in the opposite position. It is extremely important for a person who is swimming alone to be able to call the pool lift when it is in the up position so he or she will not be stranded in the water for extended periods of time awaiting assistance. The requirement for a pool lift to be independently operable does not preclude assistance from being provided.

Submerged Depth

_____H.   The lift shall be designed so that the seat will submerge to a water depth
of 18 inches (*457* mm) minimum below the stationary water level.
11B-1009.2.8 *1009.2.8* **Fig. CD-56D**

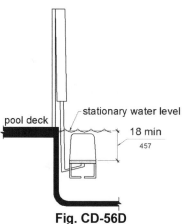

Fig. CD-56D
Pool Lift Submerged Depth
© ICC Reproduced with Permission

Lifting Capacity

_____I.   Single person pool lifts shall have a weight capacity of 300 pounds (136 kg) minimum and be capable of sustaining a static load of at least one and a half times the rated load.   **11B-1009.2.9**   *1009.2.9*

> **ADVISORY: Lifting Capacity.** *Single person pool lifts must be capable of supporting a minimum weight of 300 pounds (136 kg) and sustaining a static load of at least one and a half times the rated load. Pool lifts should be provided that meet the needs of the population they serve. Providing a pool lift with a weight capacity greater than 300 pounds (136 kg) may be advisable.*

Sloped Entries
Sloped entries shall comply with as detailed.   **11B-1009.3**   *1009.3*

> **ADVISORY: Sloped Entries.** *Personal wheelchairs and mobility devices may not be appropriate for submerging in water. Some may have batteries, motors, and electrical systems that when submerged in water may cause damage to the personal mobility device or wheelchair or may contaminate the pool water. Providing an aquatic wheelchair made of non-corrosive materials and designed for access into the water will protect the water from contamination and avoid damage to personal wheelchairs or other mobility aids.*

Sloped Entries
Sloped entries shall comply with all applicable requirements except as modified below. **11B-1009.3.1**   *1009.3.1*

> **EXCEPTION:** Where sloped entries are provided, the surfaces shall not be required to be slip resistant.

Submerged Depth

_____J.   Sloped entries shall extend to a depth of 24 inches (610 mm) minimum and 30 inches (762 mm) maximum below the stationary water level.   **11B-1009.3.2**   *1009.3.2*   **Fig. CD-56E**

_____J1.   Where landings are required by items F through L of "Landings", contained in Section 15, "RAMPS", at least one landing shall be located 24 inches (610 mm) minimum and 30 inches (762 mm) maximum below the stationary water level.   **11B-1009.3.2**   *1009.3.2*   **Fig. CD-56E**

> **EXCEPTION:** In wading pools, the sloped entry and landings, if provided, shall extend to the deepest part of the wading pool.

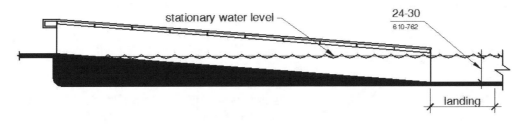

Fig. CD-56E
Sloped Entry Submerged Depth
© ICC Reproduced with Permission

Handrails

_____ K. At least two handrails complying with Section 24, "HANDRAILS", shall be provided on the sloped entry. **11B-1009.3.3** *1009.3.3* **Fig. CD-56F**

_____ K1. The clear width between required handrails shall be 33 inches (*838* mm) minimum and 38 inches (965 mm) maximum. **11B-1009.3.3** *1009.3.3* **Fig. CD-56F**

EXCEPTIONS:

1. Handrail extensions shall not be required at the bottom landing serving a sloped entry.

2. Where a sloped entry is provided for wave action pools, leisure rivers, sand bottom pools, and other pools where user access is limited to one area, the handrails shall not be required to comply with the clear width requirements of item K1, above.

3. Sloped entries in wading pools shall not be required to provide handrails complying with items K and K1, above. If provided, handrails on sloped entries in wading pools shall not be required to comply with Section 24, "HANDRAILS".

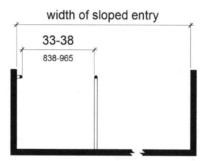

Fig. CD-56F
Handrails for Sloped Entry
© ICC Reproduced with Permission

Transfer Walls
Transfer walls shall comply as detailed. **11B-1009.4** *1009.4*

Clear Deck Space

_____ L. A clear deck space of 60 inches (*1524* mm) minimum by 60 inches (*1524* mm) minimum with a slope not steeper than 1:48 shall be provided at the base of the transfer wall. **11B-1009.4.1** *1009.4.1* **Fig. CD-56G**

_____ L1. Where one grab bar is provided, the clear deck space shall be centered on the grab bar. **11B-1009.4.1** *1009.4.1* **Fig.CD-56G**

_____L2.   Where two grab bars are provided, the clear deck space shall be centered on the clearance between the grab bars.
11B-1009.4.1   *1009.4.1*   **Fig.CD-56G**

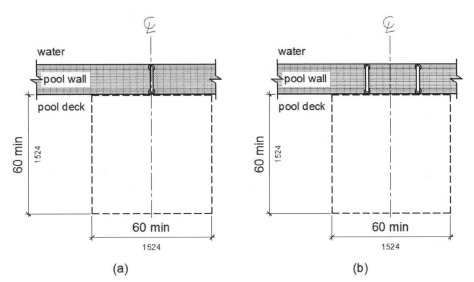

Fig. CD-56G
Clear Deck Space at Transfer Walls
© ICC Reproduced with Permission

Height

_____M.   The height of the transfer wall shall be 16 inches (*406* mm) minimum and 19 inches (*483* mm) maximum measured from the deck.   **11B-1009.4.2** *1009.4.2*   **Fig. CD-56H**

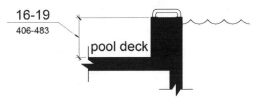

Fig. CD-56H
Transfer Wall Height
© ICC Reproduced with Permission

Wall Depth and Length

_____N.   The depth of the transfer wall shall be 12 inches (305 mm) minimum and 16 inches (*406* mm) maximum.   **11B-1009.4.3**   *1009.4.3*   **Fig. CD-56I**

_____N1.   The length of the transfer wall shall be 60 inches (*1524* mm) minimum and shall be centered on the clear deck space.
11B-1009.4.3   *1009.4.3*   **Fig. CD-56I**

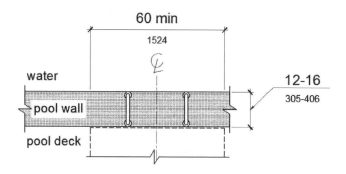

Fig. CD-56I
Depth and Length of Transfer Walls
© ICC Reproduced with Permission

Surface

_____O.   Surfaces of transfer walls shall not be sharp and shall have rounded edges. **11B-1009.4.4** *1009.4.4*

Grab Bars

_____P.   At least one grab bar complying with items E through O of Section 26H, "SHOWER COMPARTMENTS", shall be provided on the transfer wall.

_____P1.   Grab bars shall be perpendicular to the pool wall and shall extend the full depth of the transfer wall. **11B-1009.4.5** *1009.4.5* **Fig. CD-56J**

_____P2.   The top of the gripping surface shall be 4 inches (*102* mm) minimum and 6 inches (*152* mm) maximum above transfer walls. **11B-1009.4.5** *1009.4.5* **Fig. CD-56J**

_____P3.   Where one grab bar is provided, clearance shall be 24 inches (610 mm) minimum on both sides of the grab bar. **11B-1009.4.5** *1009.4.5* **Fig. CD-56J**

_____P4.   Where two grab bars are provided, clearance between grab bars shall be 24 inches (610 mm) minimum. **11B-1009.4.5** *1009.4.5* **Fig. CD-56J**

> **EXCEPTION:** Grab bars on transfer walls shall not be required to comply with the requirements for accessible grab bar height and positioning requirements.

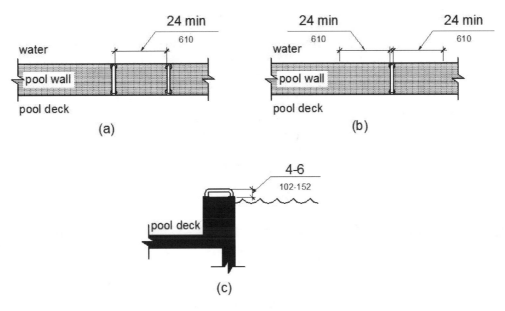

Fig. 56J
Grab Bars for Transfer Walls
© ICC Reproduced with Permission

Transfer Systems

Transfer systems shall comply as detailed.   **11B-1009.5**   *1009.5*

Transfer Platform

_____Q.   A transfer platform shall be provided at the head of each transfer system. **11B-1009.5.1**   *1009.5.*

_____Q1.   Transfer platforms shall provide 19 inches (*483* mm) minimum clear depth and 24 inches (610 mm) minimum clear width.   **11B-1009.5.1**   *1009.5.1*   **Fig. CD-56K**

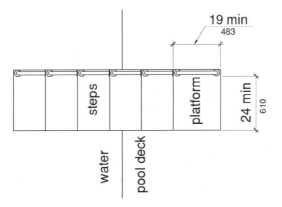

Fig. CD-56K
Size of Transfer Platform
© ICC Reproduced with Permission

Transfer Space

_____ R.   A transfer space of 60 inches (*1524* mm) minimum by 60 inches (*1524* mm) minimum with a slope not steeper than 1:48 shall be provided at the base of the transfer platform surface and shall be centered along a 24 inch (610 mm) minimum side of the transfer platform.   **11B-1009.5.2**   *1009.5.2*   **Fig. CD-56L**

_____ R1.   The side of the transfer platform serving the transfer space shall be unobstructed.   **11B-1009.5.2**   *1009.5.2*   **Fig. CD-56L**

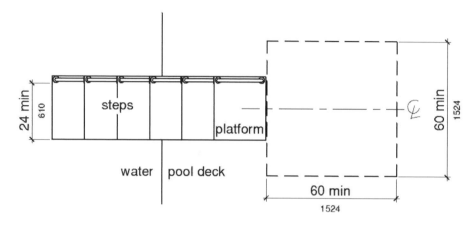

Fig. CD-56L
Clear Deck Space at Transfer Platform
© ICC Reproduced with Permission

Height

_____ S.   The height of the transfer platform shall be 16 inches (*406* mm) minimum and 19 inches (*483* mm) maximum measured from the deck. **11B-1009.4.2, 11B-1009.5.3**   *1009.4.2, 1009.5.3*

Transfer Steps

_____ T.   Transfer step height shall be 8 inches (*203* mm) maximum. **11B-1009.5.4**   *1009.5.4*   **Fig. CD-56M**

_____ T1.   The surface of the bottom tread shall extend to a water depth of 18 inches (*457* mm) minimum below the stationary water level.   **11B-1009.5.4**   *1009.5.4*   **Fig. CD-56M**

> **ADVISORY: Transfer Steps.** *Where possible, the height of the transfer step should be minimized to decrease the distance an individual is required to lift up or move down to reach the next step to gain access.*

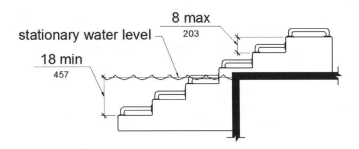

Fig. CD-56M
Transfer Steps
© ICC Reproduced with Permission

Surface

_____U.　　The surface of the transfer system shall not be sharp and shall have rounded edges.　**11B-1009.5.5**　*1009.5.5*

Size

_____V.　　Each transfer step shall have a tread clear depth of 14 inches (*356* mm) minimum and 17 inches (*432* mm) maximum and shall have a tread clear width of 24 inches (610 mm) minimum.　**11B-1009.5.6**　*1009.5.6*　**Fig. CD-56N**

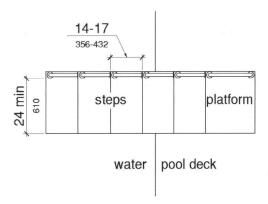

Fig. CD-56N
Size of Transfer Steps
© ICC Reproduced with Permission

Grab Bars

_____W.　　At least one grab bar on each transfer step and the transfer platform or a continuous grab bar serving each transfer step and the transfer platform shall be provided.　**11B-1009.5.7**　*1009.5.7*　**Fig. CD-56O**

_____W1.　　Where a grab bar is provided on each step, the tops of gripping surfaces shall be 4 inches (*102* mm) minimum and 6 inches (*152* mm) maximum above each step and transfer platform. **11B-1009.5.7**　*1009.5.7*　**Fig. CD-56O**

_____W2.   Where a continuous grab bar is provided, the top of the grippin surface shall be 4 inches (*102* mm) minimum and 6 inches (*152* mm) maximum above the step nosing and transfer platform.   **11B-1009.5.7** *1009.5.7*   **Fig. CD-56O**

_____W3.   Grab bars shall comply with items E through O of Section 26H, "SHOWER COMPARTMENTS", and be located on at least one side of the transfer system.   **11B-1009.5.7**   *1009.5.7*

_____W4.   The grab bar located at the transfer platform shall not obstruct transfer.   **11B-1009.5.7**   *1009.5.7*   **Fig. CD-56O**

> **EXCEPTION:** Grab bars on transfer systems shall not be required to comply with the height and positioning requirements of accessible handrails.

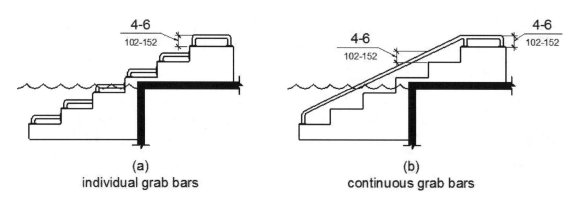

(a)
individual grab bars

(b)
continuous grab bars

Fig. CD-56O
Grab Bars
© ICC Reproduced with Permission

Pool Stairs
Pool stairs shall comply as detailed.   **11B-1009.6**   *1009.6*

Pool Stairs

_____X.   Pool stairs shall comply with Section 23, "STAIRWAYS".   **11B-1009.6.1** *1009.6.1*

> **EXCEPTION:** Pool step riser heights shall not be required to be 4 inches (*102* mm) high minimum and 7 inches (*178* mm) high maximum provided that riser heights are uniform.

Handrails

_____Y.   The width between handrails shall be 20 inches (*508* mm) minimum and 24 inches (*610* mm) maximum. Handrail extensions required by shall not be required on pool stairs.   **11B-1009.6.2**   *1009.6.2*

57. SHOOTING FACILITIES WITH FIRING POSITIONS

Shooting Facilities with Firing Positions   **11B-243**   *243*

General
Where shooting facilities with firing positions are designed and constructed at a site, at least 5 percent, but no fewer than one, of each type of firing position shall comply as detailed.   **11B-243.1**   *243.1*

Shooting Facilities with Firing Positions   **11B-1010**   *1010*

Turning Space

_____A.   A circular turning space 60 inches (*1524* mm) diameter minimum with slopes not steeper than 1:48 shall be provided at shooting facilities with firing positions.   **11B-1010.1**   *1010*

58. OUTDOOR DEVELOPED AREAS

Outdoor Developed Areas   **11B-246**

General
Outdoor developed areas shall comply as detailed.   **11B-246.1**

EXCEPTIONS:

1.   Where the enforcing agency finds that, in specific areas, the natural environment would be materially damaged by compliance with these regulations, such areas shall be subject to these regulations only to the extent that such material damage would not occur.

2.   Automobile access or accessible routes are not required when the enforcing agency determines compliance with this chapter would create an unreasonable hardship as defined (See Definitions).

> **ADVISORY: General.** *Additional information regarding accessibility best practices for outdoor occupancies can be found in the California State Parks Accessibility Guidelines available on the California State Parks website and the Draft Final Guidelines for Outdoor Developed Areas on the U.S. Access Board website.*

Camping Facilities
In camping facilities where campsites are provided, at least two campsites and one additional campsite for each 100 campsites or fraction thereof, shall be accessed by and connected to sanitary facilities by travel routes with a maximum slope of 1:12. Permanent toilet and bathing

facilities serving campsites shall comply with Section 26, "TOILET AND BATHING FACILITIES". **11B-246.2**

Beaches
Beaches shall be accessible.   **11B-246.3**

Day Use Areas and Vista Points
Day use areas, vista points, and similar areas shall be accessible.   **11B-246.4**

Picnic Areas
Where picnic tables are provided, at least one picnic table, and one additional table for each 20 tables or fraction thereof, shall be accessible and comply with Section 46, "DINING SURFACES AND WORK SURFACES".   **11B-246.5**

Parking Lots
Parking lots shall comply with Section 21, "PARKING SPACES", and shall be provided with curb cuts leading to adjacent walks, paths or trails.   **11B-246.6**

Trails and Paths
Trails, paths and nature walk areas, or portions of them, shall be constructed with gradients permitting at least partial use by wheelchair occupants. Buildings and other functional areas shall be served by paths or walks with firm and stable surfaces.   **11B-246.7**

Nature Trails
Nature trails and similar educational and informational areas shall be accessible to individuals with vision impairments by the provision of rope guidelines, raised Arabic numerals and symbols, or other similar guide and assistance devices.   **11B-246.8**

59. DEPOSITORIES, VENDING MACHINES, MAIL BOXES AND FUEL DISPENSERS

Depositories, Vending Machines, Change Machines, Mail Boxes, and Fuel Dispensers
11B-228   *228*

General
Where provided, at least one of each type of depository, vending machine, change machine, and fuel dispenser shall comply with Section 13, "OPERABLE PARTS".
11B-228.1   *228.1*

> **EXCEPTION:** Drive-up only depositories shall not be required to comply with Section 13, "OPERABLE PARTS".

> **ADVISORY: General.** *Depositories include, but are not limited to, night receptacles in banks, post offices, video stores, and libraries.*

Mail Boxes
Where mail boxes are provided in an interior location, at least 5 percent, but no fewer than one, of each type shall comply with Section 13, "OPERABLE PARTS". In residential facilities, where mail boxes are provided for each residential dwelling unit, mail boxes complying with Section 13, "OPERABLE PARTS", shall be provided for each residential dwelling unit required to provide mobility features complying with Section 43, "RESIDENTIAL FACILITIES", and adaptable features complying with Chapter 11A, Division IV.   **11B-228.2**   *228.2*

60. WINDOWS

Windows   **11B-229**   *229*

General
Where glazed openings are provided in accessible rooms or spaces for operation by occupants, at least one opening shall comply with Section 13, "OPERABLE PARTS". Each glazed opening required by an administrative authority to be operable shall comply with Section 13, "OPERABLE PARTS".   **11B-229.1**   *229.1*

EXCEPTION:

1.   Glazed openings in residential dwelling units required to comply with Section 43, "RESIDENTIAL FACILITIES" shall not be required to comply with this section.

2.   Glazed openings in guest rooms required to provide communication features and in guest rooms of transient lodging facilities where the entrances, doors and doorways providing user passage into and within the guestrooms are not required to provide compliant mobility features, shall not be required to comply with this section.

APPENDIX A

2013 CALIFORNIA CODE FORMAT

INDEX OF CBC CODE SECTIONS
REFERENCE DOCUMENT

INDEX OF CBC SECTION NUMBERS

APPENDIX B

SELECTED CALIFORNIA STATUTES
PERTAINING TO ACCESSIBILITY

TABLE OF CONTENTS

PUBLIC BUILDINGS AND FACILITIES

CONTENTS

A.   Division 5, Chapter 7. Access to Public Buildings by Physically Handicapped Persons (Government Code Sections 4450 - 4459)

B.   Division 7. Chapter 15. Facilities for Handicapped Persons (Government Code Section 7250 - Section 7252

A.   ACCESS TO PUBLIC BUILDINGS BY PHYSICALLY HANDICAPPED PERSONS

Government Code

Section 4450. Purpose; standards for access to buildings; regulations

(a) It is the purpose of this chapter to ensure that all buildings, structures, sidewalks, curbs, and related facilities, constructed in this state by the use of state, county, or municipal funds, or the funds of any political subdivision of the state shall be accessible to and usable by persons with disabilities.

(b) The State Architect shall develop and submit proposed building standards to the California Building Standards Commission for approval and adoption pursuant to Chapter 4 (commencing with Section 18935) of Part 2.5 of Division 13 of the Health and Safety Code and shall develop other regulations for making buildings, structures, sidewalks, curbs, and related facilities accessible to and usable by persons with disabilities. The regulations and building standards relating to access for persons with disabilities shall be consistent with the standards for buildings and structures that are contained in pertinent provisions of the latest edition of the selected model code, as adopted by the California Building Standards Commission, and these regulations and building standards shall contain additional requirements relating to buildings, structures, sidewalks, curbs, and other related facilities the State Architect determines are necessary to assure access and usability for persons with disabilities. In developing and revising these additional requirements, the State Architect shall consult with the Department of Rehabilitation, the League of California Cities, the California State Association of Counties, and at least one private organization representing and comprised of persons with disabilities.

(c) In no case shall the State Architect's regulations and building standards prescribe a lesser standard of accessibility or usability than provided by the Accessibility Guidelines prepared by the

---

federal Access Board as adopted by the United States Department of Justice to implement the Americans with Disabilities Act of 1990 (Public Law 101-336).

Added by Stats 1968 ch 261 Section 1. Amended by Stats 1971 ch 1301 Section 1; Stats 1974 ch 995 Section 1; Stats 1975 ch 1150 Section 1, effective September 29, 1975; Stats 1979 ch 1152 Section 10; Stats 1992 ch 913 (A.B. 1077) Section 16; Stats 1993 ch 1214 (A.B. 551) Section 4; Stats 1993 ch 1220 (A.B. 1138) Section 1.
₁42U.S.C.A. Section 12101 et seq

Section 4450.5 State architect's regulations, identification of parking spaces for handicapped

The State Architect's regulations adopted pursuant to Section 4450 shall require that all parking spaces reserved for the handicapped be identified as prescribed by Sections 22511.7 and 22511.8 of the Vehicle Code.

Added by Stats 1984 ch 484 Section 1.

Section 4451. Buildings and facilities to which chapter applicable; standards and specifications; exceptions

(a) Except as otherwise provided in this section, this chapter shall be limited in its application to all buildings and facilities stated in Section 4450 intended for use by the public, with any reasonable availability to, or usage by, persons with disabilities, including all facilities used for education and instruction, including the University of California, the California State University, and the various community college districts, that are constructed in whole or in part by the use of state, county, or municipal funds, or the funds of any political subdivision of the state.

(b) When required by federal or state law, buildings, structures, and facilities, or portions thereof, that are leased, rented, contracted, sublet, or hired by any municipal, county, or state division of government, or special district shall be made accessible to, and usable by, persons with disabilities.

(c) Except as otherwise provided by law, buildings, structures, sidewalks, curbs, and related facilities subject to the provisions of this chapter or Part 5.5 (commencing with Section 19955) of Division 13 of the Health and Safety Code shall conform to the building standards published in the California Building Standards Code relating to access for persons with disabilities and the other regulations adopted pursuant to Section 4450 that are in effect on the date of an application for a building permit. With respect to buildings, structures, sidewalks, curbs, and related facilities not requiring a building permit, building standards published in the California Building Standards Code relating to access for persons with disabilities and other regulations adopted pursuant to Section 4450, and in effect at the time construction is commenced shall be applicable.

(d) Until building standards are published in the California Building Standards Code and other regulations are developed by the State Architect and adopted by the California Building Standards Commission pursuant to Section 4450, buildings, structures, sidewalks, curbs, and related facilities subject to the provisions of this chapter or Part 5.5 (commencing with Section 19955) of Division 13 of the Health and Safety Code shall meet or exceed the requirements of Title III of Subpart D of the federal Americans with Disabilities Act of 1990.

(e) This chapter shall apply to temporary or emergency construction as well as permanent buildings.

(f) Administrative authorities, as designated under Section 4453, may grant exceptions from the literal requirements of the building standards published in the California Building Standards Code relating to access for persons with disabilities, or the other regulations adopted pursuant to this section, or permit the use of other methods or materials, but only when it is clearly evident that equivalent facilitation and protection that meets or exceeds the requirements under federal law are thereby secured.

(g) The Department of General Services shall develop, as appropriate, regulations to ensure that Braille, tactile, or visual signage for elevators, rooms, spaces, functions, and directional information is installed as required by Section 4450 and shall develop and implement an effective training program to ensure compliance with all disability access requirements.

Added by Stats 1968 ch 261 Section 1. Amended by Stats 1971 ch 1301 Section 2; Stats 1974 ch 995 Section 2; Stats 1975 ch 1150 Section 2, effective September 29, 1975; Stats 1979 ch 1152 Section 11; Stats 1983 ch 143 Section 183; Stats 1993 ch 1220 (A.B. 1138) Section 2. Amended by Stats. 2000, c. 989 (S.B. 1242), Section 2, effective January 1, 2001.

Section 4452. Minimum standards; deviation from specifications

It is the intent of the Legislature that the building standards published in the State Building Standards Code relating to access by the physically handicapped and the other regulations adopted by the State Architect pursuant to Section 4450 shall be used as minimum requirements to insure that buildings, structures and related facilities covered by this chapter are accessible to, and functional for, the physically handicapped to, through, and within their doors, without loss of function, space, or facility where the general public is concerned.

Any unauthorized deviation from such regulations or building standards shall be rectified by full compliance within 90 days after discovery of the deviation.

Added by Stats 1968 ch 261 Section 1. Amended by Stats 1971 ch 1301 Section 3; Stats 1974 ch 995 Section 3; Stats 1979 ch 1152 Section 12.

Section 4453. Responsibility for enforcement of chapter

The responsibility for enforcement of this chapter shall be as follows:

(a) By the Director of the Department of General Services where state funds are utilized for any project or where funds of counties, municipalities, or other political subdivisions are utilized for the construction of elementary, secondary, or community college projects.

(b) By the governing bodies thereof where funds of counties, municipalities, or other political subdivisions are utilized except as otherwise provided in (a) above.

Added by Stats 1968 Ch 261 Section 1. Amended by Stats 1978 ch 326 Section 1.

Section 4453.5. Inspection of state and school district buildings by physically disabled volunteers; reports; correction plan; applicability of section

(a) In addition to any other inspection requirements pertaining to building standards of state and school district buildings used by the public, the construction of which are under the jurisdiction of the Office of the State Architect in the Department of General Services, accessibility to persons with handicaps may be inspected pursuant to subdivision (b) in state and school district buildings used by the public in order to determine if the building meets minimum state standards for accessibility to handicapped persons.

(b) Inspection and approval may be made on a voluntary basis by one or more persons who have physical disabilities or who represent the interests of physically disabled persons, who are familiar with the California access laws and standards, and who have been chosen by the Department of Rehabilitation. The Department of Rehabilitation may assign these volunteers to inspect those state and school district buildings used by the public specified in subdivision (a). If the volunteer inspector finds that a building does not meet minimum state standards for accessibility to handicapped persons, the volunteer shall report this information to the Department of Rehabilitation, which shall in turn report the information to the school district if a school building is involved, to the owning agencies if a state building is involved, and to the Office of the State

Architect. When, after receipt of this information, the Office of the State Architect confirms that the building does not meet minimal state standards for accessibility to handicapped persons, the Office of the State Architect shall develop a plan to be filed with the jurisdiction owning the building that addresses the correction of the identified deficiencies.

(c) The provisions of this section shall only pertain to state and school district buildings used by the public for which building plans have been filed with the Office of the State Architect on or after January 1, 1985.

Added by Stats 1983 ch 1246 Section 2.

Section 4454. Approval of plans and specifications; filing fees; consultation
(a) Where state funds are utilized for any building or facility subject to this chapter, or where funds of counties, municipalities, or other political subdivisions are utilized for the construction of elementary school, secondary school, or community college buildings and facilities subject to this chapter, no contract shall be awarded until the Department of General Services has issued written approval stating that the plans and specifications comply with the intent of this chapter.

(b) Notwithstanding subdivision (a), for all transportation facilities, other than rail or transit stations, located within state highway rights-of-way, the Department of Transportation is authorized to issue the required written approval stating that the plans and specifications comply with intent of this chapter. If the Department of General Services, Division of the State Architect, establishes a certified access specialist program, as described in Section 4459.5, specific to standards governing access to transportation facilities, the Department of Transportation shall within 180 days of establishment of the program begin using engineers certified through that program to verify that the Department of Transportation's standards, guidelines, and design exceptions comply with the intent of this chapter.

(c) In each case the application for approval shall be accompanied by the plans and full, complete, and accurate specifications, which shall comply in every respect with any and all requirements prescribed by the Department of General Services.

(d) Except for facilities located within state highway rights-of-way, other than rail or transit stations, the application shall be accompanied by a filing fee in amounts as determined by the Department of General Services. All fees shall be deposited into the Access for Handicapped Account, which is hereby renamed the Disability Access Account as of July 1, 2001, and established in the General Fund. Not withstanding Section 13340, the account is continuously appropriated for expenditures for the use of the Department of General Services, in carrying out the department's responsibilities under this chapter.

(e) The Department of General Services shall consult with the Department of Rehabilitation in identifying the requirements necessary to comply with this chapter.

(f) The Department of General Services, Division of the State Architect, shall include the cost of carrying out the responsibilities identified in this chapter as part of the plan review costs in determining fees.

Added by Stats 1970 ch 701 Section 1. Amended by Stats 1978 ch 326 Section 2. Amended by Stats. 2000, c. 989 (S.B. 1242), Section 3. Amended by Stats, 2005, Ch, 299, Section 1, effective September 22, 2005.

Section 4455. Duties of department of rehabilitation
The Department of Rehabilitation shall be responsible for educating the public and working with officials of cities, counties, municipalities, and other political subdivisions, private architects, designers, planners, and other interested parties in order to encourage and help them make all buildings, facilities, and improved areas accessible to and usable by handicapped persons for purposes of rehabilitation, employment, business, recreation, and all other aspects of normal living.

Added by Stats 1970 ch 701 Section 2.

Section 4455.5 Elevators; Braille symbols

All new elevators in public buildings or facilities after the operative date of the act that amended this section during the first year of the 1979-80 Regular Session shall have Braille symbols and marked Arabic numerals corresponding to the numerals on the elevator buttons embossed immediately to the left thereof.

All new door casings on all elevator floors after the operative date of this section shall have the number of the floor on which the casing is located embossed in Braille symbols and marked Arabic numerals on both sides at a height of approximately 60 inches from the floor.

Added by Stats 1971 ch 1368 Section 1. Amended by Stats 1979 ch 273 Section 1.

Section 4456 Alteration of existing buildings or facilities

After the effective date of this section, any building or facility which would have been subject to this chapter but for the fact it was constructed prior to November 13, 1968, shall comply with the provisions of this chapter when alterations, structural repairs or additions are made to such building or facility. This requirement shall only apply to the area of specific alteration, structural repair or addition and shall not be construed to mean that the entire structure or facility is subject to this chapter.

Added by Stats 1971 ch 1458 Section 1.

Section 4457. Portable buildings of school district

On or after January 1, 1986, all portable buildings purchased, leased, or constructed by a school district shall meet the requirements of this chapter, except as provided in subdivision (f) of Section 4451.

Added by Stats 1985 ch 550 Section 2.

Section 4458. Violations; injunctions; district or city attorney, attorney general

The district attorney, the city attorney, the county counsel if the district attorney does not bring an action, or the Attorney General may bring an action to enjoin a violation of this chapter.

Added by Stats 1976 ch 869 Section 2. Amended by Stats 2003, Ch. 872, Section 1, effective January1, 2004.

Section 4459. State Architects adoption of minimum ADA standards

(a) The State Architect shall develop amendments for building regulations and submit them to the California Building Standards Commission for adoption to ensure that no accessibility requirements of the California Building Standards Code shall be enhanced or diminished except as necessary for (1) retaining existing state regulations that provide greater accessibility and features, or (2) meeting federal minimum accessibility standards of the federal Americans with Disabilities Act of 1990 as adopted by the United States Department of Justice, the Uniform Federal Accessibility Standards, and the federal Architectural Barriers Act.

(b) The Department of General Services shall use fees deposited in the Disability Access Account established in Section 4454 for the purposes identified in this chapter. The department shall include the cost of carrying out the responsibilities identified in this chapter as part of the plan review costs in determining fees.

(c) Notwithstanding any other provision of law, the application and scope of accessibility regulations in the California Building Standards Code shall not be less than the application and scope of accessibility requirements of the federal Americans with Disabilities Act of 1990 as adopted by the United States Department of Justice, the Uniform Federal Accessibility Standards, and the federal Architectural Barriers Act.

B.   FACILITIES FOR HANDICAPPED PERSONS

Government Code

Section 7250.   Application
Section 7251.   Toilet facilities; signs
Section 7252.   Level or ramp entrances; signs

Section 7250. Application
The provisions of this chapter apply to all buildings or other facilities owned, leased, operated or managed by the state, county, city and county, district, or other political subdivision and which are usually or regularly open to the members of the public.

Added by Stats 1968 ch 937 Section 1

Section 7251. Toilet facilities; signs
When a building contains special toilet facilities usable by a person in a wheelchair or otherwise handicapped, a sign indicating the location of such facilities shall be posted in the building directory, in the main lobby, or at any entrance specially used by handicapped persons.

Added by Stats 1968 ch 937 Section 1.

Section 7252. Level or ramp entrance; signs
When a building contains an entrance other than the main entrance which is ramped or level for use by handicapped persons, a sign showing its location shall be posted at or near the main entrance which shall be visible from the adjacent public sidewalk or way.

Added by Stats 1968 ch 937 Section 1.

PRIVATELY FUNDED BUILDINGS AND FACILITIES

CONTENTS

A.   Division 13, Part 5.3. Access to Places of Public Amusement and Resort by Physically Handicapped Persons (Health and Safety Code Sections 19952-19954)

B.   Division 13. Part 5.5. Access to Public Accommodations by Physically Handicapped Persons (Health and Safety Code Sections 19955-19959)

C.   Double Doors (Health and Safety Code Section 13011)

A.   **ACCESS TO PLACES OF PUBLIC AMUSEMENT AND RESORT BY PHYSICALLY HANDICAPPED PERSONS**

Health and Safety Code

Section 19952. Seating or accommodations in various locations within facility; removal seats; application; construction

Section 19953. Injunction; attorney's fees
Section 19954. Injunctions; persons who may bring action

Section 19952. Seating or accommodations in various locations within facility; removal seats; application; construction
(a) Any person, or public or private firm, organization, or corporation, who owns or manages places of public amusement and resort including theaters, concert halls, and stadiums shall provide seating or accommodations for physically disabled persons in a variety of locations within the facility, to the extent that this variety can be provided while meeting fire and panic safety requirements of the State Fire Marshal, so as to provide these persons a choice of admission prices otherwise available to members of the general public.

(b) Readily removable seats may be installed in wheelchair spaces when the spaces are not required to accommodate wheelchair users.

(c) The requirements of this section shall apply with respect to publicly and privately owned facilities or structures for the purposes specified in subdivision (a) for which a building permit or a building plan for new construction has been issued on or after January 1, 1985.

(d) In no case shall this section be construed to prescribe a lesser standard of accessibility or usability than provided by the Accessibility Guidelines prepared by the federal Access Board and adopted by the United States Department of Justice to implement the Americans with Disabilities Act of 1990 (Public Law 101-336).

Added by Stats 1983 ch 781 Section 1. Amended by Stats 1992 ch 913 (A.B. 1077) Section 35; Stats 1993 ch 1214 (A.B. 551) Section 6.
,42 U.S.C.A. Section 12101 et seq.

Section 19953. Injunctions; attorney fees
Any person who is aggrieved or potentially aggrieved by a violation of this part, Chapter 7 (commencing with Section 4450) of Division 5 of Title 1 of the Government Code, or Part 5.5 (commencing with Section 19955) of Division 13 of the Health and Safety Code may bring an action to enjoin the violation. The prevailing party in the action shall be entitled to recover reasonable attorney's fees.

Added by Stats 1983 ch 781 Section 1.

Section 19954. Injunctions; persons who may bring action

The district attorney, the city attorney, the county counsel if the district attorney does not bring an action, the Department of Rehabilitation acting through the Attorney General, or the Attorney General may bring an action to enjoin any violation of this part.

Added by Stats 1983 781 Section 1. Amended by Stats, 2003, Ch. 872, Section 6, effective January 1, 2004.

B.   ACCESS TO PUBLIC ACCOMMODATIONS BY PHYSICALLY HANDICAPPED PERSONS

Health and Safety Code

Section 19955. Purpose

a) The purpose of this part is to insure that public accommodations or facilities constructed in this state with private funds adhere to the provisions of Chapter 7 (commencing with Section 4450) of Division 5 of Title 1 of the Government Code. For the purposes of this part "public accommodation or facilities" means a building, structure, facility, complex, or improved area which is used by the general public and shall include auditoriums, hospitals, theaters, restaurants, hotels, motels, stadiums, and convention centers.

As used in this section, "hospitals" includes, but is not limited to, hospitals, nursing homes, and convalescent homes.

When sanitary facilities are made available for the public, clients, or employees in such accommodations or facilities, they shall be made available for the physically handicapped.

Any new requirements imposed by the amendments to this section enacted by the Legislature at its 1973-74 Regular Session shall only apply to public accommodations or facilities constructed on or after the effective date of the amendments.

Added by Stats 1969 ch 1560 Section 1, operative July 1, 1970. Amended by Stats 1971 ch 821 Section 1; Stats 1972 ch 488 Section 1; Stats 1973 ch 931 Section 1, effective September 30, 1973.

Section 19955.3. Definitions

As used in this part:

(a) "Story" means that portion of a building included between the upper surface of any floor and the upper surface of the floor next above, except that the topmost story shall be that portion of a building included between the upper surface of the topmost floor and the ceiling or roof above. If the finished floor level directly above a basement or unused under-floor space is more than six feet above grade for more than 50 percent of the total perimeter or is more than 12 feet above grade at any point, the basement or unused under-floor space shall be considered as a story.

(b) "First story" means the lowest story in a building which qualifies as a story and which provides the basic services or functions for which the building is used. A floor level in a building having only one floor level shall be classified as a first story, if the floor level is not more than four feet below grade, for more than 50 percent of the total perimeter, or more than eight feet below grade at any point.

(c) "Mezzanine" means an intermediate floor placed in any story or room. When the total area of any "mezzanine floor" exceeds 331/3 percent of the total floor area in that room, it shall be considered as constituting an additional "story." The clear height above and below a "mezzanine floor" construction shall be not less than seven feet.

(d) "Grade" means the lowest point of elevation of the finished surface of the ground, paving, or sidewalk within the area between the building and the property line or, when the property line is more than 5 feet from the building, between the building and a line 5 feet from the building.

Added by Stats 1982 ch 1416 Section 1.

Section 19955.5.   Access to passenger vehicle service stations, shopping centers, physicians' and surgeons' offices, and office buildings constructed with private funds; prospective application of section

All passenger vehicle service stations, shopping centers, offices of physicians and surgeons, and office buildings constructed in this state with private funds shall adhere to the provisions of Chapter 7 (commencing with Section 4450) of Division 5 of Title 1 of the Government Code. As used in this section, "office building" means a structure wherein commercial activity or service is performed or a profession is practiced, or wherein any combination thereof is performed or practiced in all or the majority of the building or structure.

When sanitary facilities are made available for the public, clients, or employees in these stations, centers, or buildings, they shall be made available for persons with disabilities.

Any new requirements imposed by the amendments to this section by Chapter 931 of the Statutes of 1973 shall only apply to those stations, centers, or office buildings constructed on or after September 30, 1973. Any other new requirements imposed by amendments to this section by Chapter 995 of the Statutes of 1974 shall only apply to those offices or office buildings constructed on or after January 1, 1975.

Added by Stats 1971 ch 821 Section 2. Amended by Stats 1973 ch 931 Section 2, effective Sept. 30, 1973; Stats 1974 ch 995 Section 4; Stats 1982 ch 1416 Section 2; Stats 1993 ch 1220 (A.B.1138), Section 3, effective January 1, 1994.

Section 19956.   Conformity with Government Code provisions; exceptions

All public accommodations constructed in this state shall conform to the provisions of Chapter 7 (commencing with Section 4450) of Division 5 of Title 1 of the Government Code. However, the following types of privately funded multistory buildings do not require accessibility by ramp or elevator above and below the first floor.

(a) Multistoried office buildings, other than the professional office of a health care provider, and passenger vehicle service stations less than three stories high, or less than 3,000 square feet per story.

(b) Any other privately funded multistoried building that is not a shopping center, shopping mall, or the professional office of a health care provider, and that is less than three stories high or less than 3,000 square feet per story if a reasonable portion of all facilities and accommodations normally sought and used by the public in such a building are accessible to and usable by persons with disabilities.

Added by Stats 1969 ch 1560 Section 1, operative July 1, 1970. Amended by Stats Section 1982 ch 1416 Section 3; Stats 1993 ch 1220 (A.B. 1138) Section 4.

Section 19956.5. Public curb or sidewalk constructed with private funds
Any curb or sidewalk intended for public use that is constructed in this state with private funds shall conform to the provisions of Chapter 7 (commencing with Section 4450) of Division 5 of Title 1 of the Government Code.

This section shall apply, but not be limited in application, to any curb or sidewalk which after construction with private funds will be turned over to a city or county for public use, in order to provide full and easy access to, and use of, such curb or sidewalk by the physically handicapped.

Added by Stats 1972 ch 1018 Section 1.

Section 19957. Exceptions from literal requirements of standards and specifications in hardship, etc., cases
In cases of practical difficulty, unnecessary hardship, or extreme differences, a building department responsible for the enforcement of this part may grant exceptions from the literal requirements of the standards and specifications required by this part or permit the use of other methods or materials, but only when it is clearly evident that equivalent facilitation and protection are thereby secured.

Added by Stats 1969 ch 1560 Section 1, operative July 1, 1970.

Section 19957.5. Local appeals board; jurisdiction; members; duties
(a) Every city, county, or city and county may appoint a local appeals board composed of five members to hear written appeals brought by any person regarding action taken by the building department of the city, county, or city and county in enforcement of the requirements of this part, including the exceptions contained in Section 19957.

(b) Two members of the appeals board shall be physically handicapped persons, two members shall be persons experienced in construction, and one member shall be a public member.

(c) The appeals board shall conduct hearings on written appeals made under subdivision (a) and may approve or disapprove interpretations of this part and enforcement actions taken by the building department of the city, county, or city and county. All such approvals or disapprovals shall be final and conclusive as to the building department in the absence of fraud or prejudicial abuse of discretion. The appeals board shall adopt regulations establishing procedural rules and criteria for the carrying out of its duties under this part.

Added by Stats 1976 ch 700 Section 1.

Section 19958. Enforcement; building department defined
The building department of every city, county, or city and county shall enforce this part within the territorial area of its city, county, or city and county. The responsibility for enforcing Chapter 7 (commencing with Section 4450) of Division 5 of Title 1 of the Government Code in its application under this part shall be by such building department within the territorial area of its city, county, or city and county.

"Building department" means the department, bureau, or officer charged with the enforcement of laws or ordinances regulating the erection or construction, or both the erection and construction, of buildings.

Added by Stats 1969 ch 1560 Section 1, operative July 1, 1970.

Section 19958.5. Violations; injunction; district or city attorney, attorney general
The district attorney, the city attorney, the county counsel if the district attorney does not bring an action, the Department of Rehabilitation acting through the Attorney General, or the Attorney General may bring an action to enjoin a violation of this part.

Added by Stats 1976 ch 869 Section 3, amended by Stats, 2003, Ch. 872, Section 7, effective January 1, 2004.

Section 19959. Alteration of existing public accommodations
Every existing public accommodation constructed prior to July 1, 1970, which is not exempted by Section 19956, shall be subject to the requirements of this chapter when any alterations, structural repairs or additions are made to such public accommodation. This requirement shall only apply to the area of specific alteration, structural repair or addition and shall not be construed to mean that the entire building or facility is subject to this chapter.

Added by Stats 1971 ch 1458 Section 2. Amended by Stats 1974 ch 545 Section 72.

C. DOUBLE DOORS

Health and Safety Code

Section 13011. Double doors
Both doors of any double doors designated as the public entrance to any place of business shall be kept unlocked during normal business hours.

Added by Stats 1983 ch 267 Section 1.

PARKING PRIVILEGES

CONTENTS

A. Division 3, Chapter 1. Special Plates (Section 5007)
B. Division 11. Chapter 9. Stopping-Standing-Parking
(Vehicle Code Section 22507, Section 22511, Section 22511.5, Section
22511.7, Section 22511.8, Section 22511.9, Section 22511.55.) (Vehicle Code
Section 21458)

A. SPECIAL PLATES

Vehicle Code

Section 5007. Disabled person or disabled veterans; special identification license plates; return to department on death or expiration
(a) The department shall, upon application and without additional fees, issue a special license plate or plates pursuant to procedures adopted by the department to the following:

(1) A disabled person.

(2) A disabled veteran.

(3) An organization or agency involved in the transportation of disabled persons or disabled veterans if the vehicle that will have the special license plate is used solely for the purpose of transporting those persons.

(b) The special license plates issued under subdivision (a) shall run in a regular numerical series that shall include one or more unique two-letter codes reserved for disabled person license plates or disabled veteran license plates. The International Symbol of Access adopted pursuant to Section 3 of Public Law 100-641, commonly known as the "wheelchair symbol" shall be depicted on each plate.

(c) (1) Prior to issuing a special license plate to a disabled person or disabled veteran, the department shall require the submission of a certificate, in accordance with paragraph (2), signed by the physician and surgeon, or to the extent that it does not cause a reduction in the receipt of federal aid highway funds, by a nurse practitioner, certified nurse midwife, or physician assistant, substantiating the disability, unless the applicant's disability is readily observable and uncontested. The disability of a person who has lost, or has lost use of, one or more lower extremities or one hand, for a disabled veteran, or both hands for a disabled person, or who has significant limitation in the use of lower extremities, may also be certified by a licensed chiropractor. The blindness of an applicant shall be certified by a licensed physician and surgeon who specializes in diseases of the eye or a licensed optometrist. The physician and surgeon, nurse practitioner, certified nurse midwife, physician assistant, chiropractor, or optometrist certifying the qualifying disability shall provide a full description of the illness or disability on the form submitted to the department.

(2) The physician and surgeon, nurse practitioner, certified nurse midwife, physician assistant, chiropractor, or optometrist who signs a certificate submitted under this subdivision shall retain information sufficient to substantiate that certificate and, upon request of the department, shall make that information available for inspection by the Medical Board of California or the appropriate regulatory board.

(d) A disabled person or disabled veteran issued a license plate or plates under this section shall, upon request, present to a peace officer, or person authorized to enforce parking laws, ordinances,

---

or regulations, a certification form that substantiates the eligibility of the disabled person or veteran to possess the plate or plates. The certification shall be on a form prescribed by the department and contain the name of the disabled person or disabled veteran to whom the plate or plates were issued, and the name, address, and telephone number of the medical professional described in subdivision (c) who certified the eligibility of the person or veteran for the plate or plates.

(e) The certification requirements of subdivisions (c) and (d) do not apply to an organization or agency that is issued a special license plate or plates under paragraph (3) of subdivision (a).

(f) The special license plate shall, upon the death of the disabled person or disabled veteran, be returned to the department within 60 days or upon the expiration of the vehicle registration, whichever occurs first.

(g) When a vehicle subject to paragraph (3) of subdivision (a) is sold or transferred, the special license plate or plates issued to an organization or agency under paragraph (3) of subdivision (a) for that vehicle shall be immediately returned to the department.

Added by Stats 1991 ch 893 (A.B. 274 Section 2. Amended by Status 1992 ch 785 A.B. 2289) Section 2. Amended by Stats. 2000, c. 524 (A.B. 1792), Section 4, amended by Stats, 2011, Ch. 296, Section 301, effective January 1, 2012.

B. STOPPING-STANDING-PARKING

Vehicle Code

Section 21458.	Curb markings
Section 22507.	Local regulation; preferential parking
Section 22511.	Designating stalls and off-street parking sites
Section 22511.5.	Disabled persons or disabled veterans; parking privileges
Section 22511.7.	Designation of parking for disabled persons and veterans
Section 22511.8.	Offstreet parking; designation of parking for disabled persons and veterans; removal of unauthorized vehicles
Section 22511.9.	Disabled veteran defined; parking; license plates and placards
Section 22511.55.	Distinguishing placard; application; use, display, and appearance; procedures for issuance and renewal; fee; eligibility; certificate substantiating disability; temporary disability; return upon death of disabled person
Section 22522.	Parking Near Designated Sidewalk Access Ramps

Section 21458. Curb markings
(a) Whenever local authorities enact local parking regulations and indicate them by the use of paint upon curbs, the following colors only shall be used, and the colors indicate as follows:

(1) Red indicates no stopping, standing, or parking, whether the vehicle is attended or unattended, except that a bus may stop in a red zone marked or signposted as a bus loading zone.

(2) Yellow indicates stopping only for the purpose of loading or unloading passengers or freight for the time as may be specified by local ordinance.

(3) White indicates stopping for either of the following purposes:

(A) Loading or unloading of passengers for the time as may be specified by local ordinance.

(B) Depositing mail in an adjacent mailbox.

(4) Green indicates time limit parking specified by local ordinance.

(5) Blue indicates parking limited exclusively to the vehicles of disabled persons and disabled veterans.

(b) Regulations adopted pursuant to subdivision (a) shall be effective on days and during hours or times as prescribed by local ordinances.

Added by Stats 1959 ch 3 Section 21458. Amended by Stats 1975 ch 688 Section 1; Stats 1985 ch 1041 Section 4; Stats 1992 ch 1243 (A.B. 3090) Section 88 effective September 30, 1992.

Section 22507. Local regulation; preferential parking
(a) Local authorities may, by ordinance or resolution, prohibit or restrict the stopping, parking, or standing of vehicles, including, but not limited to, vehicles that are six feet or more in height (including any load thereon) within 100 feet of any intersection, on certain streets or highways, or portions thereof, during all or certain hours of the day. The ordinance or resolution may include a designation of certain streets upon which preferential parking privileges are given to residents and merchants adjacent to the streets for their use and the use of their guests, under which the residents and merchants may be issued a permit or permits that exempt them from the prohibition or restriction of the ordinance or resolution. With the exception of alleys, the ordinance or resolution shall not apply until signs or markings giving adequate notice thereof have been placed.

A local ordinance or resolution adopted pursuant to this section may contain provisions that are reasonable and necessary to ensure the effectiveness of a preferential parking program.

(b) An ordinance or resolution adopted under this section may also authorize preferential parking permits for members of organizations, professions, or other designated groups, including, but not limited to, school personnel, to park on specified streets if the local authority determines that the use of the permits will not adversely affect parking conditions for residents and merchants in the area.

Stats 1959 ch 3 Section 22507. Amended by Stats 1963 ch 1070 Section 1; Stats 1969 ch 541 Section 1; Stats 1976 ch 1002 Section 1; Stats 1980 ch 140 Section 1; Stats 1984 ch 181 Section 2; Stats 1985 ch 912 Section 2; Stats 1987 ch 455 Section 4. Amended by Stats. 2001, c. 223 (S.B. 779), Section 1.

Section 22511. Designating stalls and off-street parking sites.

(a) Any local authority, by ordinance or resolution, and any person in lawful possession of an offstreet parking facility may designate stalls or spaces in an offstreet parking facility owned or operated by that local authority or person for the exclusive purpose of fueling and parking a vehicle that displays a valid zero-emission vehicle (ZEV) decal identification posted on the driver's side rear window or bumper of the vehicle or, notwithstanding any other provision of law, if the vehicle does not have a rear window or bumper, on the driver's side of the windshield issued by the Department of Motor Vehicles pursuant to this section. The designation shall be made by posting a sign in compliance with subdivision (d) or (e).

(b) If posted in accordance with subdivision (d) or (e), the owner or person in lawful possession of a privately owned or operated offstreet parking facility, after notifying the police or sheriff's department, may cause the removal of a vehicle from a stall or space designated pursuant to subdivision (a) in the facility to the nearest public garage if a valid ZEV decal identification issued pursuant to this section is not displayed on the vehicle.

(c) If posted in accordance with subdivision (d), the local authority owning or operating an offstreet parking facility, after notifying the police or sheriff's department, may cause the removal of a vehicle from a stall or space designated pursuant to subdivision (a) in the facility to the nearest garage, as defined in Section 340, that is owned, leased, or approved for use by a public agency if a valid ZEV decal identification issued pursuant to this section is not displayed on the vehicle.

(d) The posting required for an offstreet parking facility owned or operated either privately or by a local authority shall consist of a sign not less than 17 by 22 inches in size with lettering not less than one inch in height which clearly and conspicuously states the following: "Unauthorized vehicles not displaying valid zero-emission vehicle decal identifications will be towed away at owner's expense.

Towed vehicles may be reclaimed at

_____ or by telephoning
      (Address)

_____."
   (Telephone number of local law enforcement agency)

The sign shall be posted in either of the following locations:

(1) Immediately adjacent to, and visible from, the stall or space.

(2) In a conspicuous place at each entrance to the offstreet parking facility.

(e) If the parking facility is privately owned and public parking is prohibited by the posting of a sign meeting the requirements of paragraph (1) of subdivision (a) of Section 22658, the requirements of subdivision (b) may be met by the posting of a sign immediately adjacent to, and visible from, each stall or space indicating that a vehicle not meeting the requirements of subdivision (a) will be removed at the owner's expense and containing the telephone number of the local traffic law enforcement agency.

(f) (1) For purposes of implementing this section, the Department of Motor Vehicles shall make available for issuance, beginning July 1, 2003, for a fee determined by the Department of Motor Vehicles to be sufficient to reimburse it for actual costs incurred pursuant to this section, distinctive decals for zero-emission vehicles.

(2) The department shall design the decal, which shall be two inches by two inches, and be placed on the driver's side rear window or bumper of the vehicle, or, notwithstanding any other provision of law, if the vehicle does not have a rear window or bumper, on the driver's side of the windshield. Each decal shall display a unique number. The decal may be provided to car dealers who sell electric vehicles for distribution to ZEV purchasers.

(g) For purposes of this section, "zero-emission vehicle" means any car, truck, or any other vehicle that produces no tailpipe or evaporative emissions.

(h) Nothing in this section is intended to interfere with existing law governing the ability of local authorities to adopt ordinances related to parking programs within their jurisdiction, such as programs that provide free parking in metered areas or municipal garages for electric vehicles.
Amended by Stats, 2011, Ch. 274, Section 1, effective January 1, 2012.

Section 22511.5. Disabled persons or disabled veterans; parking privileges

(a) (1) A disabled person or disabled veteran displaying special license plates issued under Section 5007 or a distinguishing placard issued under Section 22511.55 or 22511.59 is allowed to park for unlimited periods in any of the following zones:

(A) In any restricted zone described in paragraph (5) of subdivision (a) of Section 21458 or on streets upon which preferential parking privileges and height limits have been given
pursuant to Section 22507.

(B) In any parking zone that is restricted as to the length of time parking is permitted as indicated by a sign erected pursuant to a local ordinance.

(2) A disabled person or disabled veteran is allowed to park in any metered parking space without being required to pay parking meter fees.

(3) This subdivision does not apply to a zone for which state law or ordinance absolutely prohibits stopping, parking, or standing of all vehicles, or which the law or ordinance reserves for special types of vehicles, or to the parking of a vehicle that is involved in the operation of a street vending business.

(b) A disabled person or disabled veteran is allowed to park a motor vehicle displaying a special disabled person license plate or placard issued by a foreign jurisdiction with the same parking privileges authorized in this code for any motor vehicle displaying a special license plate or a distinguishing placard issued by the Department of Motor Vehicles.

Section 2; Stats 1982 ch 975 Section 6.5; Stats 1984 ch 510 Section 1; Stats 1984 ch 1118 Section 1; Stats 1985 ch 1041 Section 7; Stats 1986 ch 351 Section 3; Stats 1988 ch 115 Section 1; Stats 1989 ch 554 Section 3; Stats 1991 ch 893 (A.B. 274) Section 4; Stats 1992 ch 785 (A.B. 2289) Section 3; Stats 1992 ch 1241 (S.B. 1615) Section 21; Stats. 1994, c. 1149 (A.B. 2878), Section 5.). Amended by Stats, 2010, Ch. 478, Section 11, effective January1, 2011.

Section 22511.7. Designation of parking for disabled persons and veterans

(a) In addition to Section 22511.8 for offstreet parking, a local authority may, by ordinance or resolution, designate onstreet parking spaces for the exclusive use of a vehicle that displays either a special identification license plate issued pursuant to Section 5007 or a distinguishing placard issued pursuant to Section 22511.55 or 22511.59.

(b) (1) Whenever a local authority so designates a parking space, it shall be indicated by blue paint on the curb or edge of the paved portion of the street adjacent to the space. In addition, the local authority shall post immediately adjacent to and visible from the space a sign consisting of a profile view of a wheelchair with occupant in white on a blue background.

(2) The sign required pursuant to paragraph (1) shall clearly and conspicuously state the following: "Minimum Fine $250." This paragraph applies only to signs for parking spaces constructed on or after July 1, 2008, and signs that are replaced on or after July 1, 2008.

(3) If the loading and unloading area of the pavement adjacent to a parking stall or space designated for disabled persons or disabled veterans is to be marked by a border and hatched lines, the border shall be painted blue and the hatched lines shall be painted a suitable contrasting color to the parking space. Blue or white paint is preferred. In addition, within the border the words "No Parking" shall be painted in white letters no less than 12 inches high. This paragraph applies only to parking spaces constructed on or after July 1, 2008, and painting that is done on or after July 1, 2008.

(c) This section does not restrict the privilege granted to disabled persons and disabled veterans by Section 22511.5.

Added by Stats 1975 ch 688 Section 2. Amended by Stats 1976 ch 1096 Section 4; Stats 1983 ch 270 Section 2, effective July 15, 1983.
(Amended by Stats. 1985, c. 1041, Section 8; Stats. 1987, c. 314, Section 2; Stats. 1989, c. 554, Section 4; Stats. 1990, c. 692 (A.B. 3398), Section 4; Stats. 1994, c. 1149 (A.B. 2878), Section 10.), Amended by Stats, 2009, Ch. 200, Section 13, effective January 1, 2010.

Section 22511.8. Offstreet parking; designation of parking for disabled persons and veterans; removal of unauthorized vehicles

(a) A local authority, by ordinance or resolution, and a person in lawful possession of an offstreet parking facility may designate stalls or spaces in an offstreet parking facility owned or operated by the local authority or person for the exclusive use of a vehicle that displays either a special license plate issued pursuant to Section 5007 or a distinguishing placard issued pursuant to Section 22511.55 or 22511.59. The designation shall be made by posting a sign as described in paragraph (1), and by either of the markings described in paragraph (2) or (3):

(1) (A) By posting immediately adjacent to, and visible from, each stall or space, a sign consisting of a profile view of a wheelchair with occupant in white on a blue background.

(B) The sign shall also clearly and conspicuously state the following: "Minimum Fine $250," This subparagraph applies only to signs for parking spaces constructed on or after July 1, 2008, and signs that are replaced on or after July 1, 2008, or as the State Architect deems necessary when renovations, structural repair, alterations, and additions occur to existing buildings and facilities on or after July 1, 2008.

(2) (A) By outlining or painting the stall or space in blue and outlining on the ground in the stall or space in white or suitable contrasting color a profile view depicting a wheelchair with occupant.

(B) The loading and unloading area of the pavement adjacent to a parking stall or space designated for disabled persons or disabled veterans shall be marked by a border and hatched lines. The border shall be painted blue and the hatched lines shall be painted a suitable contrasting color to the parking space. Blue or white paint is preferred. In addition, within the border the words "No Parking" shall be painted in white letters no less than 12 inches high. This subparagraph applies only to parking spaces constructed on or after July 1, 2008, and painting that is done on or after July 1, 2008, or as the State Architect deems necessary when renovations, structural repair, alterations, and additions occur to existing buildings and facilities on or after July 1, 2008. (3) By outlining a profile view of a wheelchair with occupant in white on a blue background, of the same dimensions as in paragraph (2). The profile view shall be located so that it is visible to a traffic enforcement officer when a vehicle is properly parked in the space.

(b) The Department of General Services under the Division of the State Architect shall develop pursuant to Section 4450 of the Government Code, as appropriate, conforming regulations to ensure compliance with subparagraph (B) of paragraph (1) of subdivision (a) and subparagraph (B) of paragraph (2) of subdivision (a). Initial regulations to implement these provisions shall be adopted as emergency regulations. The adoption of these regulations shall be considered by the Department of General Services to be an emergency necessary for the immediate preservation of the public peace, health and safety, or general welfare.

(c) If posted in accordance with subdivision (e) or (f), the owner or person in lawful possession of a privately owned or operated offstreet parking facility, after notifying the police or sheriff's department, may cause the removal of a vehicle from a stall or space designated pursuant to subdivision (a) in the facility to the nearest public garage unless a special license plate issued pursuant to Section 5007 or distinguishing placard issued pursuant to Section 22511.55 or 22511.59 is displayed on the vehicle.

(d) If posted in accordance with subdivision (e), the local authority owning or operating an offstreet parking facility, after notifying the police or sheriff's department, may cause the removal of a vehicle from a stall or space designated pursuant to subdivision (a) in the facility to the nearest public garage unless a special license plate issued pursuant to Section 5007 or a distinguishing placard issued pursuant to Section 22511.55 or 22511.59 is displayed on the vehicle.

(e) Except as provided in Section 22511.9, the posting required for an offstreet parking facility owned or operated either privately or by a local authority shall consist of a sign not less than 17 by 22 inches in size with lettering not less than one inch in height which clearly and conspicuously states the following: "Unauthorized vehicles parked in designated accessible spaces not displaying distinguishing placards or special license plates issued for persons with disabilities will be towed away at the owner's expense. Towed vehicles may be reclaimed at:

_____ or by telephoning
        (Address)
                                                            ."
    (Telephone number of local law enforcement agency)

The sign shall be posted in either of the following locations:
(1) Immediately adjacent to, and visible from, the stall or space.

(2) In a conspicuous place at each entrance to the offstreet parking facility.

(f) If the parking facility is privately owned and public parking is prohibited by the posting of a sign meeting the requirements of paragraph (1) of subdivision (a) of Section 22658, the requirements of subdivision (c) may be met by the posting of a sign immediately adjacent to, and visible from, each stall or space indicating that a vehicle not meeting the requirements of subdivision (a) will be removed at the owner's expense and containing the telephone number of the local traffic law enforcement agency.

(g) This section does not restrict the privilege granted to disabled persons and disabled veterans by Section 22511.5.

(Added by Stats 1975 ch 688 Section 3. Amended by Stats 1976 ch 1096 Section 5; Stats 1982 ch 975 Section 7; Stats 1983 ch 270 Section 3, effective July 15, 1983.)

(Amended by Stats. 1985, c. 312, Section 1; Stats. 1985, C. 1041, Section 9; Stats. 1987, c. 314, Section 3; Stats. 1989, c. 554, Section 5, Stats. 1990, c. 216 (S.B. 2510), Section 118; Stats. 1991, c. 928 (A.B. 1886), Section 27, eff. Oct. 14, 1991; Stats. 1994, c. 1149 (A.B. 2878), Section 11.) Amended by Stats, 2009, Ch. 200, Section 14, effective January 1, 2010.

Section 22511.9. Replacement signs relating to parking privileges for disabled persons; contents

Every new or replacement sign installed on or after January 1, 1992, relating to parking privileges for disabled persons shall refer to "disabled persons" rather than "physically handicapped persons" or any other similar term, whenever such a reference is required on a sign.

(Added by Stats.1991, c. 928 (A.B. 1886), Section 28, eff. October 14, 1991.)

Section 22511.55. Distinguishing placard; application; use, display, and appearance; procedures for issuance and renewal fee; eligibility; certificate substantiating disability; temporary disability; return upon death of disabled person

(a) (1) A disabled person or disabled veteran may apply to the department for the issuance of a distinguishing placard. The placard may be used in lieu of the special license plate or plates issued under Section 5007 for parking purposes described in Section 22511.5 when (A), suspended from the rearview mirror, (B) if there is no rearview mirror, when displayed on the dashboard of a vehicle, or (C) inserted in a clip designated for a distinguishing placard and installed by the manufacturer on the driver's side of the front window. It is the intent of the Legislature to encourage the use of these distinguishing placards because they provide law enforcement officers with a more readily recognizable symbol for distinguishing vehicles qualified for the parking privilege. The placard shall be the size, shape, and color determined by the department and shall bear the International Symbol of Access adopted pursuant to Section 3 of Public Law 100-641, commonly known as the "wheelchair symbol." The department shall incorporate instructions for the lawful use of a placard, and a summary of the penalties for the unlawful use of a placard, into the identification card issued to the placard owner.

(2) (A) The department may establish procedures for the issuance and renewal of the placards. The procedures shall include, but are not limited to, advising an applicant in writing on the application for the placard of the procedure to apply for a special license plate or plates, as described in Section 5007, and the fee exemptions established pursuant to Section 9105 and subdivision (a) of Section 10783 of the Revenue and Taxation Code. The placards shall have a fixed expiration date of June 30 every two years. A portion of the placard shall be printed in a contrasting color that shall be changed every two years. The size and color of this contrasting portion of the placard shall be large and distinctive enough to be readily identifiable by a law enforcement officer in a passing vehicle.

(B) As used in this section, "year" means the period between the inclusive dates of July 1 through June 30.

(C) Prior to the end of each year, the department shall, for the most current three years available, compare its record of disability placards issued against the records of the Office of Vital Statistics of the State Department of Public Health, or its successor, and withhold any renewal notices that otherwise would have been sent, for a placardholder identified as deceased.

(3) Except as provided in paragraph (4), a person shall not eligible for more than one placard at a time.

(4) Organizations and agencies involved in the transportation of disabled persons or disabled veterans may apply for a placard for each vehicle used for the purpose of transporting disabled persons or disabled veterans.

(b) (1) Except as provided in paragraph (4), prior to issuing an original distinguishing placard to a disabled person or disabled veteran, the department shall require the submission of a certificate, in accordance with paragraph (2), signed by the physician and surgeon, or to the extent that it does not cause a reduction in the receipt of federal aid highway funds, by a nurse practitioner, certified nurse midwife, or physician assistant, substantiating the disability, unless the applicant's disability is readily observable and uncontested. The disability of a person who has lost, or has lost use of, one or more lower extremities or one hand, for a disabled veteran, or both hands, for a disabled person, or who has significant limitation in the use of lower extremities, may also be certified by a licensed chiropractor. The blindness of an applicant shall be certified by a licensed physician and surgeon who specializes in diseases of the eye or a licensed optometrist. The physician and surgeon, nurse practitioner, certified nurse midwife, physician assistant, chiropractor, or optometrist certifying the qualifying disability shall provide a full description of the illness or disability on the form submitted to the department.

(2) The physician and surgeon, nurse practitioner, certified nurse midwife, physician assistant, chiropractor, or optometrist who signs a certificate submitted under this subdivision shall retain information sufficient to substantiate that certificate and, upon request of the department, shall make that information available for inspection by the Medical Board of California or the appropriate regulatory board.

(3) The department shall maintain in its records all information on an applicant's certification of permanent disability and shall make that information available to eligible law enforcement or parking control agencies upon a request pursuant to Section 22511.58.

(4) For a disabled veteran, the department shall accept, in lieu of the certificate described in paragraph (1), a certificate from the United States Department of Veterans Affairs that certifies that the applicant is a disabled veteran as described in Section 295.7.

(c) A person who is issued a distinguishing placard pursuant to subdivision (a) may apply to the department for a substitute placard without recertification of eligibility, if that placard is lost or stolen.

(d) The distinguishing placard shall be returned to the department not later than 60 days after the death of the disabled person or disabled veteran to whom the placard was issued.

(e) The department shall print on any distinguishing placard issued on or after January 1, 2005, the maximum penalty that may be imposed for a violation of Section 4461. For the purposes of this subdivision, the "maximum penalty" is the amount derived from adding all of the following:

(1) The maximum fine that may be imposed under Section 4461.

(2) The penalty required to be imposed under Section 70372 of the Government Code.

(3) The penalty required to be levied under Section 76000 of the Government Code.

(4) The penalty required to be levied under Section 1464 of the Penal Code.

(5) The surcharge required to be levied under Section 1465.7 of the Penal Code.

(6) The penalty authorized to be imposed under Section 4461.3.

Added by Stats 1991 c. 893 (A.B. 274) § 5. Amended by Stats 1991 c. 894 (S.B. 234), § 3; Stats 1993 c. 1292 (S.B. 274) § 13; Stats. 1994, c. 1149 (A.B. 2878), § 6; Stats. 1996, c. 1033 (S.B. 1498), § 1.) Amended by Stats, 2010, Ch. 491, Section 42.3, effective January 1, 2011.

Section 22522. Parking Near Designated Sidewalk Access Ramps

No person shall park a vehicle within three feet of any sidewalk access ramp constructed at, or adjacent to, a crosswalk or at any other location on a sidewalk so as to be accessible to and usable by the physically disabled, if the area adjoining the ramp is designated by either a sign or red paint.

(Added by Stats. 1974, c. 760, p. 1675, § 1.) (Amended by Stats. 1994, c. 221 (S.B. 1378), § 2, operative July 1, 1995; Stats. 1999, c. 1007 (S.B. 532), § 22.) (Amended by Stats. 1999 Ch. 1007 Sec. 22 effective January 1, 2000.).

CIVIL RIGHTS AND NON-DISCRIMINATION

CONTENTS

A.   **PART 2.5: Blind and Other Physically Disabled Persons (Civil Code 51-55.1, 1360) (Evidence Code 754)**

B.   **Discrimination (Government Code 11135-11139.5, 19230, 19231)**

C.   **Department of Fair Employment and Housing (California Government Code Section 12926 & 12926.1)**

D.   **Discrimination by Licensed Professionals (Business and Professions Code 125.6) (Insurance Code 10144-10145)**

E.   **Disabled drivers: refueling services: notice (Business and Professions Code Section 136600**

A. BLIND AND OTHER PHYSICALLY DISABLED PERSONS

Civil Code

---

§ 51. Unruh Civil Rights Act; equal rights; business establishments; violation

(a) This section shall be known, and may be cited, as the Unruh Civil Rights Act.

(b) All persons within the jurisdiction of this state are free and equal, and no matter what their sex, race, color, religion, ancestry, national origin, disability, medical condition, genetic information, marital status, or sexual orientation are entitled to the full and equal accommodations, advantages, facilities, privileges, or services in all business establishments of every kind whatsoever.

(c) This section shall not be construed to confer any right or privilege on a person that is conditioned or limited by law or that is applicable alike to persons of every sex, color, race, religion, ancestry, national origin, disability, medical condition, marital status, or sexual orientation or to persons regardless of their genetic information.

(d) Nothing in this section shall be construed to require any construction, alteration, repair, structural or otherwise, or modification of any sort whatsoever, beyond that construction, alteration, repair, or modification that is otherwise required by other provisions of law, to any new or existing establishment, facility, building, improvement, or any other structure, nor shall anything in this section be construed to augment, restrict, or alter in any way the authority of the State Architect to require construction, alteration, repair, or modifications that the State Architect otherwise possesses pursuant to other laws.

(e) For purposes of this section:

(1) "Disability" means any mental or physical disability as defined in Sections 12926 and 12926.1 of the Government Code.

(2)(A) "Genetic Information" means, with respect to any individual, information about any of the following:
(i) The individual's genetic tests.
(ii) The genetic tests of family members of the individuals
(iii) The manifestation of a disease or disorder in family members of the individual.
(B) "Genetic Information" includes any request for, or receipt of, genetic services, or participation in clinical research that includes genetic services, by an individual or any family member of the individual.
(C) "Genetic Information" does not include information about the sex or age of any individual.

(3) "Medical condition" has the same meaning as defined in subdivision (h) of Section 12926 of the Government Code.

(4) "Religion" includes all aspects of religious belief, observance, and practice.

(5) "Sex" includes, but is not limited to, pregnancy, childbirth, or medical conditions related to pregnancy or childbirth. "Sex" also includes, but is not limited to, a person's gender. "Gender" means sex, and includes a person's gender identity and gender express. "Gender expression" means a person's gender-related appearance and behavior whether or not stereotypically associated with the person's assigned sex at birth.

(6) "Sex, race, color, religion, ancestry, national origin, disability, medical condition, genetic information, marital status, or sexual orientation" includes a perception that the person has any particular characteristic or characteristics within the listed categories or that the person is associated with a person who has, or is perceived to have, any particular characteristic or characteristics within the listed categories.

(7) "Sexual orientation" has the same meaning as defined in subdivision (r) of Section 12926 of the Government Code.

(f) A violation of the right of any individual under the federal Americans with Disabilities Act of 1990 (Public Law 101-336) shall also constitute a violation of this section.

(Added by Stats. 1905, c. 413, p. 553, §. 1. Amended by Stats 1919 c. 210, p. 309, . 1; Stats 1923 ch 235, p. 485, § 1; Stats. 1959 c. 1866. P. 4424, § 1; Stats 1961 c. 1187 , p. 2920, § 1; Stats 1974 c. 1193, p. 2568, § 1.) . Amended by Stats. 1987, c. 159, Section 1; Stats. 1992, c. 913 (A.B. 1077), Section 3; Stats. 1998, c. 195 (A.B. 2702), Section 1; Stats. 2000, c. 1049 (A.B. 2222), Section 2, Amended by Stats. 2011, Ch. 719, Section 1.5 effective January 1, 2012.

§ 51.2. Age discrimination in housing prohibited; exception; intent

(a) Section 51 shall be construed to prohibit a business establishment from discriminating in the sale or rental of housing based upon age. Where accommodations are designed to meet the physical and social needs of senior citizens, a business establishment may establish and preserve that housing for senior citizens, pursuant to Section 51.3, except housing as to which Section 51.3 is preempted by the prohibition in the federal Fair Housing Amendments Act of 1988 (P.L. 100-430) and implementing regulations against discrimination on the basis of familial status. For accommodations constructed before February 8, 1982, that meet all the criteria for senior citizen housing specified in Section 51.3, a business establishment may establish and preserve that housing development for senior citizens without the housing development being designed to meet physical and social needs of senior citizens.

(b) This section is intended to clarify the holdings in Marina Point, Ltd. v. Wolfson (1982) 30 Cal. 3d 72 and O'Connor v. Village Green Owners Association (1983) 33 Cal. 3d 790.

(c) This section shall not apply to the County of Riverside.

(d) A housing development for senior citizens constructed on or after January 1, 2001, shall be presumed to be designed to meet the physical and social needs of senior citizens if it includes all of the following elements:

(1) Entryways, walkways, and hallways in the common areas of the development, and doorways and paths of access to and within the housing units, shall be as wide as required by current laws applicable to new multifamily housing construction for provision of access to persons using a standard-width wheelchair.

(2) Walkways and hallways in the common areas of the development shall be equipped with standard height railings or grab bars to assist persons who have difficulty with walking.

(3) Walkways and hallways in the common areas shall have lighting conditions which are of sufficient brightness to assist persons who have difficulty seeing.

(4) Access to all common areas and housing units within the development shall be provided without use of stairs, either by means of an elevator or sloped walking ramps.

(5) The development shall be designed to encourage social contact by providing at least one common room and at least some common open space.

(6) Refuse collection shall be provided in a manner that requires a minimum of physical exertion by residents.

(7) The development shall comply with all other applicable requirements for access and design imposed by law, including, but not limited to, the Fair Housing Act (42 U.S.C. Sec. 3601 et seq.), the Americans with Disabilities Act (42 U.S.C. Sec. 12101 et seq.), and the regulations promulgated at Title 24 of the California Code of Regulations that relate to access for persons with

disabilities or handicaps. Nothing in this section shall be construed to limit or reduce any right or obligation applicable under those laws.

(e) Selection preferences based on age, imposed in connection with a federally approved housing program, do not constitute age discrimination in housing.

(Added by Stats. 1984, c. 787, § 1. Amended by Stats. 1989, c. 501, § 1; Stats. 1993, c. 830 (S.B. 137), § 1, eff. Oct. 6, 1993; Stats. 1996, c. 1147 (S.B. 2097), § 2; Stats. 1999, c. 324 (S.B. 382), § 1. Stats. 2000, c. 1004 (S.B. 2011), Section 2. Amended by Stats, 2010, Ch. 524, Section 2, effective January 1, 2011.
[1]See Short Title note under 42 U.S.C.A. § 3601 for classification of the Act to the Code.
[2]So in chaptered copy. See, 30 Cal.3d 721, 640 P.2d 115, 180 Cal.Rptr. 496.

§ 51.3. Housing; age limitations; necessity for senior citizen housing

(a) The Legislature finds and declares that this section is essential to establish and preserve specially designed accessible housing for senior citizens. There are senior citizens who need special living environments and services, and find that there is an inadequate supply of this type of housing in the state.

(b) For the purposes of this section, the following definitions apply:

(1) "Qualifying resident" or "senior citizen" means a person 62 years of age or older, or 55 years of age or older in a senior citizen housing development.

(2) "Qualified permanent resident" means a person who meets both of the following requirements:

(A) Was residing with the qualifying resident or senior citizen prior to the death, hospitalization, or other prolonged absence of, or the dissolution of marriage with, the qualifying resident or senior citizen.

(B) Was 45 years of age or older, or was a spouse, cohabitant, or person providing primary physical or economic support to the qualifying resident or senior citizen.

(3) "Qualified permanent resident" also means a disabled person or person with a disabling illness or injury who is a child or grandchild of the senior citizen or a qualified permanent resident as defined in paragraph (2) who needs to live with the senior citizen or qualified permanent resident because of the disabling condition, illness, or injury. For purposes of this section, "disabled" means a person who has a disability as defined in subdivision (b) of Section 54. A "disabling injury or illness" means an illness or injury which results in a condition meeting the definition of disability set forth in subdivision (b) of Section 54.

(A) For any person who is a qualified permanent resident under this paragraph whose disabling condition ends, the owner, board of directors, or other governing body may require the formerly disabled resident to cease residing in the development upon receipt of six months' written notice; provided, however, that the owner, board of directors, or other governing body may allow the person to remain a resident for up to one year after the disabling condition ends.

(B) The owner, board of directors, or other governing body of the senior citizen housing development may take action to prohibit or terminate occupancy by a person who is a qualified permanent resident under this paragraph if the owner, board of directors, or other governing body finds, based on credible and objective evidence, that the person is likely to pose a significant threat to the health or safety of others that cannot be ameliorated by means of a reasonable accommodation; provided, however, that the action to prohibit or terminate the occupancy may be taken only after doing both of the following:

(i) Providing reasonable notice to and an opportunity to be heard for the disabled person whose occupancy is being challenged, and reasonable notice to the coresident parent or grandparent of that person.

(ii) Giving due consideration to the relevant, credible, and objective information provided in the hearing. The evidence shall be taken and held in a confidential manner, pursuant to a closed session, by the owner, board of directors, or other governing body in order to preserve the privacy of the affected persons. The affected persons shall be entitled to have present at the hearing an attorney or any other person authorized by them to speak on their behalf or to assist them in the matter.

(4) "Senior citizen housing development" means a residential development developed, substantially rehabilitated, or substantially renovated for, senior citizens that has at least 35 dwelling units. Any senior citizen housing development which is required to obtain a public report under Section 11010 of the Business and Professions Code and which submits its application for a public report after July 1, 2001, shall be required to have been issued a public report as a senior citizen housing development under Section 11010.05 of the Business and Professions Code. No housing development constructed prior to January 1, 1985, shall fail to qualify as a senior citizen housing development because it was not originally developed or put to use for occupancy by senior citizens.

(5) "Dwelling unit" or "housing" means any residential accommodation other than a mobilehome.
(6) "Cohabitant" refers to persons who live together as husband and wife, or persons who are domestic partners within the meaning of Section 297 of the Family Code.

(7) "Permitted health care resident" means a person hired to provide live-in, long-term, or terminal health care to a qualifying resident, or a family member of the qualifying resident providing that care. For the purposes of this section, the care provided by a permitted health care resident must be substantial in nature and must provide either assistance with necessary daily activities or medical treatment, or both. A permitted health care resident shall be entitled to continue his or her occupancy, residency, or use of the dwelling unit as a permitted resident in the absence of the senior citizen from the dwelling unit only if both of the following are applicable:

(A) The senior citizen became absent from the dwelling due to hospitalization or other necessary medical treatment and expects to return to his or her residence within 90 days from the date the absence began.

(B) The absent senior citizen or an authorized person acting for the senior citizen submits a written request to the owner, board of directors, or governing board stating that the senior citizen desires that the permitted health care resident be allowed to remain in order to be present when the senior citizen returns to reside in the development. Upon written request by the senior citizen or an authorized person acting for the senior citizen, the owner, board of directors, or governing board shall have the discretion to allow a permitted health care resident to remain for a time period longer than 90 days from the date that the senior citizen's absence began, if it appears that the senior citizen will return within a period of time not to exceed an additional 90 days.

(c) The covenants, conditions, and restrictions and other documents or written policy shall set forth the limitations on occupancy, residency, or use on the basis of age. Any such limitation shall not be more exclusive than to require that one person in residence in each dwelling unit may be required to be a senior citizen and that each other resident in the same dwelling unit may be required to be a qualified permanent resident, a permitted health care resident, or a person under 55 years of age whose occupancy is permitted under subdivision (h) of this section or under subdivision (b) of Section 51.4. That limitation may be less exclusive, but shall at least require that the persons commencing any occupancy of a dwelling unit include a senior citizen who intends to reside in the unit as his or her primary residence on a permanent basis. The application of the rules set forth in this subdivision regarding limitations on occupancy may result in less than all of the dwellings being actually occupied by a senior citizen.

(d) The covenants, conditions, and restrictions or other documents or written policy shall permit temporary residency, as a guest of a senior citizen or qualified permanent resident, by a person of less than 55 years of age for periods of time, not less than 60 days in any year, that are specified in the covenants, conditions, and restrictions or other documents or written policy.

(e) Upon the death or dissolution of marriage, or upon hospitalization, or other prolonged absence of the qualifying resident, any qualified permanent resident shall be entitled to continue his or her occupancy, residency, or use of the dwelling unit as a permitted resident. This subdivision shall not apply to a permitted health care resident.

(f) The condominium, stock cooperative, limited-equity housing cooperative, planned development, or multiple-family residential rental property shall have been developed for, and initially been put to use as, housing for senior citizens, or shall have been substantially rehabilitated or renovated for, and immediately afterward put to use as, housing for senior citizens, as provided in this section; provided, however, that no housing development constructed prior to January 1, 1985, shall fail to qualify as a senior citizen housing development because it was not originally developed for or originally put to use for occupancy by senior citizens.

(g) The covenants, conditions, and restrictions or other documents or written policies applicable to any condominium, stock cooperative, limited-equity housing cooperative, planned development, or multiple-family residential property that contained age restrictions on January 1, 1984, shall be enforceable only to the extent permitted by this section, notwithstanding lower age restrictions contained in those documents or policies.

(h) Any person who has the right to reside in, occupy, or use the housing or an unimproved lot subject to this section on January 1, 1985, shall not be deprived of the right to continue that residency, occupancy, or use as the result of the enactment of this section.

(i) The covenants, conditions, and restrictions or other documents or written policy of the senior citizen housing development shall permit the occupancy of a dwelling unit by a permitted health care resident during any period that the person is actually providing live-in, long-term, or hospice health care to a qualifying resident for compensation. For purposes of this subdivision, the term "for compensation" shall include provisions of lodging and food in exchange for care.

(j) Notwithstanding any other provision of this section, this section shall not apply to the County of Riverside.

(Added by Stats. 1984), c. 1333, § 1. Amended by Stats. 1985, c. 1505, § 2; Stats. 1989, c. 190, § 1; Stats. 1994, c. 464 (S.B. 1560), § 1; Stats. 1995, c. 147 (S.B. 332), § 1; Stats. 1996, c. 1147 (S.B. 2097), § 3; Stats. 1999, c. 324 (S.B. 382), § 2.) Stats. 2000, c. 1004 (S.B. 2011), Section 3.

§ 51.4. Exemption from special design requirement
(a) The Legislature finds and declares that the requirements for senior housing under Sections 51.2 and 51.3 are more stringent than the requirements for that housing under the federal Fair Housing Amendments Act of 1988 (P.L. 100-430) in recognition of the acute shortage of housing for families with children in California. The Legislature further finds and declares that the special design requirements for senior housing under Sections 51.2 and 51.3 may pose a hardship to some housing developments that were constructed before the decision in Marina Point, Ltd. v. Wolfson (1982) 30 Cal.3d 721. The Legislature further finds and declares that the requirement for specially designed accommodations in senior housing under Sections 51.2 and 51.3 provides important benefits to senior citizens and also ensures that housing exempt from the prohibition of age discrimination is carefully tailored to meet the compelling societal interest in providing senior housing.

(b) Any person who resided in, occupied, or used, prior to January 1, 1990, a dwelling in a senior citizen housing development that relied on the exemption to the special design requirement provided by this section prior to January 1, 2001, shall not be deprived of the right to continue that residency, occupancy, or use as the result of the changes made to this section by the enactment of Chapter 1004 of the Statutes of 2000.

(c) This section shall not apply to the County of Riverside.

(Added by Stats. 1989, c. 501, § 2. Amended by Stats. 1991, c. 59 (A.B. 125), § 1, eff. June 17, 1991; Stats. 1996, c. 1147 (S.B. 2097), § 4.).   Stats. 2000, c. 1004 (S.B. 2011(Section 4). Amended by Stats. 2006, Ch. 538, Section 37, effective January 1, 2007.
₁See Short Title note under 42 U.S.C.A. § 3601 for classification of the Act to the Code.
₂So in enrolled bill.

§ 51.5. Discrimination, boycott, blacklist, etc.; business establishments; equal rights
(a) No business establishment of any kind whatsoever shall discriminate against, boycott or blacklist, or refuse to buy from, contract with, sell to, or trade with any person in this state on account of any characteristic listed or defined in subdivision (b) or (e) of Section 51, or of the person's partners, members, stockholders, directors, officers, managers, superintendents, agents, employees, business associates, suppliers, or customers, because the person is perceived to have one or more of those characteristics, or because the person is associated with a person who has, or is perceived to have, any of those characteristics.

(b) As used in this section, "person" includes any person, firm, association, organization, partnership, business trust, corporation, limited liability company, or company.

(c) This section shall not be construed to require any construction, alteration, repair, structural or otherwise, or modification of any sort whatsoever, beyond that construction, alteration, repair, or modification that is otherwise required by other provisions of law, to any new or existing establishment, facility, building, improvement, or any other structure, nor shall this section be construed to augment, restrict, or alter in any way the authority of the State Architect to require construction, alteration, repair, or modifications that the State Architect otherwise possesses pursuant to other laws.

(Added by Stats 1976 c. 366 § 1.) (Amended by Stats 1987 c.159 § 2; Stats 1992 c. 913 (A.B. 1077) § 3.2.; Stats. 1994, c 1010 (S.B. 2053), § 28; Stats. 1998, c. 195 (A.B. 2702), § 2; Stats. 1999, c. 591 (A.B. 1670), § 2.); Stats. 2000, c. 1049 (A.B. 2222), Section 3.) Amended by Stats. 2005, Ch. 420, Section 4, effective January 1, 2006.

§ 51.7. Freedom from violence or intimidation
(a) All persons within the jurisdiction of this state have the right to be free from any violence, or intimidation by threat of violence, committed against their persons or property because of political affiliation, or on account of any characteristic listed or defined in subdivision (b) or (e) of Section 51, or position in a labor dispute, or because another person perceives them to have one or more of those characteristics. The identification in this subdivision of particular bases of discrimination is illustrative rather than restrictive.

(b) This section does not apply to statements concerning positions in a labor dispute which are made during otherwise lawful labor picketing.

(Added by Stats 1976 ch 1293 Section 2.) (Amended by Stats 1984 ch 1437 Section 1; Stats 1985 ch 497 Section 1; Stats 1987 ch 1277 Section 2; Stats. 1994, c. 407 (S.B.1595), Section 1.) Amended by Stats. 2005, Ch. 420, Section 5, effective January 1, 2006.

§ 51.8. Discrimination; franchises

(a) No franchisor shall discriminate in the granting of franchises solely on account of any characteristic listed or defined in subdivision (b) or (e) of Section 51 of the franchisee and the composition of a neighborhood or geographic area reflecting any characteristic listed or defined in subdivision (b) or (e) of Section 51 in which the franchise is located. Nothing in this section shall be interpreted to prohibit a franchisor from granting a franchise to prospective franchisees as part of a program or programs to make franchises available to persons lacking the capital, training, business experience, or other qualifications ordinarily required of franchisees, or any other affirmative action program adopted by the franchisor.

(b) Nothing in this section shall be construed to require any construction, alteration, repair, structural or otherwise, or modification of any sort whatsoever, beyond that construction, alteration, repair, or modification that is otherwise required by other provisions of law, to any new or existing establishment, facility, building, improvement, or any other structure, nor shall anything in this section be construed to augment, restrict, or alter in any way the authority of the State Architect to require construction, alteration, repair, or modifications that the State Architect otherwise possesses pursuant to other laws.

Added by Stats 1980 c. 1303 § 1. Amended by Stats 1987 c. 159 § 3; Stats 1992 c. 913 (A.B. 1077) § 3.4; Stats. 1998, c. 195 (A.B. 2702), § 3.) Amended by Stats. 2005, Ch. 420, Section 6, effective January 1, 2006.

§ 52. Denial of civil rights or discrimination; damages; civil action by people or person aggrieved; intervention; unlawful practice complaint

(a) Whoever denies, aids or incites a denial, or makes any discrimination or distinction contrary to Section 51, 51.5, or 51.6, is liable for each and every offense for the actual damages, and any amount that may be determined by a jury, or a court sitting without a jury, up to a maximum of three times the amount of actual damage but in no case less than four thousand dollars ($4,000), and any attorney's fees that may be determined by the court in addition thereto, suffered by any person denied the rights provided in Section 51, 51.5, or 51.6.

(b) Whoever denies the right provided by Section 51.7 or 51.9, or aids, incites, or conspires in that denial, is liable for each and every offense for the actual damages suffered by any person denied that right and, in addition, the following:

(1) An amount to be determined by a jury, or a court sitting without a jury, for exemplary damages.

(2) A civil penalty of twenty-five thousand dollars ($25,000) to be awarded to the person denied the right provided by Section 51.7 in any action brought by the person denied the right, or by the Attorney General, a district attorney, or a city attorney. An action for that penalty brought pursuant to Section 51.7 shall be commenced within three years of the alleged practice.

(3) Attorney's fees as may be determined by the court.

(c) Whenever there is reasonable cause to believe that any person or group of persons is engaged in conduct of resistance to the full enjoyment of any of the rights described in this section, and that conduct is of that nature and is intended to deny the full exercise of those rights, the Attorney General, any district attorney or city attorney, or any person aggrieved by the conduct may bring a civil action in the appropriate court by filing with it a complaint. The complaint shall contain the following:

(1) The signature of the officer, or, in his or her absence, the individual acting on behalf of the officer, or the signature of the person aggrieved.

(2) The facts pertaining to the conduct.

(3) A request for preventive relief, including an application for a permanent or temporary injunction, restraining order, or other order against the person or persons responsible for the conduct, as the complainant deems necessary to ensure the full enjoyment of the rights described in this section.

(d) Whenever an action has been commenced in any court seeking relief from the denial of equal protection of the laws under the Fourteenth Amendment to the Constitution of the United States on account of race, color, religion, sex, national origin, or disability, the Attorney General or any district attorney or city attorney for or in the name of the people of the State of California may intervene in the action upon timely application if the Attorney General or any district attorney or city attorney certifies that the case is of general public importance. In that action, the people of the State of California shall be entitled to the same relief as if it had instituted the action.

(e) Actions brought pursuant to this section are independent of any other actions, remedies, or procedures that may be available to an aggrieved party pursuant to any other law.

(f) Any person claiming to be aggrieved by an alleged unlawful practice in violation of Section 51 or 51.7 may also file a verified complaint with the Department of Fair Employment and Housing pursuant to Section 12948 of the Government Code.

(g) This section does not require any construction, alteration, repair, structural or otherwise, or modification of any sort whatsoever, beyond that construction, alteration, repair, or modification that is otherwise required by other provisions of law, to any new or existing establishment, facility, building, improvement, or any other structure, nor does this section augment, restrict, or alter in any way the authority of the State Architect to require construction, alteration, repair, or modifications that the State Architect otherwise possesses pursuant to other laws.

(h) For the purposes of this section, "actual damages" means special and general damages. This subdivision is declaratory of existing law.

(Added by Stats 1905 c. 413 . 2. Amended by Stats 1919 ch 210 § 2; Stats 1923 c. 235 § 2; Stats 1959 c. 1866 § 2; Stats 1974 c. 1193 §, 2; Stats 1976 c. 366 § 2; Stats 1976 c. 1293 § 2.5: Stats 1978 c. 1212 § 1; Stats 1981 c. 521 § 1, effective September 16, 1981; Stats 1986 c. 244 § 1; Stats 1987 c. 159 § 4; Stats 1989 c. 459 § 1; Stats 1991 c. 607 (S.B. 98) § 2; Stats 1991 c. 839 (A.B. 1169) § 2; Stats 1992 c. 913 (A.B. 1077) § 3.6; Stats. 1994, c. 535 (S.B. 1288), Section 1; Stats 1998, c.195 (A.B. 2702), § 4; Stats. 1999, c. 964 (A.B. 519), § 2.); Stats. 2000, c. 98 (A.B. 2719), Section 2; Stats. 2001, c. 261 (A.B. 587), Section 1.) Amended by Stats. 2005, Ch. 123, Section 1, effective January 1, 2006.

§ 53. Restrictions upon transfer or use of realty because of sex, race, color, religion, ancestry, national origin, or disability.

(a) Every provision in a written instrument relating to real property that purports to forbid or restrict the conveyance, encumbrance, leasing, or mortgaging of that real property to any person because of any characteristic listed or defined in subdivision (b) or (e) of Section 51 is void, and every restriction or prohibition as to the use or occupation of real property because of any characteristic listed or defined in subdivision (b) or (e) of Section 51 is void.

(b) Every restriction or prohibition, whether by way of covenant, condition upon use or occupation, or upon transfer of title to real property, which restriction or prohibition directly or indirectly limits the acquisition, use or occupation of that property because of any characteristic listed or defined in subdivision (b) or (e) of Section 51 is void.

(c) In any action to declare that a restriction or prohibition specified in subdivision (a) or (b) is void, the court shall take judicial notice of the recorded instrument or instruments containing the prohibitions or restrictions in the same manner that it takes judicial notice of the matters listed in Section 452 of the Evidence Code.

Added by Stats 1961 ch 1877 Section 1. Amended by Stats 1965 ch 299 Section 6, operative January 1, 1967; Stats 1974 ch 1193 Section 3; Stats 1987 ch 159 Section 5; Stats 1992 ch 913 (A.B. 1077) Section 3.8. Amended by Stats 2005, Chj. 420, Section 7, effective January 1, 2006.

Section 54. Right to streets, highways, and other public places; disability

(a) Individuals with disabilities or medical conditions have the same right as the general public to the full and free use of the streets, highways, sidewalks, walkways, public buildings, medical facilities, including hospitals, clinics, and physicians' offices, public facilities, and other public places.

(b) For purposes of this section:

(1) "Disability" means any mental or physical disability as defined in Section 12926 of the Government Code.

(2) "Medical condition" has the same meaning as defined in subdivision (h) of Section 12926 of the Government Code.

(c) A violation of the right of an individual under the Americans with Disabilities Act of 1990 (Public Law 101-336) also constitutes a violation of this section.

Added by Stats 1968 ch 461 Section 1. Amended Stats 1992 ch 913 (A.B. 1077) Section 4; Stats. 1994, c. 1257 (S.B. 1240), Section 1; Stats. 1996, c. 498 (S.B. 1687), Section 1.) Stats. 2000, c. 1049 (A.B. 2222), Section 4.)

Section 54.1. Access to public conveyances, places of public accommodation, amusement or resort, and housing accommodations

(a) (1) Individuals with disabilities shall be entitled to full and equal access, as other members of the general public, to accommodations, advantages, facilities, medical facilities, including hospitals, clinics, and physicians' offices, and privileges of all common carriers, airplanes, motor vehicles, railroad trains, motorbuses, streetcars, boats, or any other public conveyances or modes of transportation (whether private, public, franchised, licensed, contracted, or otherwise provided), telephone facilities, adoption agencies, private schools, hotels, lodging places, places of public accommodation, amusement, or resort, and other places to which the general public is invited, subject only to the conditions and limitations established by law, or state or federal regulation, and applicable alike to all persons.

(2) As used in this section, "telephone facilities" means tariff items and other equipment and services that have been approved by the Public Utilities Commission to be used by individuals with disabilities in a manner feasible and compatible with the existing telephone network provided by the telephone companies.

(3) "Full and equal access," for purposes of this section in its application to transportation, means access that meets the standards of Titles II and III of the Americans with Disabilities Act of 1990 (Public Law 101-336) and federal regulations adopted pursuant thereto, except that, if the laws of this state prescribe higher standards, it shall mean access that meets those higher standards.

(b) (1) Individuals with disabilities shall be entitled to full and equal access, as other members of the general public, to all housing accommodations offered for rent, lease, or compensation in this state, subject to the conditions and limitations established by law, or state or federal regulation, and applicable alike to all persons.

(2) "Housing accommodations" means any real property, or portion thereof, that is used or occupied, or is intended, arranged, or designed to be used or occupied, as the home, residence, or sleeping place of one or more human beings, but shall not include any accommodations included

within subdivision (a) or any single-family residence the occupants of which rent, lease, or furnish for compensation not more than one room therein.

(3) (A) Any person renting, leasing, or otherwise providing real property for compensation shall not refuse to permit an individual with a disability, at that person's expense, to make reasonable modifications of the existing rented premises if the modifications are necessary to afford the person full enjoyment of the premises. However, any modifications under this paragraph may be conditioned on the disabled tenant entering into an agreement to restore the interior of the premises to the condition existing prior to the modifications. No additional security may be required on account of an election to make modifications to the rented premises under this paragraph, but the lessor and tenant may negotiate, as part of the agreement to restore the premises, a provision requiring the disabled tenant to pay an amount into an escrow account, not to exceed a reasonable estimate of the cost of restoring the premises.

(B) Any person renting, leasing, or otherwise providing real property for compensation shall not refuse to make reasonable accommodations in rules, policies, practices, or services, when those accommodations may be necessary to afford individuals with a disability equal opportunity to use and enjoy the premises.

(4) Nothing in this subdivision shall require any person renting, leasing, or providing for compensation real property to modify his or her property in any way or provide a higher degree of care for an individual with a disability than for an individual who is not disabled.

(5) Except as provided in paragraph (6), nothing in this part shall require any person renting, leasing, or providing for compensation real property, if that person refuses to accept tenants who have dogs, to accept as a tenant an individual with a disability who has a dog.

(6) (A) It shall be deemed a denial of equal access to housing accommodations within the meaning of this subdivision for any person, firm, or corporation to refuse to lease or rent housing accommodations to an individual who is blind or visually impaired on the basis that the individual uses the services of a guide dog, an individual who is deaf or hearing impaired on the basis that the individual uses the services of a signal dog, or to an individual with any other disability on the basis that the individual uses the services of a service dog, or to refuse to permit such an individual who is blind or visually impaired to keep a guide dog, an individual who is deaf or hearing impaired to keep a signal dog, or an individual with any other disability to keep a service dog on the premises.

(B) Except in the normal performance of duty as a mobility or signal aid, nothing contained in this paragraph shall be construed to prevent the owner of a housing accommodation from establishing terms in a lease or rental agreement that reasonably regulate the presence of guide dogs, signal dogs, or service dogs on the premises of a housing accommodation, nor shall this paragraph be construed to relieve a tenant from any liability otherwise imposed by law for real and personal property damages caused by such a dog when proof of the same exists.

(C) (i) As used in this subdivision, "guide dog" means any guide dog that was trained by a person licensed under Chapter 9.5 (commencing with Section 7200) of Division 3 of the Business and Professions Code or as defined in the regulations implementing Title III of the Americans with Disabilities Act of 1990 (Public Law 101-336).

(ii) As used in this subdivision, "signal dog" means any dog trained to alert an individual who is deaf or hearing impaired to intruders or sounds.

(iii) As used in this subdivision, "service dog" means any dog individually trained to the requirements of the individual with a disability, including, but not limited to, minimal protection work, rescue work, pulling a wheelchair, or fetching dropped items.

(7) It shall be deemed a denial of equal access to housing accommodations within the meaning of this subdivision for any person, firm, or corporation to refuse to lease or rent housing accommodations to an individual who is blind or visually impaired, an individual who is deaf or hearing impaired, or other individual with a disability on the basis that the individual with a disability is partially or wholly dependent upon the income of his or her spouse, if the spouse is a party to the lease or rental agreement. Nothing in this subdivision, however, shall prohibit a lessor or landlord from considering the aggregate financial status of an individual with a disability and his or her spouse.

(c) Visually impaired or blind persons and persons licensed to train guide dogs for individuals who are visually impaired or blind pursuant to Chapter 9.5 (commencing with Section 7200) of Division 3 of the Business and Professions Code or guide dogs as defined in the regulations implementing Title III of the Americans with Disabilities Act of 1990 (Public Law 101-336), and persons who are deaf or hearing impaired and persons authorized to train signal dogs for individuals who are deaf or hearing impaired, and other individuals with a disability and persons authorized to train service dogs for individuals with a disability, may take dogs, for the purpose of training them as guide dogs, signal dogs, or service dogs in any of the places specified in subdivisions (a) and (b). These persons shall ensure that the dog is on a leash and tagged as a guide dog, signal dog, or service dog by identification tag issued by the county clerk, animal control department, or other agency, as authorized by Chapter 3.5 (commencing with Section 30850) of Division 14 of the Food and Agricultural Code. In addition, the person shall be liable for any provable damage done to the premises or facilities by his or her dog.

(d) A violation of the right of an individual under the Americans with Disabilities Act of 1990 (Public Law 101-336) also constitutes a violation of this section, and nothing in this section shall be construed to limit the access of any person in violation of that act.

(e) Nothing in this section shall preclude the requirement of the showing of a license plate or disabled placard when required by enforcement units enforcing disabled persons parking violations pursuant to Sections 22507.8 and 22511.8 of the Vehicle Code.

Added by Stats 1968 ch 461 Section 1. Amended by Stats 1969 ch 832 Section 1; Stats 1972 ch 819 Section 1; Stats 1974 ch 108 Section 1; Stats 1976 ch 971 Section 1; Stats 1976 ch 972 Section 1.5; Stats 1977 ch 700 Section 1; Stats 1978 ch 380 Section 2; Stats 1979 ch 293 Section 1; Stats 1980 773 Section 1; Stats 1988 ch 1595 Section 2; Stats 1992 ch 913 (A.B. 1077) Section 5; Stats 1993 ch 1149 (A.B. 1419) Section 4; Stats 1993 ch 1214 (A.B. 551) Section 1.5; Stats. 1994, c. 1257 (S.B. 1240), Section 2; Stats. 1996, c. 498 (S.B. 1687), Section 1.5.)   ,42 U.S.C.A. Section 12101 et seq.

Section 54.2. Guide, signal or service dogs; right to accompany individuals with a disability and trainers; damages
(a) Every individual with a disability has the right to be accompanied by a guide dog, signal dog, or service dog, especially trained for the purpose, in any of the places specified in Section 54.1 without being required to pay an extra charge or security deposit for the guide dog, signal dog, or service dog. However, the individual shall be liable for any damage done to the premises or facilities by his or her dog.

(b) Individuals who are blind or otherwise visually impaired and persons licensed to train guide dogs for individuals who are blind or visually impaired pursuant to Chapter 9.5 (commencing with Section 7200) of Division 3 of the Business and Professions Code or as defined in regulations implementing Title III of the Americans with Disabilities Act of 1990 (Public Law 101-336), and individuals who are deaf or hearing impaired and persons authorized to train signal dogs for individuals who are deaf or hearing impaired, and individuals with a disability and persons who are authorized to train service dogs for the individuals with a disability may take dogs, for the purpose of training them as guide dogs, signal dogs, or service dogs in any of the places specified in Section 54.1 without being required to pay an extra charge or security deposit for the guide dog, signal dog, or service dog. However, the person shall be liable for any damage done to the premises or facilities by his or her dog. These persons shall ensure the dog is on a leash and tagged as a guide dog, signal dog, or service dog by an identification tag issued by the county clerk, animal control department, or other agency, as authorized by Chapter 3.5 (commencing with Section 30850) of Title 14 of the Food and Agricultural Code. A violation of the right of an individual under the Americans with Disabilities Act of 1990 (Public Law 101-336) also constitutes a violation of this section, and nothing in this section shall be construed to limit the access of any person in violation of that act.

(c) As used in this section, the terms "guide dog," "signal dog," and "service dog" have the same meanings as specified in Section 54.1.

(d) Nothing in this section precludes the requirement of the showing of a license plate or disabled placard when required by enforcement units enforcing disabled persons parking violations pursuant to Sections 22507.8 and 22511.8 of the Vehicle Code.

Added by Stats 1968 ch 461 Section 1. Amended by Stats 1972 ch 819 Section 2; Stats 1979 ch 293 Section 2; Stats 1980 ch 773 Section 2; Stats 1988 ch 1595 Section 3 Stats 1992 ch 913 (A.B. 1077) Section 6; Stats. 1994, c. 1257 (S.B. 1240), Section 3; Stats. 1996, c. 498 (S.B. 1687(, Section 2. ₁42 U.S.C.A. Section 12101 et seq.

Section 54.3. Violations; liability

(a) Any person or persons, firm or corporation who denies or interferes with admittance to or enjoyment of the public facilities as specified in Sections 54 and 54.1 or otherwise interferes with the rights of an individual with a disability under Sections 54, 54.1 and 54.2 is liable for each offense for the actual damages and any amount as may be determined by a jury, or the court sitting without a jury, up to a maximum of three times the amount of actual damages but in no case less than one thousand dollars ($1,000), and attorney' s fees as may be determined by the court in addition thereto, suffered by any person denied any of the rights provided in Sections 54, 54.1, and 54.2. "Interfere," for purposes of this section, includes, but is not limited to, preventing or causing the prevention of a guide dog, signal dog, or service dog from carrying out its functions in assisting a disabled person.

(b) Any person who claims to be aggrieved by an alleged unlawful practice in violation of Section 54, 54.1, or 54.2 may also file a verified complaint with the Department of Fair Employment and Housing pursuant to Section 12948 of the Government Code. The remedies in this section are nonexclusive and are in addition to any other remedy provided by law, including, but not limited to, any action for injunctive or other equitable relief available to the aggrieved party or brought in the name of the people of this state or of the United States.

(c) A person may not be held liable for damages pursuant to both this section and Section 52 for the same act or failure to act.

(Added by Stats 1968 ch 461 Section 1. Amended by Stats 1976 ch 971 Section 2; Stats 1976 ch 972 Section 2.5; Stats 1977 ch 881 Section 1; Stats 1981 ch 395 Section 1.)
(Amended by Stats 1992 c 913 (A.B. 1077) Section 7; Stats 1994, c. 1257 (S.B. 1240), Section 4; Stats. 1996, c. 498 (S.B. 1687), Section 2.3.)

Section 54.4. Blind pedestrian; failure to carry white cane

A blind or otherwise visually impaired pedestrian shall have all of the rights and privileges conferred by law upon other persons in any of the places, accommodations, or conveyances specified in Sections 54 and 54.1, notwithstanding the fact that the person is not carrying a predominantly white cane (with or without a red tip), or using a guide dog. The failure of a blind or otherwise visually impaired person to carry such a cane or to use such a guide dog shall not constitute negligence per se.

(Added by Stats 1968 ch 461 Section 1.)
(Amended by Stats 1994, c. 1257 (S.B. 1240), Section 5.)

Section 54.5. White cane safety day; proclamation
Each year, the Governor shall publicly proclaim October 15 as White Cane Safety Day. He or she shall issue a proclamation in which:

(a) Comments shall be made upon the significance of this chapter.

(b) Citizens of the state are called upon to observe the provisions of this chapter and to take precautions necessary to the safety of disabled persons.

(c) Citizens of the state are reminded of the policies with respect to disabled persons declared in this chapter and he urges the citizens to cooperate in giving effect to them.

(d) Emphasis shall be made on the need of the citizenry to be aware of the presence of disabled persons in the community and to keep safe and functional for the disabled the streets, highways, sidewalks, walkways, public buildings, public facilities, other public places, places of public accommodation, amusement and resort, and other places to which the public is invited, and to offer assistance to disabled persons upon appropriate occasions.

(e) It is the policy of this state to encourage and enable disabled persons to participate fully in the social and economic life of the state and to engage in remunerative employment.

(Added by Stats 1968 ch 461 Section 1.) (Amended by Stats. 1994, c. 1257 (S.B. 1240), Section 6.)

Section 54.6. Visually impaired
As used in this part, "visually impaired" includes blindness and means having central visual acuity not to exceed 20/200 in the better eye, with corrected lenses, as measured by the Snellen test, or visual acuity greater than 20/200, but with a limitation in the field of vision such that the widest diameter of the visual field subtends an angle not greater than 20 degrees.

(Added by Stats 1968, c. 461, p. 1092, Section 1.) Amended by Stats 1994, c. 1257 (S.B. 1240), Section 7.)
₁So in chaptered copy. Amewnded by Stats. 2006, Ch. 538, Section 38, effective January 1, 2007.

Section 54.7. Zoos or wild animal parks; facilities for guide, service or signal dogs accompanying individuals with a disability
(a) Notwithstanding any other provision of law, the provisions of this part shall not be construed to require zoos or wild animal parks to allow guide dogs, signal dogs, or service dogs to accompany individuals with a disability in areas of the zoo or park where zoo or park animals are not separated from members of the public by a physical barrier. As used in this section, "physical barrier" does not include an automobile or other conveyance.

(b) Any zoo or wild animal park that does not permit guide dogs, signal dogs, or service dogs to accompany individuals with a disability therein shall maintain, free of charge, adequate kennel facilities for the use of guide dogs, signal dogs, or service dogs belonging to these persons. These facilities shall be of a character commensurate with the anticipated daily attendance of individuals with a disability. The facilities shall be in an area not accessible to the general public, shall be equipped with water and utensils for the consumption thereof, and shall otherwise be safe, clean, and comfortable.

(c) Any zoo or wild animal park that does not permit guide dogs to accompany blind or visually impaired persons therein shall provide free transportation to blind or visually impaired persons on any mode of transportation provided for members of the public. Each zoo or wild animal park that does not permit service dogs to accompany individuals with a disability shall provide free transportation to individuals with a disability on any mode of transportation provided for a member of the public in cases where the person uses a wheelchair and it is readily apparent that the person is unable to maintain complete or independent mobility without the aid of the service dog.

(d) Any zoo or wild animal park that does not permit guide dogs to accompany blind or otherwise visually impaired persons therein shall provide sighted escorts for blind or otherwise visually impaired persons if they are unaccompanied by a sighted person.

(e) As used in this section, "wild animal park" means any entity open to the public on a regular basis, licensed by the United States Department of Agriculture under the Animal Welfare Act as an exhibit, and operating for the primary purposes of conserving, propagating, and exhibiting wild and exotic animals, and any marine, mammal, or aquatic park open to the general public.

(Added by Stats 1979 ch 525 Section 1.)
(Amended by Stats 1988 ch 1595 Section 4; Stats 1994, c. 1257 (S.B. 1240), Section 8.)

Section 54.8. Hearing impaired persons; assertive listening systems in civil or criminal proceedings; notice of need; availability; use in proceedings
(a) In any civil or criminal proceeding, including, but not limited to, traffic, small claims court, family court proceedings and services, and juvenile court proceedings, in any court-ordered or court-provided alternative dispute resolution, including mediation and arbitration, or in any administrative hearing of a public agency, where a party, witness, attorney, judicial employee, judge, juror, or other participant who is hearing impaired, the individual who is hearing impaired, upon his or her request, shall be provided with a functioning assistive listening system or a computer-aided transcription system. Any individual requiring this equipment shall give advance notice of his or her need to the appropriate court or agency at the time the hearing is set or not later than five days before the hearing.

(b) Assistive listening systems include, but are not limited to, special devices which transmit amplified speech by means of audio-induction loops, radio frequency systems (AM or FM), or infrared transmission. Personal receivers, headphones, and neck loops shall be available upon request by individuals who are hearing impaired.

(c) If a computer-aided transcription system is requested, sufficient display terminals shall be provided to allow the individual who is hearing impaired to read the real-time transcript of the proceeding without difficulty.

(d) A sign shall be posted in a prominent place indicating the availability of, and how to request, an assistive listening system and a computer-aided transcription system. Notice of the availability of the systems shall be posted with notice of trials.

(e) Each superior court shall have at least one portable assistive listening system for use in any court facility within the county. When not in use, the system shall be stored in a location determined by the court.

(f) The Judicial Council shall develop and approve official forms for notice of the availability of assistive listening systems and computer-aided transcription systems for individuals who are hearing impaired. The Judicial Council shall also develop and maintain a system to record utilization by the courts of these assistive listening systems and computer-aided transcription systems.

(g) If the individual who is hearing impaired is a juror, the jury deliberation room shall be equipped with an assistive listening system or a computer-aided transcription system upon the request of the juror.

(h) A court reporter may be present in the jury deliberating room during a jury deliberation if the services of a court reporter for the purpose of operating a computer-aided transcription system are required for a juror who is hearing impaired.

(i) In any of the proceedings referred to in subdivision (a), or in any administrative hearing of a public agency, in which the individual who is hearing impaired is a party, witness, attorney, judicial employee, judge, juror, or other participant, and has requested use of an assistive listening system or computer-aided transcription system, the proceedings shall not commence until the system is in place and functioning.

(j) As used in this section, "individual who is hearing impaired" means an individual with a hearing loss, who, with sufficient amplification or a computer-aided transcription system, is able to fully participate in the proceeding.

(k) In no case shall this section be construed to prescribe a lesser standard of accessibility or usability than that provided by Title II of the Americans with Disabilities Act of 1990 (Public Law 101-336) and federal regulations adopted pursuant to that act.

Added by Stats 1989 ch 1002 Section 1. Amended Stats 1992 ch 913 (A.B. 1077) Section 8; Stats 1993 ch 1214 (A.B. 551) Section 2; Stats. 2001, c. 824 (A.B. 1700), Section 1.)
₁42 U.S.C.A. Section 12101 et seq.

Section 55. Violations; injunction; action by person actually or potentially aggrieved
Any person who is aggrieved or potentially aggrieved by a violation of Section 54 or 54.1 of this code, Chapter 7 (commencing with Section 4450) of Division 5 of Title 1 of the Government Code, or Part 5.5 (commencing with Section 19955) of Division 13 of the Health and Safety Code may bring an action to enjoin the violation. The prevailing party in the action shall be entitled to recover reasonable attorney's fees.

Added by Stats 1974 ch 1443 Section 1.

Section 55.1. Violations; injunctions; district or city attorney, attorney general
In addition to any remedies available under the federal Americans with Disabilities Act of 1990, Public Law 101-336 (42 U.S.C. Sec. 12102), or other provisions of law, the district attorney, the city attorney, the Department of Rehabilitation acting through the Attorney General, or the Attorney General may bring an action to enjoin any violation of Section 54 or 54.1.

(Added by Stats 1976, c. 869, p. 1979, Section 1.)
(Amended by Stats 1994, c. 1257 (S.B. 1240) Section 9.)

Section 1360. Modification of unit by owner; facilitation of access for handicapped; approval by project association
(a) Subject to the provisions of the governing documents and other applicable provisions of law, if the boundaries of the separate interest are contained within a building, the owner of the separate interest may do the following:

(1) Make any improvements or alterations within the boundaries of his or her separate interest that do not impair the structural integrity or mechanical systems or lessen the support of any portions of the common interest development.

(2) Modify a unit in a condominium project, at the owner's expense, to facilitate access for persons who are blind, visually handicapped, deaf, or physically disabled, or to alter conditions which could be hazardous to these persons. These modifications may also include modifications of the route from the public way to the door of the unit for the purposes of this paragraph if the unit is on the ground floor or already accessible by an existing ramp or elevator. The right granted by this paragraph is subject to the following conditions:

(A) The modifications shall be consistent with applicable building code requirements.

(B) The modifications shall be consistent with the intent of otherwise applicable provisions of the governing documents pertaining to safety or aesthetics.

(C) Modifications external to the dwelling shall not prevent reasonable passage by other residents, and shall be removed by the owner when the unit is no longer occupied by persons requiring those modifications who are blind, visually handicapped, deaf, or physically disabled.

(D) Any owner who intends to modify a unit pursuant to this paragraph shall submit his or her plans and specifications to the association of the condominium project for review to determine whether the modifications will comply with the provisions of this paragraph. The association shall not deny approval of the proposed modifications under this paragraph without good cause.

(b) Any change in the exterior appearance of a separate interest shall be in accordance with the governing documents and applicable provisions of law.

Added by Stats 1985 ch 874 Section 14.

EVIDENCE CODE

Section 754. Deaf or hearing impaired persons; interpreters; qualifications; guidelines; compensation; questioning; use of statements
(a) As used in this section, "individual who is deaf or hearing impaired" means an individual with a hearing loss so great as to prevent his or her understanding language spoken in a normal tone, but does not include an individual who is hearing impaired provided with, and able to fully participate in the proceedings through the use of, an assistive listening system or computer-aided transcription equipment provided pursuant to Section 54.8 of the Civil Code.

(b) In any civil or criminal action, including, but not limited to, any action involving a traffic or other infraction, any small claims court proceeding, any juvenile court proceeding, any family court proceeding or service, or any proceeding to determine the mental competency of a person, in any court-ordered or court-provided alternative dispute resolution, including mediation and arbitration, or any administrative hearing, where a party or witness is an individual who is deaf or hearing impaired and the individual who is deaf or hearing impaired is present and participating, the proceedings shall be interpreted in a language that the individual who is deaf or hearing impaired understands by a qualified interpreter appointed by the court or other appointing authority, or as agreed upon.

(c) For purposes of this section, "appointing authority" means a court, department, board, commission, agency, licensing or legislative body, or other body for proceedings requiring a qualified interpreter.

(d) For the purposes of this section, "interpreter" includes, but is not limited to, an oral interpreter, a sign language interpreter, or a deaf-blind interpreter, depending upon the needs of the individual who is deaf or hearing impaired.

(e) For purposes of this section, "intermediary interpreter" means an individual who is deaf or hearing impaired, or a hearing individual who is able to assist in providing an accurate interpretation between spoken English and sign language or between variants of sign language or between American Sign Language and other foreign languages by acting as an intermediary between the individual who is deaf or hearing impaired and the qualified interpreter.

(f) For purposes of this section, "qualified interpreter" means an interpreter who has been certified as competent to interpret court proceedings by a testing organization, agency, or educational institution approved by the Judicial Council as qualified to administer tests to court interpreters for individuals who are deaf or hearing impaired.

(g) In the event that the appointed interpreter is not familiar with the use of particular signs by the individual who is deaf or hearing impaired or his or her particular variant of sign language, the court or other appointing authority shall, in consultation with the individual who is deaf or hearing impaired or his or her representative, appoint an intermediary interpreter.

(h) Prior to July 1, 1992, the Judicial Council shall conduct a study to establish the guidelines pursuant to which it shall determine which testing organizations, agencies, or educational institutions will be approved to administer tests for certification of court interpreters for individuals who are deaf or hearing impaired. It is the intent of the Legislature that the study obtain the widest possible input from the public, including, but not limited to, educational institutions, the judiciary, linguists, members of the State Bar, court interpreters, members of professional interpreting organizations, and members of the deaf and hearing-impaired communities. After obtaining public comment and completing its study, the Judicial Council shall publish these guidelines. By January 1, 1997, the Judicial Council shall approve one or more entities to administer testing for court interpreters for individuals who are deaf or hearing impaired. Testing entities may include educational institutions, testing organizations, joint powers agencies, or public agencies. Commencing July 1, 1997, court interpreters for individuals who are deaf or hearing impaired shall meet the qualifications specified in subdivision (f).

(i) Persons appointed to serve as interpreters under this section shall be paid, in addition to actual travel costs, the prevailing rate paid to persons employed by the court to provide other interpreter services unless such service is considered to be a part of the person's regular duties as an employee of the state, county, or other political subdivision of the state. Accept as provided in subdivision (j), payment of the interpreter's fee shall be a charge against the court. Payment of the interpreter's fee in administrative proceedings shall be a charge against the appointing board or authority.

(j) Whenever a peace officer or any other person having a law enforcement or prosecutorial function in any criminal or quasi-criminal investigation or proceeding questions or otherwise interviews an alleged victim or witness who demonstrates or alleges deafness or hearing impairment, a good faith effort to secure the services of an interpreter shall be made, without any unnecessary delay unless either the individual who is deaf or hearing impaired affirmatively indicates that he or she does not need or cannot use an interpreter, or an interpreter is not otherwise required by Title II of the Americans with Disabilities Act of 1990 (Public Law 101-336) and federal regulations adopted thereunder. Payment of the interpreter's fee shall be charged against the county, or other political subdivision of the state, in which the action is pending.

(k) No statement, written or oral, made by an individual who the court finds is deaf or hearing impaired in reply to a question of a peace officer, or any other person having a law enforcement or prosecutorial function in any criminal or quasi-criminal investigation or proceeding, may be used against that individual who is deaf or hearing impaired unless the question was accurately interpreted and the statement was made knowingly, voluntarily, and intelligently and was accurately interpreted, or the court makes special findings that either the individual could not have used an interpreter or an interpreter was not otherwise required by Title II of the Americans with Disabilities Act of 1990 (Public Law 101-336) and federal regulations adopted thereunder and that the statement was made knowingly, voluntarily, and intelligently.

(l) In obtaining services of an interpreter for purposes of subdivision (j) or (k), priority shall be given to first obtaining a qualified interpreter.

(m) Nothing in subdivision (j) or (k) shall be deemed to supersede the requirement of subdivision (b) for use of a qualified interpreter for individuals who are deaf or hearing impaired participating as parties or witnesses in a trial or hearing.

(n) In any action or proceeding in which an individual who is deaf or hearing impaired is a participant, the appointing authority shall not commence proceedings until the appointed interpreter is in full view of and spatially situated to assure proper communication with the participating individual who is deaf or hearing impaired.

(o) Each superior court shall maintain a current roster of qualified interpreters certified pursuant to subdivision (f).

Amended by Stats 1977 ch 1182 Section 1; Stats 1984 ch 768 Section 2; Stats 1989 ch 1002 Section 2; Stats 190 ch 1450 (S.B. 2046) Section 2; Stats 1991 ch 883 (S.B. 585) Section 1; Stats 1992 ch 118 (S.B. 16) Section 1, effective July 7, 1992; Stats 1992 ch 913 (A.B. 1077) Section 14. (Amended by Stats. 1995, c. 143 (A.B. 1833), Section 1, eff. July 18, 1995.). Amended by Stats, 2012, Ch. 470, Section 11, effective January 1, 2013.
₁42 U.S.C.A. Section 12101 et seq.

B. DISCRIMINATION

Government Code

Section 11135.	Programs or activities funded by state; discrimination on basis of ethnic group identification, religion, age, sex, color, or disability; federal act; definition
Section 11136.	Notice or contractor, grantee or local agency by state agency; probable cause to believe violation of statute or regulation; hearing
Section 11137.	Action to curtail state funding upon determination of violation.
Section 11138.	Rules and regulations.
Section 11139.	Prohibitions and sanctions; construction of article.
Section 11139.5	Standards and guidelines; establishment; assistance
Section 19230.	Legislative declaration; state policy
Section 19231.	Definitions; judging undue hardship on a department program

Section 11135. Programs or activities funded by state; discrimination on basis of ethnic group identification, religion, age, sex, color, or disability; federal act; definition.
(a) No person in the State of California shall, on the basis of race, national origin, ethnic group identification, religion, age, sex, sexual orientation, color, genetic information or disability, be unlawfully denied full and equal access to the benefits of, or be unlawfully subjected to discrimination under, any program or activity that is conducted, operated, or administered by the state or by any state agency, is funded directly by the state, or receives any financial assistance from the state. Notwithstanding Section 11000, this section applies to the California State University.

(b) With respect to discrimination on the basis of disability, programs and activities subject to subdivision (a) shall meet the protections and prohibitions contained in Section 202 of the federal Americans with Disabilities Act of 1990 (42 U.S.C. Sec. 12132), and the federal rules and regulations adopted in implementation thereof, except that if the laws of this state prescribe stronger protections and prohibitions, the programs and activities subject to subdivision (a) shall be subject to the stronger protections and prohibitions.

(c) (1) As used in this section, "disability" means any mental or physical disability, as defined in Section 12926.

(2) The Legislature finds and declares that the amendments made to this act are declarative of existing law. The Legislature further finds and declares that in enacting Senate Bill 105 of the 2001-02 Regular Session (Chapter 1102 of the Statutes of 2002), it was the intention of the Legislature to apply subdivision (d) to the California State University in the same manner that subdivisions (a), (b), and (c) of this section already applied to the California State University, notwithstanding Section 11000. In clarifying that the California State University is subject to paragraph (2) of subdivision (d), it is not the intention of the Legislature to increase the cost of developing or procuring electronic and information technology. The California State University shall, however, in determining the cost of developing or procuring electronic or information technology, consider whether technology that meets the standards applicable pursuant to paragraph (2) of subdivision (d) will reduce the long-term cost incurred by the California State University in providing access or accommodations to future users of this technology who are persons with disabilities, as required by existing law, including this section, Title II of the Americans with Disabilities Act of 1990 (42 U.S.C. Sec. 12101 and following), and Section 504 of the Rehabilitation Act of 1973 (29 U.S.C. Sec. 794).

(d) (1) The Legislature finds and declares that the ability to utilize electronic or information technology is often an essential function for successful employment in the current work world.

(2) In order to improve accessibility of existing technology, and therefore increase the successful employment of individuals with disabilities, particularly blind and visually impaired and deaf and

hard-of-hearing persons, state governmental entities, in developing, procuring, maintaining, or using electronic or information technology, either indirectly or through the use of state funds by other entities, shall comply with the accessibility requirements of Section 508 of the Rehabilitation Act of 1973, as amended (29 U.S.C. Sec. 794d), and regulations implementing that act as set forth in Part 1194 of Title 36 of the Federal Code of Regulations.

(3) Any entity that contracts with a state or local entity subject to this section for the provision of electronic or information technology or for the provision of related services shall agree to respond to, and resolve any complaint regarding accessibility of its products or services that is brought to the attention of the entity.

(e) As used in this section, "sex" and "sexual orientation" have the same meanings as those terms are defined in subdivisions (p) and (q) of Section 12926.

(f) As used in this section, "race, national origin, ethnic group identification, religion, age, sex, sexual orientation, color, or disability" includes a perception that a person has any of those characteristics or that the person is associated with a person who has, or is perceived to have, any of those characteristics.

(g) As used in this section, "genetic information", has the same definition as in paragraph (2), of subdivision(e) of section 51 of the Civil Code.

Added by Stats 1977 ch 972 Section 1. Amended by Stats 1992 ch 913 (A.B. 1077) Section 18; Stats. 1994, c. 146 (A.B. 3601), Section 66; Stats. 2001, c. 708 (A.B. 677, Section 1.). Amended by Stats. 2011, Ch. 261, Section 6, effective January 1, 2012.

Section 11136. Notice to contractor, grantee or local agency by state agency; probable cause to believe violation of statute or regulation; hearing

Whenever a state agency that administers a program or activity that is funded directly by the state or receives any financial assistance from the state, has reasonable cause to believe that a contractor, grantee, or local agency has violated the provisions of Section 11135, or any regulation adopted to implement such section, the head of the state agency shall notify the contractor, grantee, or local agency of such violation and shall, after considering all relevant evidence, determine whether there is probable cause to believe that a violation of the provisions of Section 11135, or any regulation adopted to implement such section, has occurred. In the event that it is determined that there is probable cause to believe that the provisions of Section 11135, or any regulation adopted to implement such section, have been violated, the head of the state agency shall cause to be instituted a hearing conducted pursuant to the provisions of Chapter 5 (commencing with Section 11500) of this part to determine whether a violation has occurred.

Added by Stats 1977 ch 972 Section 1.

Section 11137. Action to curtail state funding upon determination of violation

If it is determined that a contractor, grantee, or local agency has violated the provisions of this article, the state agency that administers the program or activity involved shall take action to curtail state funding in whole or in part to such contractor, grantee, or local agency.

Added by Stats 1977 ch 972 Section 1.

Section 11138. Rules and regulations

Each state agency that administers a program or activity that is funded directly by the state or receives any financial assistance from the state and that enters into contracts for the performance of services to be provided to the public in an aggregate amount in excess of one hundred thousand dollars ($100,000) per year shall, in accordance with the provisions of Chapter 4.5 (commencing with Section 11371) of this part, adopt such rules and regulations as are necessary to carry out the purpose and provisions of this article.

Added by Stats 1977 ch 972 Section 1.

Section 11139. Prohibitions and sanctions; construction and enforcement of article

The prohibitions and sanctions imposed by this article are in addition to any other prohibitions and sanctions imposed by law. This article shall not be interpreted in a manner that would frustrate its purpose. This article shall not be interpreted in a manner that would adversely affect lawful programs which benefit the disabled, the aged, minorities, and women. This article and regulations adopted pursuant to this article may be enforced by a civil action for equitable relief, which shall be independent of any other rights and remedies.

Added by Stats. 1977 c. 972 § 1. (Amended by Stats. 1999, c. 591 (A.B. 1670), § 3; Stats 2001, c. 708 (A.B. 677), Section 2.)

Section 11139.5. Standards and guidelines; establishment; assistance

The Secretary of the Health and Welfare Agency, with the advice and concurrence of the Fair Employment and Housing Commission, shall establish standards for determining which persons are protected by this article and standards for determining what practices are discriminatory. The secretary, with the cooperation of the Fair Employment and Housing Commission, shall assist state agencies in coordinating their programs and activities and shall consult with such agencies, as necessary, so that consistent policies, practices, and procedures are adopted with respect to the enforcement of the provisions of the article.

Added by Stats 1977 ch 972 Section 1. Amended by Stats 1980 ch 992 Section 1; Stats 1982, ch 1270 Section 24. Amended by stats. 2012, Ch. 46, Section 18, effective June 27, 2012, operative January 1, 2013, by section 140 of Ch. 46.

Section 19230. Legislative declaration; state policy

(a) It is the policy of this state to encourage and enable individuals with a disability to participate fully in the social and economic life of the state and to engage in remunerative employment.

(b) It is the policy of this state that qualified individuals with a disability shall be employed in the state service, the service of the political subdivisions of the state, in public schools, and in all other employment supported in whole or in part by public funds on the same terms and conditions as the nondisabled, unless it is shown that the particular disability is job related.

(c) It is the policy of this state that a department, agency, or commission shall make reasonable accommodation to the known physical or mental limitations of an otherwise qualified applicant or employee who is an individual with a disability, unless the hiring authority can demonstrate that the accommodation would impose an undue hardship on the operation of its program. A department shall not deny any employment opportunity to a qualified applicant or employee who is an individual with a disability if the basis for the denial is the need to make reasonable accommodation to the physical or mental limitations of the applicant or employee.

Added by Stats 1977 ch 1196 Section 2. Amended by Stats 1987 ch 292 Section 1; Stats 1992 ch 913 (A.B 1077) Section 27.

Section 19231. Definitions; judging undue hardship on a department program

As used in this article, "individual with a disability" means any individual who has a physical or mental disability as defined in Section 12926.

Added by Stats 1977 ch 1196 Section 2. Amended by Stats 1987 ch 292 Section 2; Stats 1992 ch 913 (A.B.1077) Section 28 (Amended by Stats. 2000, c. 1048 (S.B. 2025), Section 1; Stats. 2000, c. 1049 (A.B. 2222), Section 9.)

C.   DEPARTMENT OF FAIR EMPLOYMENT AND HOUSING

California Government Code

Section 12926. Additional definitions.
Section 12926.1. Legislative findings and declarations; disability, mental disability, and medical condition; broad coverage under state law; interaction in determining reasonable accommodation

Section 12926. Additional definitions.

As used in this part in connection with unlawful practices, unless a different meaning clearly appears from the context:

(a) "Affirmative relief" or "prospective relief" includes the authority to order reinstatement of an employee, awards of backpay, reimbursement of out-of-pocket expenses, hiring, transfers, reassignments, grants of tenure, promotions, cease and desist orders, posting of notices, training of personnel, testing, expunging of records, reporting of records, and any other similar relief that is intended to correct unlawful practices under this part.

(b) "Age" refers to the chronological age of any individual who has reached his or her 40th birthday.

(c) "Employee" does not include any individual employed by his or her parents, spouse, or child, or any individual employed under a special license in a nonprofit sheltered workshop or rehabilitation facility.

(d) "Employer" includes any person regularly employing five or more persons, or any person acting as an agent of an employer, directly or indirectly, the state or any political or civil subdivision of the state, and cities, except as follows: "Employer" does not include a religious association or corporation not organized for private profit.

(e) "Employment agency" includes any person undertaking for compensation to procure employees or opportunities to work.

(f) "Essential functions" means the fundamental job duties of the employment position the individual with a disability holds or desires. "Essential functions" does not include the marginal functions of the position.

(1) A job function may be considered essential for any of several reasons, including, but not limited to, any one or more of the following:

(A) The function may be essential because the reason the position exists is to perform that function.

(B) The function may be essential because of the limited number of employees available among whom the performance of that job function can be distributed.

(C) The function may be highly specialized, so that the incumbent in the position is hired for his or her expertise or ability to perform the particular function.

(2) Evidence of whether a particular function is essential includes, but is not limited to, the following:

(A) The employer's judgment as to which functions are essential.

(B) Written job descriptions prepared before advertising or interviewing applicants for the job.

(C) The amount of time spent on the job performing the function.

(D) The consequences of not requiring the incumbent to perform the function.

(E) The terms of a collective bargaining agreement.

(F) The work experiences of past incumbents in the job.

(G) The current work experience of incumbents in similar jobs.

(g) (1) "Genetic information" means, with respect to any individual, information about any of the following:

(A) The individual's genetic tests.

(B) The genetic tests of family members of the individual.

(C) The manisfestation of a disease or disorder in family members of the individual.

(2) "Genetic information" includes any request for, or receipt of, genetic services, or participation in clinical research that includes genetic services, by an individual or any family member of the individual.

(3) "Genetic information" does not include information about the sex or age of any individual.

(h) "Labor organization" includes any organization that exists and is constituted for the purpose, in whole or in part, of collective bargaining or of dealing with employers concerning grievances, terms or conditions of employment, or of other mutual aid or protection.

(i) "Medical condition" means either of the following:

(1) Any health impairment related to or associated with a diagnosis of cancer or a record or history of cancer.

(2) Genetic characteristics. For purposes of this section, "genetic characteristics" means either of the following:

(A) Any scientifically or medically identifiable gene or chromosome, or combination or alteration thereof, that is known to be a cause of a disease or disorder in a person or his or her offspring, or that is determined to be associated with a statistically increased risk of development of a disease or disorder, and that is presently not associated with any symptoms of any disease or disorder.

(B) Inherited characteristics that may derive from the individual or family member, that are known to be a cause of a disease or disorder in a person or his or her offspring, or that are determined to be associated with a statistically increased risk of development of a disease or disorder, and that are presently not associated with any symptoms of any disease or disorder.

(j) "Mental disability" includes, but is not limited to, all of the following:

(1) Having any mental or psychological disorder or condition, such as mental retardation, organic brain syndrome, emotional or mental illness, or specific learning disabilities, that limits a major life activity. For purposes of this section:

(A) "Limits" shall be determined without regard to mitigating measures, such as medications, assistive devices, or reasonable accommodations, unless the mitigating measure itself limits a major life activity.

(B) A mental or psychological disorder or condition limits a major life activity if it makes the achievement of the major life activity difficult.

(C) "Major life activities" shall be broadly construed and shall include physical, mental, and social activities and working.

(2) Any other mental or psychological disorder or condition not described in paragraph (1) that requires special education or related services.

(3) Having a record or history of a mental or psychological disorder or condition described in paragraph (1) or (2), which is known to the employer or other entity covered by this part.

(4) Being regarded or treated by the employer or other entity covered by this part as having, or having had, any mental condition that makes achievement of a major life activity difficult.

(5) Being regarded or treated by the employer or other entity covered by this part as having, or having had, a mental or psychological disorder or condition that has no present disabling effect, but that may become a mental disability as described in paragraph (1) or (2).

"Mental disability" does not include sexual behavior disorders, compulsive gambling, kleptomania, pyromania, or psychoactive substance use disorders resulting from the current unlawful use of controlled substances or other drugs.

(k) "Military and veteran status" means a member or veteran of the United States Armed Forces, United States Armed Forces Reserve, the United States National Guard, and the California National Guard.

(l) "On the bases enumerated in this part" means or refers to discrimination on the basis of one or more of the following: race, religious creed, color, national origin, ancestry, physical disability, mental disability, medical condition, genetic information, marital status, sex, age, or sexual orientation.

(m) "Physical disability" includes, but is not limited to, all of the following:

(1) Having any physiological disease, disorder, condition, cosmetic disfigurement, or anatomical loss that does both of the following:

(A) Affects one or more of the following body systems: neurological, immunological, musculoskeletal, special sense organs, respiratory, including speech organs, cardiovascular, reproductive, digestive, genitourinary, hemic and lymphatic, skin, and endocrine.

(B) Limits a major life activity. For purposes of this section:

(i) "Limits" shall be determined without regard to mitigating measures such as medications, assistive devices, prosthetics, or reasonable accommodations, unless the mitigating measure itself limits a major life activity.

(ii) A physiological disease, disorder, condition, cosmetic disfigurement, or anatomical loss limits a major life activity if it makes the achievement of the major life activity difficult.

(iii) "Major life activities" shall be broadly construed and includes physical, mental, and social activities and working.

(2) Any other health impairment not described in paragraph (1) that requires special education or related services.

(3) Having a record or history of a disease, disorder, condition, cosmetic disfigurement, anatomical loss, or health impairment described in paragraph (1) or (2), which is known to the employer or other entity covered by this part.

(4) Being regarded or treated by the employer or other entity covered by this part as having, or having had, any physical condition that makes achievement of a major life activity difficult.

(5) Being regarded or treated by the employer or other entity covered by this part as having, or having had, a disease, disorder, condition, cosmetic disfigurement, anatomical loss, or health impairment that has no present disabling effect but may become a physical disability as described in paragraph (1) or (2).

(6) "Physical disability" does not include sexual behavior disorders, compulsive gambling, kleptomania, pyromania, or psychoactive substance use disorders resulting from the current unlawful use of controlled substances or other drugs.

(n) Notwithstanding subdivisions (j) and (n), if the definition of "disability" used in the federal Americans with Disabilities Act of 1990 (Public Law 101-336) would result in broader protection of the civil rights of individuals with a mental disability or physical disability, as defined in subdivision (j) or (m), or would include any medical condition not included within those definitions, then that broader protection or coverage shall be deemed incorporated by reference into, and shall prevail over conflicting provisions of, the definitions in subdivisions (j) and (m).

(o) "Race, religious creed, color, national origin, ancestry, physical disability, mental disability, medical condition, genetic information, marital status, sex, age, sexual orientation or military and veteran status" includes a perception that the person has any of those characteristics or that the person is associated with a person who has, or is perceived to have, any of those characteristics.

(p) "Reasonable accommodation" may include either of the following:

(1) Making existing facilities used by employees readily accessible to, and usable by, individuals with disabilities.

(2) Job restructuring, part-time or modified work schedules, reassignment to a vacant position, acquisition or modification of equipment or devices, adjustment or modifications of examinations, training materials or policies, the provision of qualified readers or interpreters, and other similar accommodations for individuals with disabilities.

(q) "Religious creed," "religion," "religious observance," "religious belief," and "creed" include all aspects of religious belief, observance, and practice, including religious dress and grooming practices. "Religious dress practice" shall be construed broadly to include the wearing or carrying of religious clothing, head or face coverings, jewelry, artifacts, and any other item that is part of the observance by an individual of his or her religious creed. "Religious grooming practice" shall be construed broadly to include all forms of head, facial, and body hair that are part of the observance by an individual of his or her religious creed.

(r)(1) "Sex" includes, but is not limited to, the following;

(A) Pregnancy or medical conditions related to pregnancy.

(B) Childbirth, or medical conditions related to childbirth.

(C) Breastfeeding or medical conditions related to breastfeeding.

(2) "Sex" also includes, but is not limited to, a person's gender. "Gender" means sex, and includes a person's gender identity and gender expression. "Gender expression" means a person's gender-related appearance and behavior whether or not stereotypically associated with the persons's assigned sex at birth.

(s) "Sexual orientation" means heterosexuality, homosexuality, and bisexuality.

(t) "Supervisor" means any individual having the authority, in the interest of the employer, to hire, transfer, suspend, lay off, recall, promote, discharge, assign, reward, or discipline other employees, or the responsibility to direct them, or to adjust their grievances, or effectively to recommend that action, if, in connection with the foregoing, the exercise of that authority is not of a merely routine or clerical nature, but requires the use of independent judgment.

(u) "Undue hardship" means an action requiring significant difficulty or expense, when considered in light of the following factors:

(1) The nature and cost of the accommodation needed.

(2) The overall financial resources of the facilities involved in the provision of the reasonable accommodations, the number of persons employed at the facility, and the effect on expenses and resources or the impact otherwise of these accommodations upon the operation of the facility.

(3) The overall financial resources of the covered entity, the overall size of the business of a covered entity with respect to the number of employees, and the number, type, and location of its facilities.

(4) The type of operations, including the composition, structure, and functions of the workforce of the entity.

(5) The geographic separateness, administrative, or fiscal relationship of the facility or facilities.

(Added by Stats.1980, c. 992 Section 4. Amended by Stats.1985, c.1151, Section 1; Stats.1990, c. 15 (S.B. 1027), Section 1.) (Amended by Stats.1992, c. 911 (A.B. 311), Section 3; Stats. 1992, c. 912 (A.B. 1286), Section 3; Stats.1992, c. 913 (A.B.1077), Section 21.3; Stats. 1993, c. 1214 (A.B.551), Section 5; Stats.1998, c.99 (S.B.654), Section 1; Stats.1999, c. 311 (S.B. 1185), Section 2; Stats.1999, c. 591 (A.B. 1670), Section 5.1; Stats.1999, c. 592 (A.B.1001), Section 3.7; Stats.2000, c. 1049 (A.B.2222), Section 5.). Amended by Stats. 2013, Ch. 691, Section 3, effective January 1, 2014.

Section 12926.1. Legislative findings and declarations; disability, mental disability, and medical condition; broad coverage under state law; interaction in determining reasonable accommodation
The Legislature finds and declares as follows:

(a) The law of this state in the area of disabilities provides protections independent from those in the federal Americans with Disabilities Act of 1990 (Public Law 101-336). Although the federal act provides a floor of protection, this state's law has always, even prior to passage of the federal act, afforded additional protections.

(b) The law of this state contains broad definitions of physical disability, mental disability, and medical condition. It is the intent of the Legislature that the definitions of physical disability and mental disability be construed so that applicants and employees are protected from discrimination due to an actual or perceived physical or mental impairment that is disabling, potentially disabling, or perceived as disabling or potentially disabling.

(c) Physical and mental disabilities include, but are not limited to, chronic or episodic conditions such as HIV/AIDS, hepatitis, epilepsy, seizure disorder, diabetes, clinical depression, bipolar disorder, multiple sclerosis, and heart disease. In addition, the Legislature has determined that the definitions of "physical disability" and "mental disability" under the law of this state require a "limitation" upon a major life activity, but do not require, as does the federal Americans with Disabilities Act of 1990, a "substantial limitation." This distinction is intended to result in broader coverage under the law of this state than under that federal act. Under the law of this state, whether a condition limits a major life activity shall be determined without respect to any mitigating measures, unless the mitigating measure itself limits a major life activity, regardless of federal law under the Americans with Disabilities Act of 1990. Further, under the law of this state, "working" is a major life activity, regardless of whether the actual or perceived working limitation implicates a particular employment or a class or broad range of employments.

(d) Notwithstanding any interpretation of law in Cassista v. Community Foods (1993) 5 Cal.4th 1050, the Legislature intends (1) for state law to be independent of the federal Americans with Disabilities Act of 1990, (2) to require a "limitation" rather than a "substantial limitation" of a major life activity, and (3) by enacting paragraph (4) of subdivision (j) and paragraph (4) of subdivision (l) of Section 12926, to provide protection when an individual is erroneously or mistakenly believed to have any physical or mental condition that limits a major life activity.

(e) The Legislature affirms the importance of the interactive process between the applicant or employee and the employer in determining a reasonable accommodation, as this requirement has been articulated by the Equal Employment Opportunity Commission in its interpretive guidance of the federal Americans with Disabilities Act of 1990.

(Added by Stats.2000, c. 1049 (A.B.2222), Section 6.) (Amended by Stats. 2011, Ch. 261, Sec. 10. Effective January 1, 2012.)

D.   DISCRIMINATION BY LICENSED PROFESSIONALS

Business and Professions Code

Section 125.6.   Refusal to perform licensed activity; aiding or inciting refusal of performance by another licensee; discrimination or restriction in performance; race, color, sex, religion, ancestry, disability, marital status, or national origin

Insurance Code

Section 10144.   Physically or mentally impaired person; coverage and rates
Section 10145.   Blindness or partial blindness

Section 125.6. Refusal to perform licensed activity; aiding or inciting refusal of performance by another licensee; discrimination or restriction in performance; race, color, sex, religion, ancestry, disability, marital status, or national origin
(a) (1) With regard to an applicant, every person who holds a license under the provisions of this code is subject to disciplinary action under the disciplinary provisions of this code applicable to that person if, because of any characteristic listed or defined in subdivision (b) or (e) of Section 51 of the Civil Code, he or she refuses to perform the licensed activity or aids or incites the refusal to perform that licensed activity by another licensee, or if, because of any characteristic listed or defined in subdivision (b) or (e) of Section 51 of the Civil Code, he or she makes any discrimination, or restriction in the performance of the licensed activity.

(2) Nothing in this section shall be interpreted to prevent a physician or health care professional licensed pursuant to Division 2 (commencing with Section 500) from considering any of the

characteristics of a patient listed in subdivision (b) or (e) of Section 51 of the Civil Code if that consideration is medically necessary and for the sole purpose of determining the appropriate diagnosis or treatment of the patient.

(3) Nothing in this section shall be interpreted to apply to discrimination by employers with regard to employees or prospective employees, nor shall this section authorize action against any club license issued pursuant to Article 4 (commencing with Section 23425) of Chapter 3 of Division 9 because of discriminatory membership policy.

(4) The presence of architectural barriers to an individual with physical disabilities that conform to applicable state or local building codes and regulations shall not constitute discrimination under this section.

(b) (1) Nothing in this section requires a person licensed pursuant to Division 2 (commencing with Section 500) to permit an individual to participate in, or benefit from, the licensed activity of the licensee where that individual poses a direct threat to the health or safety of others. For this purpose, the term "direct threat" means a significant risk to the health or safety of others that cannot be eliminated by a modification of policies, practices, or procedures or by the provision of auxiliary aids and services.

(2) Nothing in this section requires a person licensed pursuant to Division 2 (commencing with Section 500) to perform a licensed activity for which he or she is not qualified to perform.

(c) (1) "Applicant," as used in this section, means a person applying for licensed services provided by a person licensed under this code.

(2) "License," as used in this section, includes "certificate," "permit," "authority," and "registration" or any other indicia giving authorization to engage in a business or profession regulated by this code.

Added by Stats 1974 ch 1350 Section 1. Amended by Stats 1977 ch 293 Section 1; Stats 1980 ch 191 Section 1; Stats 1992 ch 913 (A.B. 1077) Section 2. Amended by Stats 2007, Ch. 568, Sec. 2. Effective January 1, 2008.

Insurance Code

Section 10144. Physically or mentally impaired person; coverage and rates
No insurer issuing, providing, or administering any contract of individual or group insurance providing life, annuity, or disability benefits applied for and issued on or after January 1, 1984, shall refuse to insure, or refuse to continue to insure, or limit the amount, extent, or kind of coverage available to an individual, or charge a different rate for the same coverage solely because of a physical or mental impairment, except where the refusal, limitation or rate differential is based on sound actuarial principles or is related to actual and reasonably anticipated experience. "Physical or mental impairment" means any physical, sensory, or mental impairment which substantially limits one or more of that person's major life activities.

Added by Stats 1980 ch 352 Section 2; Amended by Stats 1982 ch 620 Section 1, operative January 1, 1984; Stats 1985 ch 971 Section 1.5.

Section 10145. Blindness or partial blindness
No insurer issuing, providing, or administering any contract of individual or group insurance providing life, annuity, or disability benefits applied for and issued on or after January 1, 1986, shall refuse to insure, or refuse to continue to insure, or limit the amount, extent, or kind of coverage available to an individual, or charge a different rate for the same coverage solely because of blindness or partial blindness. "Blindness or partial blindness" means central visual acuity of not more than 20/200 in the better eye, after correction, or visual acuity greater than 20/200 but with a

limitation in the fields of vision so that the widest diameter of the visual field subtends an angle no greater than 20 degrees, certified by a licensed physician and surgeon who specializes in diseases of the eye or a licensed optometrist.

Added by Stats 1985 ch 971 Section 2.

E.   SERVICE STATION REFUELING SERVICE FOR DISABLED PERSONS; EXCEPTIONS; VIOLATIONS; PUNISHMENT; NOTICE.

Business and Professions Code

§ 13660. (a) Every person, firm, partnership, association, trustee, or corporation that operates a service station shall provide, upon request, refueling service to a disabled driver of a vehicle that displays a disabled person's plate or placard, or a disabled veteran's plate, issued by the Department of Motor Vehicles. The price charged for the motor vehicle fuel shall be no greater than that which the station otherwise would charge the public generally to purchase motor vehicle fuel without refueling service.

(b) Any person or entity specified in subdivision (a) that operates a service station shall be exempt from this section during hours when:

(1) Only one employee is on duty.

(2) Only two employees are on duty, one of whom is assigned exclusively to the preparation of food. As used in this subdivision, the term "employee" does not include a person employed by an unrelated business that is not owned or operated by the entity offering motor vehicle fuel for sale to the general public.

(c) (1) Every person, firm, partnership, association, trustee, or corporation required to provide refueling service for persons with disabilities pursuant to this section shall post the following notice, or a notice with substantially similar language, in a manner and single location that is conspicuous to a driver seeking refueling service:

"Service to Disabled Persons

Disabled individuals properly displaying a disabled person's plate or placard, or a disabled veteran's plate, issued by the Department of Motor Vehicles, are entitled to request and receive refueling service at this service station for which they may not be charged more than the self-service price."

(2) If refueling service is limited to certain hours pursuant to an exemption set forth in subdivision (b), the notice required by paragraph (1) shall also specify the hours during which refueling service for persons with disabilities is available.

(3) Every person, firm, partnership, association, trustee, or corporation that, consistent with subdivision (b), does not provide refueling service for persons with disabilities during any hours of operation shall post the following notice in a manner and single location that is conspicuous to a driver seeking refueling service:

"No Service for Disabled Persons"
This service station does not provide refueling service for disabled individuals.

(4) The signs required by paragraphs (1) and (3) shall also include a statement indicating that drivers seeking information about enforcement of laws related to refueling services for persons with

disabilities may call one or more toll-free telephone numbers specified and maintained by the Department of Rehabilitation. By January 31, 1999, the Director of the Department of Rehabilitation shall notify the State Board of Equalization of the toll-free telephone number or numbers to be included on the signs required by this subdivision. At least one of these toll-free telephone numbers shall be accessible to persons using telephone devices for the deaf. The State Board of Equalization shall publish information regarding the toll-free telephone numbers as part of its annual notification required by subdivision (i). In the event that the toll-free telephone number or numbers change, the Director of the Department of Rehabilitation shall notify the State Board of Equalization of the new toll-free telephone number or numbers to be used.

(d) During the county sealer's normal petroleum product inspection of a service station, the sealer shall verify that a sign has been posted in accordance with subdivision (c). If a sign has not been posted, the sealer shall issue a notice of violation to the owner or agent. The sealer shall be reimbursed, as prescribed by the department, from funds provided under Chapter 14. If substantial, repeated violations of subdivision (c) are noted at the same service station, the sealer shall refer the matter to the appropriate local law enforcement agency.

(e) The local law enforcement agency shall, upon the verified complaint of any person or public agency, investigate the actions of any person, firm, partnership, association, trustee, or corporation alleged to have violated this section. If the local law enforcement agency determines that there has been a denial of service in violation of this section, or a substantial or repeated failure to comply with subdivision (c), the agency shall levy the fine prescribed in subdivision (f).

(f) Any person who, as a responsible managing individual setting service policy of a service station, or as an employee acting independently against the set service policy, acts in violation of this section is guilty of an infraction punishable by a fine of one hundred dollars ($100) for the first offense, two hundred dollars ($200) for the second offense, and five hundred dollars ($500) for each subsequent offense.

(g) In addition to those matters referred pursuant to subdivision

(e), the city attorney, the district attorney, or the Attorney General, upon his or her own motion, may investigate and prosecute alleged violations of this section. Any person or public agency may also file a verified complaint alleging violation of this section with the city attorney, district attorney, or Attorney General.

(h) Enforcement of this section may be initiated by any intended beneficiary of the provisions of this section, his or her representatives, or any public agency that exercises oversight over the service station, and the action shall be governed by Section 1021.5 of the Code of Civil Procedure.

(i) An annual notice setting forth the provisions of this section shall be provided by the State Board of Equalization to every person, firm, partnership, association, trustee, or corporation that operates a service station.

(j) A notice setting forth the provisions of this section shall be printed on each disabled person's placard issued by the Department of Motor Vehicles on and after January 1, 1999. A notice setting forth the provisions of this section shall be provided to each person issued a disabled person's or disabled veteran's plate on and after January 1, 1998.

(k) For the purposes of this action "refueling service" means the service of pumping motor vehicle fuel into the fuel tank of a motor vehicle.

(Amended by Stats. 1998, Ch. 879, Sec. 26.4 Effective January 1, 1999)

TELECOMMUNICATIONS DEVICES

Civil Code 54.8.

Section 54.8. Hearing impaired persons; assistive listening systems in civil or criminal proceedings; notice of need; availability; use in proceedings

(a) In any civil or criminal proceeding, including, but not limited to, traffic, small claims court, family court proceedings and services, and juvenile court proceedings, in any court-ordered or court-provided alternative dispute resolution, including mediation and arbitration, or in any administrative hearing of a public agency, where a party, witness, attorney, judicial employee, judge, juror, or other participant who is hearing impaired, the individual who is hearing impaired, upon his or her request, shall be provided with a functioning assistive listening system or a computer-aided transcription system. Any individual requiring this equipment shall give advance notice of his or her need to the appropriate court or agency at the time the hearing is set or not later than five days before the hearing.

(b) Assistive listening systems include, but are not limited to, special devices which transmit amplified speech by means of audio-induction loops, radio frequency systems (AM or FM), or infrared transmission. Personal receivers, headphones, and neck loops shall be available upon request by individuals who are hearing impaired.

(c) If a computer-aided transcription system is requested, sufficient display terminals shall be provided to allow the individual who is hearing impaired to read the real-time transcript of the proceeding without difficulty.

(d) A sign shall be posted in a prominent place indicating the availability of, and how to request, an assistive listening system and a computer-aided transcription system. Notice of the availability of the systems shall be posted with notice of trials.

(e) Each superior court shall have at least one portable assistive listening system for use in any court facility within the county. When not in use, the system shall be stored in a location determined by the court.

(f) The Judicial Council shall develop and approve official forms for notice of the availability of assistive listening systems and computer-aided transcription systems for individuals who are hearing impaired. The Judicial Council shall also develop and maintain a system to record utilization by the courts of these assistive listening systems and computer-aided transcription systems.

(g) If the individual who is hearing impaired is a juror, the jury deliberation room shall be equipped with an assistive listening system or a computer-aided transcription system upon the request of the juror.

(h) A court reporter may be present in the jury deliberating room during a jury deliberation if the services of a court reporter for the purpose of operating a computer-aided transcription system are required for a juror who is hearing impaired.

(i) In any of the proceedings referred to in subdivision (a), or in any administrative hearing of a public agency, in which the individual who is hearing impaired is a party, witness, attorney, judicial employee, judge, juror, or other participant, and has requested use of an assistive listening system or computer-aided transcription system, the proceedings shall not commence until the system is in place and functioning.

(j) As used in this section, "individual who is hearing impaired" means an individual with a hearing loss, who, with sufficient amplification or a computer-aided transcription system, is able to fully participate in the proceeding.

(k) In no case shall this section be construed to prescribe a lesser standard of accessibility or usability than that provided by Title II of the Americans with Disabilities Act of 1990 (Public Law 101-336) and federal regulations adopted pursuant to that act.

Added by Status 1989 ch 1002 Section 1. Amended by Status 1992 ch 913 (A.B. 1007) Section 8; Stats 1993 ch 1214 (A.B. 551) Section 2; Stats. 2001, c. 824 9A.B. 1700), Section 1.) ₁42 U.S.C.A. Section 12101 et seq.

TRANSPORTATION

Government Code Section 4500.

Section 4500. Contracts for rapid transit equipment or structures; ready access for individuals with disabilities

(a) Notwithstanding the provisions of any statute, rule, regulation, decision, or pronouncement to the contrary, other than subdivision (b), every state agency, board, and department, every local governmental subdivision, every district, every public and quasi-public corporation, every local public agency and public service corporation, and every city, county, city and county and municipal corporation, whether incorporated or not and whether chartered or not, in awarding contracts for operations, equipment, or structures shall be obligated to require that all fixed-route transit equipment and public transit structures shall be so built that individuals with disabilities shall have ready access to, from and in such equipment and structures.

(b) Notwithstanding any other provision of law, public transit facilities and operations, whether operated by or under contract with a public entity, shall meet the applicable standards of Titles II and III of the federal Americans with Disabilities Act of 1990 (Public Law 101-336) and the federal regulations adopted pursuant thereto, subject to the exceptions provided in that act. However, if the laws of this state in effect on December 31, 1992, prescribe higher standards than the Americans with Disabilities Act of 1990 (Public Law 101-336) and federal regulations adopted pursuant thereto, then those public transit facilities and operations shall meet the higher standards.

(Added Stats 1971 ch 444 Section 1. Amended by Stats 1992 ch 913 (A.B. 1077) Section 17.)
₁42 U.S.C.A. Section 12101 et seq.

STATE HISTORIC BUILDING CODE
Division 13, Part 2.7

Health and Safety Code

Section 18951. Purpose

It is the purpose of this part to provide alternative regulations and standards for the rehabilitation, preservation, restoration (including related reconstruction), or relocation of qualified historical buildings or structures, as defined in Section 18955. These alternative standards and regulations are intended to facilitate the rehabilitation, restoration, or change of occupancy so as to preserve their original or restored architectural elements and features, to encourage energy conservation and a cost-effective approach to preservation, and to provide for the safety of the building occupants.

Added by Stats 1975 ch 906 Section 1. Amended by Stats 1977 ch 707 Section 1, effective September 8, 1977; Stats 1979 ch 1152 Section 164. Amended by Stats 200, Ch. 504, Sec. 1 Effective January 1, 2004.

Section 18953. Intent

It is the intent of this part to provide means for the preservation of the historical value of qualified historical buildings or structures and, concurrently, to provide reasonable safety from fire, seismic forces or other hazards for occupants of these buildings or structures, and to provide reasonable availability to and usability by, the disabled.

Added by Stats 1975 ch 906 Section 1. Amended by Stats 1981 ch 598 Section 1. Amended by Stats. 2003, Ch. 504, Sec. 3, Effective January 1, 2004.

Section 18954. Repairs, alterations, and additions; application of building standards and building regulations; physically handicapped accessibility standards

Repairs, alterations, and additions necessary for the preservation, restoration, rehabilitation, moving, or continued use of a qualified historical building or structure may be made if they conform to this part. The building department of every city or county or other local agency that has jurisdiction over the enforcement of code within its legal authority shall apply the alternative standards and regulations adopted pursuant to Section 18959.5 in permitting repairs, alterations, and additions necessary for the preservation, restoration, rehabilitation, safety, moving, or continued use of a qualified historical building or structure. A state agency shall apply the alternative regulations adopted pursuant to Section 18959.5 in permitting repairs, alterations, and additions necessary for the preservation, restoration, rehabilitation, safety, moving, or continued use of a qualified historical building or structure.

The application of any alternative standards for the provision of access to the disabled or exemption from access requirements shall be done on a case-by-case and item-by-item basis, and

---

shall not be applied to an entire qualified historical building or structure without individual consideration of each item, and shall not be applied to related sites or areas except on an item-by-item basis.

Added by Stats 1975 ch 906 Section 1. Amended by Stats 1979 ch 1152 Section 165; Stats 1981 ch 598 Section 2; Stats 1984 ch 1314 Section 1, operative July 1, 1985; Stats 1990 ch 625 (S.B. 2775) Section 1. Amended by Stats. 2003, Ch. 504, Sec. 4 Effective January 1, 2004.

Section 18955. Qualified historical building or structure

For the purposes of this part, a qualified historical building or structure is any structure or property, collection of structures, and their related sites deemed of importance to the history, architecture, or culture of an area by an appropriate local or state governmental jurisdiction. This shall include historical buildings or structures on existing or future national, state or local historical registers or official inventories, such as the National Register of Historic Places, State Historical Landmarks, State Points of Historical Interest, and city or county registers or inventories of historical or architecturally significant sites, places, historic districts, or landmarks. This shall also include places, locations, or sites identified on these historical registers or official inventories and deemed of importance to the history, architecture, or culture of an area by an appropriate local or state governmental jurisdiction.

Added by Stats 1975 ch 906 Section 1. Amended by Stats 1977 ch 707 Section 2, effective September 8, 1977. Amended by Stats. 2003, Ch. 504 Sec. 5, Effective January 1, 2004.

Section 19856. Application of Government Code, Public Resources Code, and other statutes and regulations

The application of the provisions of Part 5.5 (commencing with Section 19955) of Division 13 of this code, Chapter 7 (commencing with Section 4450) of Division 5 of Title 1 of the Government Code, Division 15 (commencing with Section 25000) of the Public Resources Code, and of any other statute or regulation, as they may apply to qualified historical buildings or structures, shall be governed by this part.

Added by Stats 1975 ch 906 Section 1. Amended by Stats 1976 ch 192 Section 1, effective May 28, 1976; Stats 1978 ch 555 Section 1, effective August 25, 1978; Stats 1980 ch 676 Section 169.

Section 18958. Additional agencies authorized to adopt rules and regulations

Except as provided in Section 18930, the following state agencies, in addition to the State Historical Building Safety Board, shall have the authority to adopt rules and regulations pursuant to the State Historical Building Code governing the rehabilitation, preservation, restoration, related reconstruction, safety, or relocation of qualified historical buildings and structures within their jurisdiction:

(a) The Division of the State Architect.
(b) The State Fire Marshal.
(c) The State Building Standards Commission, but only with respect
    to approval of building standards.
(d) The Department of Housing and Community Development.
(e) The Department of Transportation.
(f) Other state agencies that may be affected by this part.

Added by Stats 1975 ch 906 Section 1. Amended by Stats 1979 ch 1152 Section 166; Stats 1981 ch 598 Section 3; Stats 1984 ch 1314 Section 2, operative July 1, 1985; Stats 1990 ch 625 (S.B. 2775) Section 3. Amended by Stats. 2003, Ch. 504, Sec 6 Effective January 1, 2004.

Section 18959. Administration and enforcement

(a) Except as otherwise provided in Part 2.5 (commencing with Section 18901), all state agencies shall administer and enforce this part with respect to qualified historical buildings or structures under their respective jurisdiction.

(b) Except as otherwise provided in Part 2.5 (commencing with Section 18901), all local authorities shall, within their legal authority, administer and enforce this part with respect to qualified historical buildings or structures under their respective jurisdictions where applicable.

(c) The State Historical Building Safety Board shall coordinate and consult with the other applicable state agencies affected by this part and, except as provided in Section 18943, disseminate provisions adopted pursuant to this part to all local building authorities and state agencies at cost.

(d) Regulations adopted by the State Fire Marshal pursuant to this part shall be enforced in the same manner as regulations are enforced under Sections 13145, 13146, and 13146.5.

(e) Regular and alternative building standards published in the California Building Standards Code shall be enforced in the same manner by the same governmental entities as provided by law.

(f) When administering and enforcing this part, each local agency may make changes or modifications in the requirements contained in the California Historical Building Code, as described in Section 18944.7, as it determines are reasonably necessary because of local climatic, geological, seismic, and topographical conditions. The local agency shall make an express finding that the modifications or changes are needed, and the finding shall be available as a public record. A copy of the finding and change or modification shall be filed with the State Historical Building Safety Board. No modification or change shall become effective or operative for any purpose until the finding and modification or change has been filed with the board.

Added by Stats 1975 ch 906 Section 1. Amended by Stats 1977 ch 707 Section 3, effective September 6, 1977; Stats 1979 ch 1152 Section 167; Stats 1981 ch 598 Section 4; Stats 1984 ch 1314 Section 3; operative July 1, 1985; Stats 1985 ch 106 Section 84; Stats 1990 ch 625 (S.B. 2775) Section 5. Amended by Stats. 2003, Ch. 504, Sec. 7. Effective January 1, 2004.

Section 18959.5. Alternative building standards, rules and regulations; historical buildings code board

Subject to the applicable provisions of Part 2.5 (commencing with Section 18901) of this division, the State Historical Building Safety Board shall adopt and submit alternative building standards for approval pursuant to Chapter 4 (commencing with Section 18935) of Part 2.5 of this division and may adopt, amend, and repeal other alternative rules and regulations under this part which the board has recommended for adoption under subdivision (b) of Section 18960 by the State Architect or other appropriate state agencies.

Added by Stats 1977 ch 707 Section 4, effective September 8, 1977. Amended by Stats 1979 ch 1152 Section 168; Stats 1984 ch 1314 Section 4, operative July 1, 1985; Stats 1990 ch 625 (S.B. 2775) Section 7.

Section 18960. State historical building code board

(a) A State Historical Building Safety Board is hereby established as a unit within the Division of the State Architect. The board shall be composed of qualified experts in their respective fields who shall represent various state and local public agencies, professional design societies and building and preservation oriented organizations.

(b) This board shall act as a consultant to the State Architect and to the other applicable state agencies for purposes of this part. The board shall recommend to the State Architect and the other applicable state agencies rules and regulations for adoption pursuant to this part.

(c) The board shall also act as a review body to state and local agencies with respect to interpretations of this part as well as on matters of administration and enforcement of it. The board's decisions shall be reported in printed form.

(1) Notwithstanding subdivision (b) of Section 18945, if any local agency administering and enforcing this part or any person adversely affected by any regulation, rule, omission,

interpretation, decision, or practice of this agency representing a building standard wishes to appeal the issue for resolution to the State Historical Building Safety Board, these parties may appeal to the board. The board may accept the appeal only if it determines that issues involved in the appeal have statewide significance.

(2) The State Historical Building Safety Board shall, upon making a decision on an appeal pursuant to paragraph (1), send a copy to the State Building Standards Commission.

(3) Requests for interpretation by local agencies of the provisions of this part may be accepted for review by the State Historical Building Safety Board. A copy of an interpretation decision shall be sent to the State Building Standards Commission in the same manner as paragraph (2).

(4) The State Historical Building Safety Board may charge a reasonable fee, not to exceed the cost of the service, for requests for copies of their decisions and for requests for reviews by the board pursuant to paragraph (1) or (3). All funds collected pursuant to this paragraph shall be deposited in the State Historical Building Code Fund, which is hereby established, for use by the State Historical Building Safety Board. The State Historical Building Code Fund and the fees collected therefor, and the budget of the State Historical Building Safety Board, shall be subject to annual appropriation in the Budget Act.

(5) Local agencies may also charge reasonable fees not to exceed the cost for making an appeal pursuant to paragraph (1) to persons adversely affected as described in that appeal.

(6) All other appeals involving building standards under this part shall be made as set forth in subdivision (a) of Section 18945.

(d) The board shall be composed of representatives of state agencies and public and professional building design, construction, and preservation organizations experienced in dealing with historic buildings.   Unless otherwise indicated, each named organization shall appoint its own representatives. Each of the following shall have one member on the board who shall serve without pay, but shall receive actual and necessary expenses incurred while serving on the board:

(1) The Division of the State Architect.
(2) The State Fire Marshal.
(3) The State Historical Resources Commission.
(4) The California Occupational Safety and Health Standards Board.
(5) California Council, American Institute of Architects.
(6) Structural Engineers Association of California.
(7) A mechanical engineer, Consulting Engineers and Land Surveyors
    of California.
(8) An electrical engineer, Consulting Engineers and Land
    Surveyors of California.
(9) California Council of Landscape Architects.
(10) The Department of Housing and Community Development.
(11) The Department of Parks and Recreation.
(12) The California State Association of Counties.
(13) League of California Cities.
(14) The Office of Statewide Health Planning and Development.
(15) The Department of Rehabilitation.
(16) The California Chapter of the American Planning Association.
(17) The Department of Transportation.
(18) The California Preservation Foundation.
(19) The Seismic Safety Commission.
(20) The California Building Officials.
(21) The Building Owners and Managers Association of California.

The 21 members listed above shall select a building contractor as a member of the board, who shall serve without pay, but shall receive actual and necessary expenses incurred while serving on the board.

Each of the appointing authorities shall appoint, in the same manner as for members, an alternate in addition to a member. The alternate member shall serve in place of the member at the meetings of the board that the member is unable to attend. The alternate shall have all of the authority that the member would have when the alternate is attending in the place of the member. The board may appoint, from time to time, as it deems necessary, consultants who shall serve without pay but shall receive actual and necessary expenses as approved by the board.

(e) The term of membership on the board shall be for four years, with the State Architect's representative serving continually until replaced. Vacancies on the board shall be filled in the same manner as original appointments. The board shall annually select a chairperson from among the members of the board.

Added by Stats 1975 ch 906 Section 1. Amended by Stats 1977 ch 1001 Section 1; Stats 1981 ch 598 Section 5; Stats 1984 1314 Section 5, operative July 1, 1985; Stats 1990 ch 625 (S.B. 2775) Section 9. Amended by Stats. 2007, Ch. 55, Sec. 1 Effective January 1, 2008.

Section 18961. Review, enforcement and administration of variances, appeals procedure; agency to consider alternative provisions of this part and obtain review

All state agencies that enforce and administer approvals, variances, or appeals procedures or decisions affecting the preservation or safety of the historical aspects of qualified historical buildings or structures shall use the alternative provisions of this part and shall consult with the State Historical Building Safety Board to obtain its review prior to undertaking action or making decisions on variances or appeals that affect qualified historical buildings or structures.

Added by Stats 1982 ch 1358 Section 3. Amended by Stats 1984 ch 1314 Section 6, operative July 1, 1985; Stats 1990 h 625 (S.B. 2775) Section 11. Amended by Stats 2003, Ch 504, Sec 9, Effective January 1, 2004.

POLLING PLACES

Elections Code

Section 2050.	Short title
Section 2051.	"Visually impaired" defined
Section 2052.	Legislative intent
Section 2053.	Advisory board; membership; duties; compensation
Section 13303.	Sample ballots for other elections; printing; mailing; notice of polling place
Section 13304.	Notice of polling place; accessibility by physically handicapped
Section 14282.	Aid to disabled voters; access by physically handicapped

Section 2050. Short Title
This article shall be known, and may be cited, as the Visually Impaired Voter Assistance Act of 1989.

Added by Stats 1989 ch 774 Section 1. Enacted by Stats. 1994, Ch. 920, Sec. 2.

Section 2051. "Visually Impaired" defined
As used in this article, "visually impaired" means a person having central visual acuity not to exceed 20-200 in the better eye, with corrected lenses, or visual acuity greater than 20-200, but

with a limitation in the field of vision such that the widest diameter of the visual field subtends at an angle of not greater than 20 degrees.

Added by Stats 1989 ch 774 Section 1. Enacted by Stats. 1994, Ch. 920, Sec. 2.

Section 2052. Legislative Intent
It is the intent of the Legislature to promote the fundamental right to vote of visually impaired individuals, and to make efforts to improve public awareness of the availability of ballot pamphlet audio recordings and improve their delivery to these voters.

Added by Stats 1989 ch 774 Section 1. Amended by Stats. 2009, Ch. 88, Sec. 30, Effective January 1, 2010.

Section 2053. Advisory board; membership; duties; compensation
(a) The Secretary of State shall establish a Visually Impaired Voter Assistance Advisory Board. This board shall consist of the Secretary of State or his or her designee and the following membership, appointed by the Secretary of State:

(1) A representative from the State Advisory Council on Libraries.

(2) One member from each of three private organizations. Two of the organizations shall be representative of organizations for blind persons in the state.

(b) The board shall do all of the following:

(1) Establish guidelines for reaching as many visually impaired persons as practical.

(2) Make recommendations to the Secretary of State for improving the availability and accessibility of ballot pamphlet audio recordings and their delivery to visually impaired voters. The Secretary of State may implement the recommendations made by the board.

(3) Increase the distribution of public service announcements identifying the availability of ballot pamphlet audio recordings at least 45 days before any federal, state, and local election.

(4) Promote the Secretary of State's toll-free voter registration telephone line for citizens needing voter registration information, including information for those who are visually handicapped, and the toll-free telephone service regarding the California State Library and regional library service for the visually impaired.

(c) No member shall receive compensation, but each member shall be reimbursed for his or her reasonable and necessary expenses in connection with service on the board.

Added by Stats 1989 ch 774 Section 1. Amended by Stats. 2009, Ch. 88, Sec. 31. Effective January 1, 2010.

(END OF VISUALLY IMPAIRED VOTER ASSISTANCE ACT OF 1989.)

Section 13303. Sample ballots for other elections; printing; mailing; notice of polling place
(a) For each election, each appropriate elections official shall cause to be printed, on plain white paper or tinted paper, without watermark, at least as many copies of the form of ballot provided for use in each voting precinct as there are voters in the precinct. These copies shall be designated "sample ballot" upon their face and shall be identical to the official ballots used in the election, except as otherwise provided by law. A sample ballot shall be mailed, postage prepaid, not more than 40 nor less than 21 days before the election to each voter who is registered at least 29 days prior to the election.

(b) The elections official shall send notice of the polling place to each voter with the sample ballot. Only official matter shall be sent out with the sample ballot as provided by law.

(c) The elections official shall send notice of the polling place to each voter who registered after the 29th day prior to the election and is eligible to participate in the election. The notice shall also include information as to where the voter can obtain a sample ballot and a ballot pamphlet prior to the election, a statement indicating that those documents will be available at the polling place at the time of the election, and the address of the Secretary of State's website and, if applicable, of the county website where a sample ballot may be viewed.

Added by Stats 1979 ch 494 Section 1. Amended by Stats 1990 ch 106 (A.B. 211) Section 1.
(Stats. 1994, c. 920 (S.B. 1547), Section 2.) (Amended by Stats. 2000, c. 899 (A.B. 1094), Section 12.) (Amended by Stats. 2005, Ch. 72, Sec. 2, Effective January 19, 2005.)

Section 13304. Notice of polling place; accessibility by physically handicapped

The notice of the polling place which is sent to each voter as provided in Section 13303 may, at the option of the local elections official, inform the voter as to whether the polling place is accessible to the physically handicapped. In addition, this notice may inform the voter of his or her rights under Section 14282, if applicable.

Added by Stats 1979 ch 494 Section 2. Amended by Stats. 2005, Ch. 72, Sec. 2, Effective January 19, 2005.

Section 14282. Aid to disabled voters; access by physically handicapped

(a) When a voter declares under oath, administered by any member of the precinct board at the time the voter appears at the polling place to vote, that the voter is then unable to mark a ballot, the voter shall receive the assistance of not more than two persons selected by the voter, other than the voter's employer, an agent of the voter's employer, or an officer or agent of the union of which the voter is a member.

(b) No person assisting a voter shall divulge any information regarding the marking of the ballot.

(c) In those polling places that are inaccessible under the guidelines promulgated by the Secretary of State for accessibility by the physically handicapped, a physically handicapped person may appear outside the polling place and vote a regular ballot. The person may vote the ballot in a place that is as near as possible to the polling place and that is accessible to the physically handicapped. A precinct board member shall take a regular ballot to that person, qualify that person to vote, and return the voted ballot to the polling place. In those precincts in which it is impractical to vote a regular ballot outside the polling place, vote by mail ballots shall be provided in sufficient numbers to accommodate physically handicapped persons who present themselves on election day. The vote by mail ballot shall be presented to and voted by a physically handicapped person in the same manner as a regular ballot may be voted by that person outside the polling place.

Added by Stats 1976 ch 220 Section 6. Amended by Stats 1985 ch 277 Section 1; Stats 1990 ch 106 (A.B. 211) Section 2. Amended by Stats. 2007, Ch. 508, Sec. 82, Effective January 1, 2008.

RECREATION

A. Division 5, Chapter 1, California Recreation Trails
(Public Resources Code 5070.5, 5071, 5071.7, 5073.1, 5075.3.)

B. Division 5, Chapter 2.6. Public Playground Equipment
(Public Resources Code 5410, 5411.)

C. Division 1. Chapter 2. Fish and Game Code 217.5.

A. CALIFORNIA RECREATION TRAILS

Public Resources Code

Section 5070.5. Declaration of policy
Section 5071.   Contents of plan
Section 5071.7. Consultations; signs along heritage corridors; segments oriented to waterways;
              off-highway motor vehicle use; physically disabled persons
Section 5073.1. Repealed by Stats 1990 ch 1495 (A.B. 4044) Section 3.
Section 5075.3. Standards and criteria for trail routes and complementary facilities

Section 5070.5. Declaration of policy
Unless the context otherwise requires, the following definitions shall govern construction of this article:

(a) "Affirmative access area" means an area of already existing disability access improvements along a heritage corridor.

(b) "Committee" means the California Recreational Trails Committee.

(c) "Heritage corridor" means a regional, state, or nationwide alignment of historical, natural, or conservation education significance, with roads, state and other parks, greenways, or parallel recreational trails, intended to have guidebooks, signs, and other features to enable self-guiding tourism, and environmental conservation education along most of its length and of all or some of the facilities open to the public along its length, with an emphasis on facilities whose physical and interpretive accessibility meet "whole-access" goals.

(d) "Heritage corridors access map" means a 1:500,000 publicly distributed map combining listings and locations of parks, trails, museums, and roadside historical and natural access points, including disability and interpretive access data, along designated heritage corridors.

(e) "Plan" means the California Recreational Trails System Plan.

(f) "System" means the California Recreational Trails System.

(g) "Whole-access" means a general level of trail and human accessibility that includes not only disabled persons but all others making up the "easy-access" majority of the public. This level of accessibility may also benefit from amplified concepts of natural terrain accessibility and cooperation with volunteer and nonprofit accessibility groups.

Added by Stats 1974 ch 1461 Section 2. Amended by Stats 1977 ch 946 Section 1, effective September 21, 1977; Stats 1979 ch 944 Section 2; Status 1982 ch 994 Section 8.

Section 5071. Contents of plan
The plan shall contain, but shall not be limited to, the
following elements:

(a) Pedestrian trails.
(b) Bikeways.
(c) Equestrian trails.
(d) Boating trails.
(e) Trails and areas suitable for use by physically disabled persons, the elderly, and others in
   need of graduated trails, especially along designated heritage corridors.
(f) Cross-country skiing trails.
(g) Heritage corridors.

Added by Stats 1974 ch 1461 Section 2. Amended by Stats 1977 ch 946 Section 2, effective September 21,
1977; Stats 1979 ch 844 Section 3; Stats 1982 ch 994 Section 9.

**Section 5071.7.   Consultations; signs along heritage corridors; segments oriented to
waterways; off-highway motor vehicle use; physically disabled persons**
(a) (1) In planning the system, the director shall consult with and seek the assistance of the
Department of Transportation. The Department of Transportation shall plan and design those trail
routes that are in need of construction contiguous to state highways and serve both a
transportation and a recreational need.

(2) The Department of Transportation shall install or supervise the installation of signs along
heritage corridors consistent with the plan element developed pursuant to this section and Section
5073.1; provided, however, that it shall neither install nor supervise the installation of those signs
until it determines that it has available to it adequate volunteers or funds, or a combination thereof,
to install or supervise the installation of the signs, or until the Legislature appropriates sufficient
funds for the installation or supervision of installation, whichever occurs first.

(b) The element of the plan relating to boating trails and other segments of the system which are
oriented to waterways shall be prepared and maintained by the Department of Boating and
Waterways pursuant to Article 2.6 (commencing with Section 68) of Chapter 2 of Division 1 of the
Harbors and Navigation Code. Those segments shall be integrated with the California Protected
Waterways Plan developed pursuant to Chapter 1273 of the Statutes of 1968, and shall be planned
so as to be consistent with the preservation of rivers of the California Wild and Scenic Rivers
System, as provided in Chapter 1.4 (commencing with Section 5093.50) of this division.

(c) Any element of the plan relating to trails and areas for the use of off/highway motor vehicles
shall be prepared and maintained by the Division of Off/Highway Motor Vehicle Recreation
pursuant to Chapter 1.25 (commencing with Section 5090.01).

(d) In planning the system, the director shall consult with and seek the assistance of the
Department of Rehabilitation, representatives of its California Access Network volunteers, and
nonprofit disability access groups to assure that adequate provision is made for publicizing the
potential use of recreational trails, including heritage corridors by physically disabled persons.

Added by Stats 1974 ch 1461 Section 2. Amended by Stats 1977 ch 946 Section 3, effective September 21,
1977; Stats 1978 ch 365 Section 15; Stats 1979 ch 844 Section 4; Stats 1980 ch 595 Section 1; Stats 1982 ch
994 Section 10; Stats 1990 ch 1495 (A.B. 4044) Section 2.

Section 5073.1. Repealed by Stats 1990 ch 1495 (A.B. 4044) Section 3.

Section 5075.3. Standards and criteria for trail routes and complementary facilities
In specifying criteria and standards for the design and construction of trail routes and complementary facilities as provided in subdivisions (b) and (c) of Section 5071.3, the director shall include the following:

(a) The following routes shall be given priority in the allocation of funds:

(1) Routes which are in proximity or accessible to major urban areas of the state.

(2) Routes which are located on lands in public ownership.

(3) Routes which provide linkage or access to natural, scenic, historic, or recreational areas of clear statewide significance.

(4) Routes which are, or may be, the subject of agreements providing for participation of other public agencies, cooperating volunteer trail associations, or any combination of those entities, n state trail acquisition, development, or maintenance.

(b) Where feasible, trail uses may be combined on routes within the system; however, where trail use by motor vehicles is incompatible with other trail uses, separate areas and facilities should be provided.

(c) Trails should be located and managed so as to restrict trail users to established routes and to aid in effective law enforcement.

(d) Trails should be located so as to avoid severance of private property and to minimize impact on adjacent landowners and operations. The location of any trail authorized by this article shall, if the property owner so requests, be placed as nearly as physically practicable to the boundary lines of the property traversed by the trail, as such boundary lines existed as of January 1, 1975.

(e) Insofar as possible, trails should be designed and maintained to harmonize with, and complement, established forest, agricultural, and resource management plans. No trail, or property acquisition therefor, shall interfere with a landowner's water rights or his right to access to the place of exercise of such water rights.

(f) Trails should be planned as a system and each trail segment should be part of the overall system plan.

(g) Trails should be appropriately signed to provide identification, direction, and information.

(h) Rest areas, shelters, sanitary facilities, or other conveniences should be designed and located to meet the needs of trail users, including physically handicapped persons, and to prevent intrusion into surrounding areas.

(i) The department shall erect fences along any trail when requested to do so by the owner of adjacent land, or with the consent of the owner of such land when the department determines it will be in the best interests of the users of the trail and adjoining property owners, and shall place gates in such fences when necessary to afford proper access and at each point of intersection with existing roads, trails, or at used points of access to or across such trail. The department shall maintain such fences and gates in good condition.

(j) A landowner's right to conduct agricultural, timber harvesting, or mining activities on private lands adjacent to, or in the vicinity of, a trail shall not be restricted because of the presence of the trail.

Added by Stats 1974 ch 1461 Section 2. Amended by Stats 1975 ch 945 Section 1; Stats 1977 ch 946 Section 4, effective September 21, 1977; Stats 1982 ch 903 Section 3, effective September 13, 1982.

B. PUBLIC PLAYGROUND EQUIPMENT

Public Resources Code

Section 5410.   Legislative findings and declarations
Section 5411.   Purchase of equipment usable by physically handicapped persons

Section 5410. Legislative findings and declarations
The Legislature hereby finds and declares that playground equipment accessible to and usable by all persons regardless of physical condition is becoming increasingly available. The legislature further finds and declares that it is a matter of statewide interest and concern that all public agencies providing playgrounds utilize such facilities to the maximum extent possible, and, where feasible, provide playground equipment usable by both able-bodied and physically disabled persons so as to integrate able-bodied and physically disabled persons within such playgrounds.

Added by Stats 1978 ch 1006 Section 1.

Section 5411. Purchase of equipment usable by physically handicapped persons
All public agencies operating playgrounds, including state agencies, cities, counties, a city and county, school districts and other districts, shall, with respect to new playground equipment purchased on or after January 1, 1979, provide that a portion of the equipment within the playground is accessible and usable by all persons regardless of their physical condition whenever such equipment is available at a quality and cost which is comparable to the quality and cost of standard equipment.

Added by Stats 1978 ch 1006 Section 1.

C. FISH AND GAME CODE SECTION 217.5.

Section 217.5. Sport fishing areas accessible to disabled persons; notation in booklet of regulations

(a) The department shall identify property it owns or manages that includes areas for sport fishing which are accessible to disabled persons.

(b) Commencing with the booklet of sport fishing regulations published by the commission in 1986, the availability of sport fishing areas, identified by the department as accessible to disabled persons under subdivision (a), shall be noted in the booklet of regulations, together with telephone numbers and instructions for obtaining a list of those areas from regional department offices.

Added by Stats 1984 ch 1148 Section 1.

APPENDIX C

RESOURCES FOR FURTHER INFORMATION;
ORGANIZATIONS AND AGENCIES THAT
PROVIDE MATERIALS AND SERVICES
REGARDING COMPLIANCE FOR
DISABLED ACCESSIBILITY

REFERENCE SOURCE

ADA TELEPHONE INFORMATION SERVICES

This list contains the telephone numbers of federal and state agencies and other organizations that provide information about the Americans with disabilities Act (ADA) and informal guidance in understanding and complying with different provisions of the ADA and state laws and building codes. These agencies and organizations are not, and should not be viewed as, sources for obtaining legal advice or legal opinions about your rights or responsibilities under the accessibility laws.

Department of Justice offers technical assistance on the ADA Standards for Accessible Design and other ADA provisions applying to businesses, non-profit service agencies, and state and local government programs; also provides information on how to file ADA complaints.
ADA Information Line for documents, questions, and referrals
800-514-0301 (voice)   800-514-0383 (TDD)

Access Board, or Architectural and Transportation Barriers Compliance Board, offers technical assistance on the ADA Accessibility Guidelines.
ADA documents and questions
800-872-2253 (voice)   800-993-2822 (TDD)
Electronic bulletin board
202-272-5448

Pacific Disability and Business Technical Assistance Center (DBTAC), provides individualized responses to information requests, and a range of written and audio-visual materials on specific aspects of ADA compliance.   Provides referrals to local sources of expertise in all aspects of ADA compliance, provides training on ADA provisions and disability awareness.
ADA questions and documents
800-949-4232 (voice and TDD)

Equal Employment Opportunity Commission offers technical assistance on the ADA provisions applying to employment; also provides information on how to file ADA complaints.
Employment questions
800-669-4000 (voice)   800-669-6820 (TDD)
Employment documents
800-669-3362 (voice)   800-800-3302 (TDD)

California Division of the State Architect, Office of Universal Design
Provides technical assistance to projects under DSA jurisdiction. Publishes California Accessibility Reference Manual.
(916) 445-5827
www.dsa.dgs.ca.gov

Department of Housing and Community Development (HCD)
For information pertaining to the accessibility provisions contained in the California Building Code regarding covered multifamily residential housing.
(916) 445-9471

California Department of Rehabilitation
Provides technical assistance and documents on compliance with state and federal disabled accessibility compliance, and other information relating to specific types of disabilities.
(916) 322-8614

Fair Housing Act, for questions, publications, and referrals on housing access issues, call:
HUD Distribution Center
800-767-7468

California Council of the Blind
In California: (800) 221-6359
Direct: (916) 441-2100

INDEX

A

D

The CalDAG - California Disabled Accessibility Guidebook   ©2014 Michael P. Gibbens

"COMPETENT ACCESS CONSULTING DEMANDS KNOWING BOTH WHAT IS AND WHAT IS NOT NOT REQUIRED IN A PARTICULAR SITUATION"

Mike Gibbens' experience and expertise is unmatched in the field of disabled Accessibility compliance. Call for CV.

- CASp Certified #054 (California Certified Access Specialist
- ICC Certified Accessibility/Usability Specialist
- ICC Certified Accessibility Inspector/Plans Examiner
- Author of 16 Books, Including the California Disabled Accessibility Guidebook (The CalDAG™), the State's Leading Resource for Combined ADA/Title 24 Access Compliance
- 17 Year Member of the California State Building Standards Commission Accessibility Code Advisory Committee (5 Years as Chairman)
- Court Approved Expert/Court Appointed Expert
- Expert Witness/Consultant (1000+ cases). ADA, ADAAG, Title 24, ANSI, FHA, UFAS, HCD, Section 504, and IBC/State Specific Standards.
- Property Inspections/Compliance Reviews (Large Portfolio Compliance Survey Projects a Specialty).
- Barrier Removal and Transition Plan Services
- Training and Education Services, Seminars
- Plan Check and Related Services for Designers, Architects, Contractors, etc.

Gibbens & Associates, LLC

4258 Avenida Prado
Thousand Oaks, CA 91360
Phone: (805) 870-0900
Web: Http://www.theaccessguy.com
Email: Info@theaccessguy.com

INTERNATIONAL CODE COUNCIL®

People Helping People Build a Safer World®

Testing for the 2013 California Codes *Now Available*

The new California Codes are the path to building a safe, sustainable, and resilient California—an ICC Certification in the 2013 codes demonstrates your commitment to this mission. If you need help preparing, ICC sells reference materials at shop.iccsafe.org

Don't forget, ICC also offers National Special Inspector Certifications in seven categories:

- Reinforced Concrete
- Structural Masonry
- Spray-applied Fireproofing
- Prestressed Concrete
- Soils
- Structural Steel and Bolting
- Structural Welding

Learn more about the Special Inspector exams or see a list of all ICC National Certification exams at www.iccsafe.org/exams

Register for California code exams at www.iccsafe.org/CAexams